Investing into North African Solar Power

Robert Gömmel

Investing into North African Solar Power

A Legal Framework for Risk Management and Prospects for Arbitration

Springer

Robert Gömmel
Engelsbach
Germany

Inaugural-Dissertation zur Erlangung des Doktorgrades am Fachbereich Rechtswissenschaft der Justus-Liebig-Universität Gießen
Eingereicht von Robert Gömmel
Betreuer: Prof. Dr. Thilo Marauhn (M.Phil.)
Zweitkorrektor: Prof. Dr. Christoph Benicke
Tag der Disputation: 18.12.2013

ISBN 978-3-319-15755-9 ISBN 978-3-319-15756-6 (eBook)
DOI 10.1007/978-3-319-15756-6

Library of Congress Control Number: 2015936187

Springer Cham Heidelberg New York Dordrecht London

Printed on acid-free paper

Springer International Publishing AG Switzerland is part of Springer Science+Business Media (www.springer.com)

For my parents, wife and daughters.
Thank you for all your support and the joy you have brought into my life.

Acknowledgements

I would like to thank everyone who has helped me over the past few years to realize this project.

First of all, I owe a special thanks to my supervisor, Prof. Dr. Thilo Marauhn (M.Phil.). This thesis would not have eventuated without his support of the topic and help during the writing process and the financing of the project.

Furthermore, I am extremely grateful to the University of Giessen for providing a postgraduate scholarship while I was writing the main part of my thesis. The financial support gave me the freedom to focus on this project.

I also want to thank the SEPA (solar energy partnership with Africa) working group of the University of Giessen/ZEU. In addition to extending me a warm welcome and making me feel a part of your group, the exchange of ideas, opinions, and suggestions was a crucial part of my thesis and helped me to assess problems from different angles and perspectives.

I also owe special thanks to Mrs. Joanna Niederer for all the work she put into this project by proofreading my thesis many times and giving valuable feedback and suggestions.

The seminar for doctoral students hosted by Prof. Dr. Thilo Marauhn (M.Phil.) was also a great help as it gave me a further opportunity to talk about my thesis and my ideas. It was always a pleasure to organize these events, and I thank you all for your valuable feedback on my thesis after numerous presentations and discussions. I always had a great time with all of you.

In addition, I also want to thank all of my friends, no matter where I met them (Braunschweig, Gießen, Düsseldorf, Frankfurt, Usingen, Auckland, Darmstadt...). You helped me to switch off from my thesis from time to time, which allowed me to relax and recharge my batteries.

Finally, I want to thank my parents Rainer and Gerlinde, my wife Maike, and my daughters Marie and Amelie. Your lifelong support and love mean the world to me.

Contents

List of Figures

List of Tables

Chapter 1
Introduction

Since the beginning of the commercial production of energy, the world has changed. Society's energy need is constantly growing and is threatening to outpace the electricity that we generate. Therefore, we must search for new methods of satisfying this energy need. Since recent studies have pointed out that there may be a strong correlation between CO_2 and the greenhouse effect, almost all fossil energy sources are no longer an option, as they are limited and will cause further problems in the future. Nuclear energy is often seen as a possible alternative because nuclear power plants produce CO_2-free energy. Besides the risk of a meltdown or abuse by terrorist organizations, studies have revealed that it would take over 13,000 nuclear plants just to replace all the fossil fueled plants.[1] Although uranium deposits are not as low as often predicted,[2] this number of power plants would be an unpredictable threat.

From 2003 to 2005, global energy demands were predicted to grow by 1.2 % per year, whereby the growth rate of the OECD countries is slightly higher.[3] Experts assume that the energy demands of the EU member states will rise from 3,200 to 4,200 TWh by 2020.[4] Up to the present day, the energy demand has been covered by conventional power plants.[5] The International Energy Agency (IEA) predicts that CO_2 emissions will have risen by 50 % by 2030 compared to today.[6] In order to comply with the energy demands of the future and the need to replace the old infrastructure, at a rate of 2 % Gross Domestic Product (GDP), a total investment of

---

[1] Düren (2008), 6 (7).

[2] Nuclear power plants as medium-acting CO_2 relief: Deutsche Physikalische Gesellschaft e.V. (2005), 68–69.

[3] Lewiner (2010), 41 (41); Energy growth by 25 % by 2030, in: Renner and Jasper (2010), 91 (94); Growth rate at 1.6 % per year, in: Stigler and Bachhiesl (2010), 183 (186).

[4] Renner and Jasper (2010), 91 (96).

[5] Renner and Jasper (2010), 91 (94).

[6] Renner and Jasper (2010), 91 (95); Raise by 60 % by 2030, in: Pfeiffer (2010), 131 (132); Importance of coal, in: Stigler and Bachhiesl (2010), 183 (183) and (195).

R. Gömmel, *Investing into North African Solar Power*,
DOI 10.1007/978-3-319-15756-6_1

$22,000 billion dollars is needed between 2006 and 2030, half of it within developing countries alone.[7] In the past, there have been crises related to conventional energies, caused by non-EU states and EU states, which have affected the EU.[8] Thus, it is important for the EU to replace oil and gas completely in order to secure its energy demand.[9] Since 50 % of Europe's energy is imported and this might rise to 70 % by 2030, the EU should look for alternative solutions.[10]

The sun is a powerful energy source, producing around 1.6 billion terawatt hours (TWh) per year.[11] To put this amount into perspective, one day of sunlight equals the annual energy needs of the world's population.[12] This means that sunlight surpasses the capacities of all other renewable energies.[13] Nowadays, technology provides us with the means to make use of sunlight, such as with photovoltaics and solar thermal power.

These facts, combined with the need to rearrange energy production, opened the door for a new project called the Desertec Concept. In 2003, this concept started with the aim of building solar thermal power plants in the Middle East and North Africa (MENA) region. Due to the great amount of sunshine and the opportunities provided by solar thermal technology, there is a chance to produce "clean" energy for MENA countries and Europe. So far, the Desertec Concept has found a lot of support within society and politics. Even the MENA states view it as a good alternative to their current energy production.

Realization of the Desertec Concept entails numerous risks and problems. This thesis mainly focuses on the issue of investment security. The purpose of this thesis is to offer a guideline to assess the main legal problems connected to investment in a solar thermal project. The introduction will address questions concerning possible contracting parties and the contractual setup. The choice of the contracting parties as well as the contractual setup is based on legal and economic evaluations. The assessment will deal predominantly with Public Private Partnerships as a form of cooperation, funding and risk assessment, and allocation.

This chapter will address the question of dispute resolution. To successfully implement such a large scale project, it is necessary to work out how possible disputes arising from the contract will be handled. There are several options for dispute resolution, e.g. national, international courts or arbitration. This chapter will also explain why international commercial arbitration is the best dispute resolution tool for the Desertec project.

[7] Lewiner (2010), 41 (42).

[8] Heun-Rehn and Dratwa (2010), 101 (105–106).

[9] Heun-Rehn and Dratwa (2010), 101 (116); Countries which export frequently use commodities as tools of foreign policy, in: Pfeiffer (2010), 131 (136).

[10] Pfeiffer (2010), 131 (132); cf. Stigler and Bachhiesl (2010), 183 (183).

[11] Schäfer (2008), 11 (12).

[12] Schäfer (2008), 11 (12).

[13] Schäfer (2008), 11 (12).

Chapter 2 will discuss the issues underpinning investment arbitration. As investment arbitration is distinct from commercial arbitration, other rules and procedures apply. In the case of a solar thermal project, questions about the definition of investment may arise and thus need further assessment. Therefore, Chap. 2 will introduce the system of investment arbitration according to the International Centre for the Settlement of Investment Disputes Convention (ICSID Convention) and assess the meaning of investment in the context of a solar thermal power plant in the MENA region.

1.1 Contracting Parties

Before discussing a possible contract, it is important to know who can be a party to it. This project involves two sides: the European side and the MENA region. It is important to assess all the possible contracting parties as this will have an influence on questions of commercial and investment arbitration.

1.1.1 The European Side

The EU cannot be a contracting party because it lacks the necessary competence. A violation of competence leads to an immediate invalidity *ultra vires* according to Art. 46 of the Vienna Convention on the Law of Treaties (VCLT).[14] Due to the inaccurate wording and a lack of clarity with regard to interactions with Directive 2003/54/EC,[15] neither the TFEU[16] nor Directive 2009/28/EC[17] create a legal basis for EU action. Another way the EU cooperates with the MENA region is through the European Neighborhood Policy (ENP),[18] which includes initiatives like the Neighborhood Investment Facility (NIF).[19] The ENP offers some interesting tools, but so far it has failed to live up to expectations,[20] so the EU is consequently "limited" to playing the role of mediator. In addition, the Euro-Mediterranean Partnership (EMP) and the launch of the Union of Mediterranean (UM) and its

[14] Pache and Bielitz (2006), 316 (320).

[15] Klinski (2005), 207 (209–210) and (212).

[16] As this provision was quite important under the ECT, it is argued that it lost some of its relevance due to the stipulation in the TfEU and not the EUT, in: Bungenberg (2009), 205 (211); Baur and Blask (2002), 636 (640).

[17] Ringel and Bitsch (2009), 807 (810); Lehnert and Vollprecht (2009), 307 (308–309).

[18] Werenfels and Westphal (2010), 1 (21).

[19] One practical example of NIF in action is a feasibility study for a Concentrated Solar Power (CSP) plant in Tunisia, in: European Commission (2010c) NIF—Projects in the Southern Region.

[20] Werenfels and Westphal (2010), 1 (21).

Mediterranean Solar Plan would not be suitable hosts for this project. So far the MSP has been purely political and thus can only create the legal framework for industrial initiatives,[21] and the UM has not yet established the Secretariat.[22] Finally, European states are not capable of being or are not willing to be an active actor within the concept. This is due to political animosity between different states and their reluctance to spend their own money and time on this project.

1.1.1.1 Private Sector

The solution would be to include a private investor.[23] To date, there are several companies interested in developing a solar thermal power plant in the MENA region. However, there is no investor at the moment actively trying to implement a solar thermal project within the MENA region. The last serious attempt to implement a solar thermal project in the MENA region was made by Dii, but it failed by the end of 2014. Until 2014, Desertec almost served as an epitome for solar thermal projects. It is important to differ between the three different terms connected with the word Desertec as only Dii failed. First of all, there is the "Desertec Foundation", secondly the "Desertec Concept" and, finally, Dii.

1.1.1.1.1 Organization and Intentions of Desertec

The "Desertec Concept" started in 2003 and was designed by the Club of Rome and Trans-Mediterranean Renewable Energy Cooperation (TREC).[24] One of the main purposes of the Desertec Concept is the construction of a MENA-wide solar thermal network.[25] The Desertec Concept does not necessarily limit its focus to supplying Europe. Instead it looks towards the energy needs of the host state.[26] Between 2004 and 2007, the German Aerospace Center (DLR) released three feasibility studies concerning the Desertec Concept and approved its practicability.[27] In this examination the term "Desertec Concept" is used as an example of a solar thermal power plant project. Up-to-date the Desertec Concept is the most sophisticated solar thermal project.

---

[21] Ruchser (2010).

[22] Werenfels and Westphal (2010), 1 (21).

[23] Deutscher Bundestag (2009), 8; Europäische Kommission (2006), 7.

[24] Desertec Foundation (2009b); Desertec-UK (2010); The Club of Rome was founded in 1968 by 100 people, each from different cultural backgrounds and representing the five continents, in: The Club of Rome (2014).

[25] Desertec Foundation (2009c); Desertec Foundation (2014b).

[26] While Algeria interest is more towards exporting additional energy, Morocco has a big energy demand itself and would rather use the energy itself, in: Knies (2008).

[27] Bundesumweltministerium (2005); German Aerospace Center (DLR) et al. (2005); German Aerospace Center (DLR) et al. (2006); German Aerospace Center (DLR) et al. (2007).

The Desertec Foundation was founded in 2008 and is based in Berlin. Its main purpose is to coordinate the Desertec network, to establish new alliances and create a greater awareness of the whole program.[28] At the moment, the Desertec network is widespread, involving almost every region of the world.[29] The main aim of the Desertec Foundation is to encourage research into new energy solutions and to work as a mediator for interested commercial organizations, governments and NGOs.[30]

1.1.1.1.2 Organization and Intentions of Dii

Dii on the other hand is an independent company established as a German limited liability company according to the German limited liability company act in 2009.[31] Dii was founded as part of an initiative between the Desertec Foundation and Munich Re (one of the biggest reinsurance companies in the world).[32] Furthermore, a memorandum of understanding was signed in Munich by the Desertec Foundation and 12 different companies.[33] The number of shareholders is not limited to the current amount as it is envisaged that Dii will incorporate companies from all over the MENA region.[34] Dii has since grown from its original 12–44 shareholders and partners, encompassing partners from all over the world with different specifications.[35] Over the past 2 years, Dii has faced several problems with major partners leaving,[36] but has also managed to win the State Grid Corporation of China (SGCC) as a new partner.[37] Due to disagreements between the Desertec Foundation and Dii, the Foundation pulled out as a shareholder of Dii in 2013.[38] The Desertec Foundation based this decision on irreconcilable differences concerning future alignments of Dii.[39] Overall, Dii was unsuccessful and its failure was announced in late 2014.[40] Nevertheless, the idea behind Dii remains valid for possible investment in North African solar power. A financially powerful and highly experienced investor is

---

[28] Desertec Foundation (2009b).

[29] Desertec Foundation (2014a).

[30] Desertec Foundation (2009a).

[31] Desertec Industries (2010a).

[32] Deutsch and Hobohm (2010), 54 (54); Werenfels and Westphal (2010), 1 (7).

[33] Dii founding companies: ABB, Abengoa Solar, Cevital, DESERTEC Foundation, Deutsche Bank, E.ON, HSH Nordbank, MAN Solar Millennium, Munich Re, M+W Zander, RWE, SCHOTT Solar and Siemens, in: Desertec Industries (2014).

[34] An amount of 25 companies is targeted: Werenfels and Westphal (2010), 1 (8).

[35] Desertec Industries (2014).

[36] FAZ.Net (2014b, c).

[37] Wetzel (2013).

[38] Desertec Foundation (2013).

[39] Desertec Foundation (2013).

[40] FAZ.Net (2014d).

necessary to implement a solar thermal power plant. A cooperation of several companies still seems to be a suitable solution in this case.

1.1.1.1.3 Conclusion

Although there was discord between Dii and the Desertec Foundation, this does not mean that the whole concept failed. The concept and the foundation are still in place and offer a great think tank and an interesting idea to facilitate a solar project in the MENA region. The discord between Dii and the Desertec Foundation was not really surprising as there have been differences in the past. One example is their differing opinions on the primary supply of energy (MENA or Europe), which will be discussed later on in this thesis. In hindsight, the discord was the first sign that Dii was going to fail as it was not able to combine, accommodate or mediate between all the different shareholders' concerns.

Admittedly, the MENA region's electricity demand will be the same as Europe's demand today by 2050,[41] but it would be wrong to believe that the European demand will remain at today's level. It is more likely that it will increase as more of our daily life is "electrified". Dii's homepage supported this by stating that:

> [F]rom the very beginning, the aim will be to start supplying electricity to the European Union and to generate sufficient power to meet the needs of the producer countries as soon as possible. The aim is to supply around 15 % of Europe's electricity by 2050.[42]

The aforementioned statement might also illustrate one of the main areas of disagreement between the Desertec Foundation and Dii—Dii's primary goal was to supply Europe. This goal was not Dii's alone as any investor might have focused on the European market to recoup their investment. However, there was always the risk that the needs of the MENA region would be sidelined in favor of those of the primary customer—the European market. Despite its failure, Dii still serves as a good example of an international investor for the following examination. Finally, any investor that tries to realize the Desertec Concept can potentially become a "Transnational Company" (TNC). A TNC is best understood as a company that operates across borders, with operations contained in its host state and in at least one other state.[43] Another main feature of a TNC is that there is a parent company which has power over each of its branches.[44] An investor could be considered a TNC from the moment of cooperation as it may operate in several countries, and as an investor would retain majority control over all its branches.

---

[41] Schüssler (2008), 221 (222).

[42] Desertec Industries (2010b).

[43] Schöbener et al. (2010) 60, Kap. 1 § 5 para 196.

[44] Weh (2008), 60; This also includes the majority of shares, in: Weh (2008), 62.

1.1.1.2 Other Solar Thermal Projects

Although "Desertec" represents almost the whole industry dealing with solar energy, there are other companies as well. One company that has the ability to become a serious competitor is Transgreen. Transgreen is a new solar project and was founded in 2010. From the very beginning, there was a major difference between Transgreen and Dii. While Transgreen is a political initiative of the French government, Dii was an economic initiative in which national governments have no direct involvement.[45]

To date, Transgreen has not established a homepage or implemented any of its projects. It has almost completely disappeared from the newspapers, meaning that the latest news stems from 2011.

1.1.2 The MENA Side

It is necessary to remember that a CSP can only exist within the "sun belt", which encompasses most of the MENA region and only small parts of Europe.[46]

1.1.2.1 The Middle East and North Africa

The Middle East is currently unable to host a large scale solar thermal project. The application of the Shari'a is not a problem in itself, but leads to reluctance on the part of many Western investors due to their limited knowledge of it. Furthermore, the Israel–Palestine Conflict and the Syrian civil war are ongoing security issues within the area. Another problem that this region deals with is terrorism. All these issues are also reflected within the Organization for Economic Co-operation and Development (OECD) Country Risk Classification, where countries from the Middle East, such as Iraq, Iran, Lebanon, Jordan, Syria and Israel, are ranked higher than countries located in North Africa.[47]

What is more, neither the African Union (AU) nor any Regional Economic Community (REC) can be an active partner for the implementation of a solar thermal power plant as well. Besides the internal problems,[48] still prevalent African

---

[45] FAZ.Net (2014a).

[46] Schäfer (2008), 11 (12 and 14); Deutsche Physikalische Gesellschaft e.V. (2005), 80; Schüssler (2008), 221 (223).

[47] The OECD ranking goes from 0 to 7, whereas 0 is no risk and 7 very high risk: Middle East countries: Iraq 7, Iran 7, Lebanon 7, Jordan 5, Israel (0 without prejudice to the status of the Golan Heights, East Jerusalem and Israeli settlements in the West Bank under the terms of international law/Rating of 3 in 2010) on the other hand North African states: Morocco 3, Algeria 3, Tunisia 4 (Germany 0), in: OECD (2014).

[48] Makinda and Okumu (2008), 6; Soest (2008), 2; Schmidt (2008), 5.

resentments[49] make cooperation impossible at present. Most of the RECs are not integrated regionally and are hampered by internal conflicts.[50]

1.1.2.2 Morocco

At the moment, only Morocco appears to be a suitable partner for the implementation of a solar thermal project. Besides its good relations with the EU,[51] Morocco is frequently described as the most dynamic Arab state within a conflict-ridden region.[52] Morocco is a constitutional monarchy and an Islamic country,[53] with King Mohammed VI leading the country since 1999.[54] Over the past few years, there has been progress towards the modernization and democratization of Morocco, especially in the field of human rights.[55] This is also underlined by the recent referendum concerning a reform of the constitution to restrict the king's power. A total of 98.49 % of the population voted for the reform, which was initially organized by the king himself.[56] Morocco is also very committed to the fight against terrorism, which is reflected in the 2003 Terrorism Act.[57] Morocco's laws are influenced more by the West than other Islamic states and are, in fact, very similar to French laws.[58] Due to its experience with PPPs[59] and the existing solar thermal power plants in Morocco,[60] it is a suitable partner for the abovementioned project.

---

[49] Muchie et al. (2006), 3 (16).

[50] Integration, in: Brenton et al. (2006), 3–5; Conflicts, in: Wippel (2005), 5 (7); West Sahara Conflict, in: European Neighbourhood and Partnership Instrument (2007), 9; Werenfels and Westphal (2010), 1 (33).

[51] Statute avancé, in: European Commission (2010b); Werenfels and Westphal (2010), 1 (21); Many European states value the stabilizing factor of Morocco within North Africa, in: Kausch (2009), 165 (166) and (171–175).

[52] Maghraoui (2009), 143 (143); Some authors mention that the good reputation of Morocco only result from their very good PR work, in: Kausch (2009), 165 (170).

[53] European Neighbourhood and Partnership Instrument (2007), 8; European Commission (2004a), 5.

[54] Joffé (2009), 151 (154); Kausch (2009), 165 (166); European Commission (2004a), 8.

[55] So far democracy is not reality in Morocco, in: Maghraoui (2009), 143 (144); European Neighbourhood and Partnership Instrument (2007), 8.

[56] n-tv.de (2011).

[57] European Commission (2004a), 10.

[58] Deutsche Industrie- und Handelskammer in Marokko (2008), 9; Kutty (2006), 565 (595).

[59] BOTs, in: Badawi (2003), 3.

[60] Jaouad (2010).

1.2 Solar Thermal Energy

In solar thermal power plants, sunlight is focused onto an absorber by giant mirrors, which then transform the light into thermal energy.[61] By intensifying the sunlight's concentration, a suitable fluid is heated up.[62] Thermo oil would be an example of a fluid that can reach a temperature of 400 °C.[63] The heat is then used to run a thermal engine, e.g. a steam engine to produce electricity.[64] As illustrated by the picture below, there are three different types of solar thermal power plants: firstly, the solar tower; secondly, the parabolic channel; and finally, the paraboloid.[65]

The main advantage of solar thermal power plants is that they allow thermal energy (heat storage) to be stored more easily and more efficiently than electricity.[66] Studies have illustrated that the efficiency of solar thermal power plants increases from 20 to 30 % through heat storage. Such a plant could therefore operate as a base load power station.[67] The possibility of storing the thermal energy would also allow the power plant to operate at all times.[68] CSP's functionality has also been proven in various plants all over the world, and recently also in Africa.[69] Modern transmission lines (HVDC) enable the energy to be transported to Europe from North Africa, with only a loss of 10–15 %.[70]

Calculations predict a total investment of 400 billion Euros within the next 40 years.[71] To achieve this aim, the EU is backing the construction of undersea cables in the Mediterranean Sea and other possible subsidies.[72] Facilities in Spain had production costs of 14 cents/kWh, but due the greater solar radiation within North Africa the price might be reduced to 9.5 cents/kWh.[73] Further reductions could take place because of night storage and connected desalination of 3–5 cents.[74] Transport costs are around 2 cents/kWh to Europe.[75] In Spain, the feed-in compensation was around 18–22 cents/kWh (including possible state investments).[76]

---

[61] Deutsch and Hobohm (2010), 54 (55); Düren (2008), 6 (10).

[62] Deutsche Physikalische Gesellschaft e.V. (2005), 80; Deutsch and Hobohm (2010), 54 (55).

[63] Deutsche Physikalische Gesellschaft e.V. (2005), 81.

[64] Schäfer (2008), 11 (15); cf. Deutsch and Hobohm (2010), 54 (55); Deutsche Physikalische Gesellschaft e.V. (2005), 80.

[65] Deutsch and Hobohm (2010), 54 (55); Deutsche Physikalische Gesellschaft e.V. (2005), 80.

[66] Schäfer (2008), 11 (14).

[67] Schäfer (2008), 11 (14).

[68] Schüssler (2008), 221 (221); Düren (2008), 6 (10).

[69] Schüssler (2008), 221 (225).

[70] Düren (2008), 6 (10).

[71] Biobay.de (2009).

[72] Biobay.de (2009).

[73] Winker and Preußner (2008), 34 (35); Starting costs are higher than operating costs later on, in: Berger (2003), 65 (65).

[74] Winker and Preußner (2008), 34 (35).

[75] Winker and Preußner (2008), 34 (36).

[76] Winker and Preußner (2008), 34 (35).

However, these are all predictions and are subject to change, e.g. increased plant construction costs.[77] Many questions still remain unanswered, such as the cost of maintenance, payments for ground usage and the price of energy.[78]

In the long term, it is predicted that 60 billion kWh could be imported between 2020 and 2025, and there are suggestions that an extension to 700 billion kWh/year could be achieved from 2050 onwards.[79] All predictions concerning the economic feasibility of the Desertec Concept are based on predicted European import rates, as there are no studies concerning the energy demands of the MENA region. However, due to the fact that the European market is of great relevance for the successful implementation of the Desertec Concept, these figures may be useful.

These statistics (Tables 1.1 and 1.2) reveal that a concession (based on the export) should last for at least 30 years if a price of 22 cents/kWh is charged for the duration. Some predictions suggest that a possible and realistic price for the MENA region could be 16 cents/kWh,[80] and that it would take around 41.7 years to "repay".

It is also important to compare the Moroccan price for energy with the expected price of the "desert sun energy" to ascertain whether the Desertec Concept is compatible. If it is to be successfully implemented into the MENA region, the solar thermal power plant must fit into the local system.[81] Hence, the price of the produced energy is an indicator of possible market integration. At the moment, the costs of energy in Morocco are as follows (see Table 1.3)[82].

Other authors mention an average cost of 13 cents/kWh in Morocco.[83] The cost of energy in Morocco is therefore the highest in the MENA region (aside from Jordan, with which it is on a par).[84] Concerning the application of photovoltaics, there are cost predictions for Morocco of between 25 and 30 cents/kWh.[85] Some estimate that the costs of CSP will be 16 cents/kWh and that this price will decrease by 2.5–5 % per year within the whole MENA region.[86]

---

[77] Winker and Preußner (2008), 34 (36).

[78] n-tv.de (2009).

[79] Deutsches Zentrum für Luft- und Raumfahrt (2010).

[80] Deutsch and Hobohm (2010), 54 (56–57).

[81] cf. Hunter (2007), 165 (166).

[82] Morris (2009).

[83] Deutsch and Hobohm (2010), 54 (56); RAL (2010), 53.

[84] Deutsch and Hobohm (2010), 54 (56); Other countries: Egypt 7 cent/kWh, Algeria 5 cent/kWh, Israel 10 cent/kWh, Lebanon 10 cent/kWh, Syria 7 cent/kWh, Tunisia 11 cent/kWh, Turkey 12 cent/kWh, West Bank/Gaza 11 cent/kWh, in: Deutsch and Hobohm (2010), 54 (56).

[85] RAL (2010), 52.

[86] RAL (2010), 52.

Table 1.1 Calculation of revenue based on predicted European import

Time frame and total (kWh)	Proposed price (22 cents/kWh)	Spain (14 cents/kWh)	MENA (7 cents/kWh)[a]	MENA (10.5 cents/kWh)[b]
2020–2025 (60 billion kWh)	13.2 billion Euro per year	8.4 billion Euro per year	4.2 billion Euro per year	6.3 billion Euro per year
From 2050 (700 billion kWh)	154 billion Euro per year	98 billion Euro per year	49 billion Euro per year	73.5 billion Euro per year

[a]Composition of the price 9.5 cents/kWh predicted costs −4.5 cents/kWh predicted relief +2 cents/kWh transport costs
[b]The first MENA price is the pure production price; since Spain produced for 14 cents/kWh and charged a total of 18–22 cents/kWh, this is an increase of approximately 50 %. Therefore, 50 % was added to the initial MENA price

Table 1.2 Worst-case scenario to raise 400 billion Euro by only 60 billion kWh/year

Time to raise 400 billion Euro	22 cents/kWh	14 cents/kWh	7 cents/kWh	10.5 cents/kWh
	30 years	47 years	95 years	63 years

Table 1.3 Moroccan energy costs

0–100 kWh per person (pp)	101–200 kWh (pp)	201–500 kWh (pp)	>500 kWh (pp)
8 cents/kWh	9 cents/kWh	9 cents/kWh	13 cents/kWh

1.3 Contract Design: Public Private Partnership

The kind of contract is important to assess whether commercial and investment arbitration is applicable. Furthermore, only a few cooperation models are applicable. The best way to realize such a large scale project would be to make use of PPPs as they offer numerous forms and types of cooperation between the private and public sector.[87] Internationally, the utilization of PPPs is quite common.[88] As long as both parties do not choose a national law to govern the PPP, it is most likely that it will be categorized as an international contract, despite the inclusion of private parties.[89]

Due to the wide variety of cooperation forms, it is not possible to define a PPP.[90] However, there are certain features which distinguish a PPP from other forms of cooperation. These attributes are: (1) the long duration of the contract; (2) the

[87] Noltensmeier (2009), 26; Nickel and Kopf (2004); Yescombe (2007), 4.

[88] Deutscher Bundestag (2005), 1; Spread of PPPs, in: Bremer (2005), 1; Yescombe (2007), 22.

[89] Herdegen (1994), 41 (48).

[90] Metje (2008), 5; Corbacho and Schwartz (2008), 85 (86); cf. Seidel/Mertens in Dauses (2010), H. IV. Öffentliches Auftragswesen para 205; Jonas and Paulsen (2007), 13 (14); Höfler (2007), 35 (37); Koman (2004).

parties include one private and one public party; (3) they are privately funded projects, and (4) there is an allocation of risks.

The main reason why the public sector frequently decides to join a PPP is because of insufficient money. The main idea is to use the money of the private sector to realize the project. At the same time, the public sector always seeks to minimize risks and thus to transfer them to the private sector. The allocation of risks is often a contentious issue between both partners, so it needs to be assessed further.

1.3.1 Contract Period

The average contract period ranges between 15 and 30 years.[91] On an international scale, there are some PPP agreements with a lifespan of between 50 and 99 years.[92] If both parties desire comparable, long-term cooperation, the law that governs the PPP is significant as national law can prohibit excessive contract periods.

1.3.2 Contractual Partner

There are primary project participants, namely the public authority and the private partner, and secondary participants like banks, subcontractors, consumers, consultants and possible operators.[93] Regardless of whether one is talking about the role of primary or secondary project participants, it is always important that no one party has too much power within a PPP (e.g. public intervention).[94]

1.3.3 Project Funding

Normally, the project is not financed solely by the private operator, but rather by secondary participants (e.g. banks).[95] The private sector tries to keep the amount of equity as small as possible.[96] There are different proposals concerning the necessary amount of equity, as it should be around 30 %.[97] In the case of the Desertec Concept, this would be 120 billion Euros. In 2005, commercial banks provided

---

[91] Corner (2006), 37 (41); cf. Noltensmeier (2009), 36; Knütel in Siebel et al. (2008), para 1272.

[92] Knütel in Siebel et al. (2008), para 1309.

[93] Metje (2008), 17.

[94] Wolfs and Woodroffe (2002), 101 (102).

[95] Wahl (1994), 15 (16).

[96] Höfler (2007), 35 (46).

[97] Schoeltzke and Claus (2007), 133 (138).

$140 billion dollars or, in other words, 85 % of all project-finance funding for PPPs.[98]

Direct agreements' between banks and the public sector,[99] credit agreements,[100] cash-flow agreements[101] between banks and the private partner and possible central bank guarantees[102] complete the complex contractual setup for successful project funding.

1.3.4 Types of Risk

All the risks can be separated into two categories: namely, procedural risks and project risks.[103] Procedural risks encompass everything from execution of the contract to political and legal acceptance, while project risks encompass the construction, operation and transfer phases.[104]

1.3.4.1 Procedural Risks

To avoid any risks concerning the execution of the contract, a 'Service Level Agreement' (SLA) should be concluded in advance, which sets standards and formulates contractual penalties for disobeying these standards.[105] In terms of procedural risks, the two other major issues that a project faces if it is situated in a foreign country are political and legal risks.

Political risks include practically all governmental activities, such as an unexpected change to the political system[106] and the risk of any kind of state interference.[107] However, political commitment and legal stability are meaningless if the agreements are not adequately enforced by the government.[108] One way to ensure the security of the project would be to apply the 'material adverse government

---

98 Yescombe (2007), 126; cf. Knütel in Siebel et al. (2008), para 1254.

99 Yescombe (2007), 210.

100 Dorner (1994), 75 (85–86).

101 Harries (1996), 19 (21); "Escrow Accounts" should be suited in countries of the Western hemisphere, in: Schoeltzke and Claus (2007), 133 (138).

102 Verveniotis (1996), 115 (125); Wahl (1994), 15 (19–20).

103 Höfler (2007), 35 (39).

104 Höfler (2007), 35 (39).

105 Höfler (2007), 35 (39); Knütel in Siebel et al. (2008), para 1415; Bertelsmann Stiftung et al. (2011), 50–51.

106 Berger (2003), 65 (65); cf. Joffé (2000), 33 (36); HyunChan et al. (1997), 93 (100); Kumaraswamy and Zhang (2001), 195 (198); Palmer (2009), 14; Metje (2008), 54; cf. Happ (2002), 39 (40).

107 Metje (2008), 53.

108 Christiansen (2008), 143 (150).

action' (MAGA) rules, which provide for readjustment in case of political interference, alternatively for termination of the contract.

Legal risks relate to the change of laws and thus the legal framework of the project.[109] Here the support of the host government is just as essential as in the case of political risk.[110] While discriminatory changes are covered by MAGA rules, other changes should be addressed separately. The state should agree to keep all necessary impacts as small as possible[111] and, if this is not possible, additional costs should be split between the host state and the investor.[112]

1.3.4.2 Project Risks

Construction risks encompass the conditions on the ground, the failure to stay on schedule and the general default of obligations by the state (e.g. permission) or construction partners.[113]

Other project risks include financial risks, demand risks, availability risks, *force majeure* and residual risks.[114] The risk of changes in consumer preference, a decrease in public solvency or the appearance of a competitive product (demand risk),[115] and the guarantee of availability, capacity and efficiency during the contract period (operating risk)[116] are frequent issues for any large scale project. Inflation, political reluctance towards change, applicable currency of payment and any negative influence of the cash flow are considered financial risks.[117] Environmental risks include potentially negative influences on the environment.[118] Social

[109] HyunChan et al. (1997), 93 (100); cf. Palmer (2009), 18; Höfler (2007), 35 (43–44); Legal risks exists in long-term projects, in: Joffé (2000), 33 (36).

[110] Metje (2008), 18.

[111] Knütel in Siebel et al. (2008), para 1325; In cases where the change of law is adversely to the private operator, he can demand adjustment to be put into the same economic position prior to the changes, in: Wang and Tiong (2000), 69 (74).

[112] Roquette/Butcher in Roquette and Otto (2005), F. Public–Private Partnership III paras 70 and 72–73; Knütel in Siebel et al. (2008), para 1329.

[113] Wagner (1996), 3 (11); cf. HyunChan et al. (1997), 93 (100); Palmer (2009), 14; cf. Nicklisch (2007), 3 (6); cf. Höfler (2007), 35 (43); Subcontractor should be chosen wisely, and facts like the subcontractor's average size projects must be examined, in: Yescombe (2007), 89 and 256; Risks related to the planning and constructions of the project are normally borne by the private sector, in: Jonas and Paulsen (2007), 13 (28); And permission granting, in: Höfler (2007), 35 (43) and (46).

[114] Corbacho and Schwartz (2008), 85 (88); Construction risk, in: Hesse (1994), 31 (35–36); Market risk (similar to demand risk), in: Pahl (1994), 23 (29).

[115] cf. Yescombe (2007), 19; OECD (2008), 53; Nicklisch (2007), 3 (7); Jonas and Paulsen (2007), 13 (30–31).

[116] Wagner (1996), 3 (11); cf. HyunChan et al. (1997), 93 (100).

[117] Estache et al. (2007), 14; Verveniotis (1996), 115 (124); Nicklisch (2007), 3 (7); HyunChan et al. (1997), 93 (100); cf. Kumaraswamy and Zhang (2001), 195 (198); Metje (2008), 51; Including risk of inflation, in: Metje (2008), 52.

[118] Palmer (2009), 14; HyunChan et al. (1997), 93 (100); Consideration of environmental issues, in: Kumaraswamy and Zhang (2001), 195 (198).

risks include the local population and/or government no longer supporting the project.[119] PPPs often include an obligation to transfer the asset back to the state at the end of the contract period,[120] which might cause problems with regard to the 'should-be state' of the asset. Finally, vandalism and hardship are further project risks which a solar thermal power plant might face.

1.3.4.3 Conclusion

There is a connection between risk and reward.[121] If the private operator is able to manage the transferred risk, the reward will be equivalent to the amount of risk assumed.[122] However, the bigger the project, the greater the risk is that it will fail.[123] In the case of risk allocation, it is essential to properly assign each risk.[124] For example, the government will be more adept at handling political risks (problems created by government actions); while the private sector is better equipped to deal with construction risks (design and schedule problems).[125] Best coverage means that the appropriate party manages those risks cost-effectively and can therefore influence the outcome to a certain extent.[126] The amount of risk transferred to the public sector depends on the PPP model chosen, so it can differ tremendously.[127] The amount of risk which can be transferred increases for the private sector from a simple PPP, such as Design-Build, to a concession.[128] A Value-for-Money (VFM) will work best if the risk allocation process is thorough and balanced.[129]

Finally, the state must be aware that the amount of influence on the project decreases by the amount of risk that it transfers to the private sector. It is still a common misbelieve that the state can divest itself of all risk but retain all control of the project. In return for the investment made by the private sector, the state must

---

[119] Kumaraswamy and Zhang (2001), 195 (198); HyunChan et al. (1997), 93 (100).

[120] Höfler (2007), 35 (44); Jonas and Paulsen (2007), 13 (22).

[121] Corner (2006), 37 (44).

[122] Corner (2006), 37 (49); The more the risk, the more owners' equity in a BOT, in: Wolfgang Straßburg (1996), 29 (32); cf. Wiwen-Nilsson (1996), 91 (104).

[123] Blanchard (1998), 417 (468).

[124] Wiwen-Nilsson (1996), 91 (97–99); Becher (2007), 55 (58); Corner (2006), 37 (46); OECD (2008), 48 and 235–237; Ahadzi and Bowles (2004), 967 (968); Davies and Eustice (2011), 20; Corbacho and Schwartz (2008), 85 (89); Christiansen (2008), 143 (149); cf. Hemming (2008), 235 (237); Hathorn (2008), 245 (248); Bremer (2005), 87–88; Palmer (2009), 15; Nicklisch (2007), 3 (7); Höfler (2007), 35 (47); Metje (2008), 48.

[125] Corbacho and Schwartz (2008), 85 (88–89); cf. Christiansen (2008), 143 (149); cf. Yescombe (2007), 19; Risks concerning building or contaminated sites are usually located by the government, in: Bremer (2005), 88.

[126] Corner (2006), 37 (46–47).

[127] Christiansen (2008), 143 (149).

[128] OECD (2008), 51.

[129] Corner (2006), 37 (46); OECD (2008), 48.

accept that it is going to "lose" part of its influence. Otherwise, no private investor will be willing to invest in any kind of PPP.

1.3.5 Additional PPP Regulations

In addition to the four main characteristics of a PPP, there are other contractual aspects which must be kept in mind. Despite the private operator's independence, the government should also include rights of inspection.[130] The term "partnership" with a PPP should not belie the need to monitor the private operator,[131] as public control (to a certain extent) is advisable since big scale projects often affect core public interests. However, it would be wrong to assume that the private partner will automatically and purposely violate any given law purely because it is legally beneficial.[132]

In cases of termination, there should be specific rules on how a transfer of the asset will take place.[133] In addition, PPPs need to address the means by which the value of the company will be calculated.[134] PPPs should also include a "solution manual" on how to deal with breaches of contract.[135] In the event of a one-sided termination that has resulted from a contractual breach on the part of the terminating party, there is a need to negotiate possible compensation.[136] This is also known as '(...) compensation on termination (...).'[137] There is also the issue of what happens in the event that the project is destroyed. Some PPPs contain a regulation which, if the project is destroyed, requires the private sector to rebuild the whole project despite any insurance and the question of whether the project can be insured at all (e.g. all-risk insurance).[138]

Finally, the importance of international dispute resolution processes, such as international arbitration, is frequently emphasized in large scale projects.[139] All contracts (independent of the drafting language) should take care to regulate a consistent dispute settlement process.[140] Nevertheless, regular consultation can help to avoid conflicts and strengthen the public–private cooperation.[141]

---

[130] Knütel in Siebel et al. (2008), paras 1311–1312 and 1420–1421; Roquette/Butcher in Roquette and Otto (2005), F. Public–Private Partnership III para 46.

[131] Davis (2005), 439 (439–440).

[132] Knütel in Siebel et al. (2008), para 1313.

[133] Werner and Fiedler (2007), 91 (104); Roquette (2007), 147 (157).

[134] Roquette (2007), 147 (157).

[135] Roquette/Butcher in Roquette and Otto (2005), F. Public–Private Partnership III paras 85–87.

[136] Werner and Fiedler (2007), 91 (103–104); Roquette/Butcher in Roquette and Otto (2005), F. Public–Private Partnership III paras 88–90.

[137] Roth (2007), 107 (114).

[138] Jonas and Paulsen (2007), 13 (31).

[139] Wahl (1994), 15 (21); cf. Christiansen (2008), 143 (158).

[140] Ruchay (1996), 69 (84).

[141] Christiansen (2008), 143 (155).

1.3.6 *Disadvantages of PPPs*

PPP-specific regulations do not exist in many countries. A long negotiation phase and increasing costs are often connected with PPPs.[142] This creates uncertainties in relation to the regulatory framework and, due to the lengthy contract period, it can be subject to law changes enacted by a newly elected government.[143]

The drive for economic efficiency may affect the quality and performance of the project, and weaken democracy in the region, because all decisions concerning the service are made in the private company's headquarters, far away from the affected region.[144] In terms of project finance, the private sector has higher borrowing costs than the government.[145] It therefore comes down to a comparison between the efficiency of the private sector operating a PPP and the higher cost of borrowing money.[146] Corruption may also occur at almost every stage of a PPP: from the consultation phase, to the tender phase, to the execution phase.[147] Moreover, corruption may lead to public skepticism and subsequently to doubts as to the integrity of the whole PPP process.[148]

Although off-balance sheet treatment is convenient for the local state because it means that the long-term PPP project does not appear in its budget,[149] this may lead to the government purposely choosing a PPP to avoid any appearance of costs within its annual budget.[150] As this only works if the state has no costs at all, off-balance sheet treatments are not advisable, since the private sector cannot guarantee the success of the project.[151] The host state should be aware that a PPP will limit the state's future budget flexibility.[152]

---

142 Ahadzi and Bowles (2004), 967 (971); cf. Davies and Eustice (2011), 30; cf. Roquette/Butcher in Roquette and Otto (2005), F. Public–Private Partnership III para 24.

143 Straßburg (1996), 29 (38–39).

144 Pfaffhausen (2008), 95 (97 and 101); Importance to maintain the level of service, in: Davies and Eustice (2011), 25.

145 Jonas and Paulsen (2007), 13 (21); Roquette/Butcher in Roquette and Otto (2005), F. Public–Private Partnership III para 21; Corbacho and Schwartz (2008), 85 (88); Yescombe (2007), 18.

146 Irwin (2008), 105 (107).

147 Christiansen (2008), 143 (151); OECD (2008), 124.

148 Christiansen (2008), 143 (151).

149 Corbacho and Schwartz (2008), 85 (85); Davies and Eustice (2011), 26; cf. Hemming (2008), 235 (235); Schoeltzke and Claus (2007), 133 (134).

150 Corbacho and Schwartz (2008), 85 (89).

151 Possible state liabilities, in: Corbacho and Schwartz (2008), 85 (89); Possible state guarantees, in: Hemming (2006), 9; Hemming (2008), 235 (239); Kumaraswamy and Zhang (2001), 195 (200).

152 Irwin (2008), 105 (106).

1.3.7 Advantages of PPPs and Solutions to Problems

Despite the negative aspects of PPP usage, there are also benefits. Initially, it is helpful to establish Public Sector Comparators (PSC) to determine whether the market is suitable for a PPP, or whether another form of cooperation will lead to a better VFM.[153] A PSC is an estimate of the project's cost assuming it had been performed by conventional public procurement.[154] If a PPP is best, the advantage is that the private operator undertakes most of the capital expenses, such as the design, building, financing and operating costs.[155] Although a special PPP framework is not necessary for the "smooth running" of a PPP, it is useful to have a clear and comprehensive legal framework.[156] It is very important that both parties accept that a PPP cannot regulate every possible event. To avoid inflexibility, it is necessary to include renegotiation or adjustment clauses.[157]

There is no doubt that there is chance of creating a monopoly. Time limited contracts such as concessions can help to prevent this. The contract should include certain provisions pertaining to the government's ability to check the costs relative to the prices, and to adjust the price if necessary.[158] It is beneficial to involve the private sector because it is better able to operate and administer a large scale project.[159] Increased efficiency may have a negative effect on the service, which means that it essential to choose the correct partner and to include a 'no service/no pay' agreement in order to facilitate PPPs.[160] Furthermore, a contract should include sections dealing with the issue of a change of control, for example in the event of a shareholder replacement.[161]

To avoid off-balance-sheet treatment, it would be best to be as transparent as possible from the beginning.[162] PPPs can help states with budget constraints, because they are not bearing the costs (and some of the risks) and are thus not

[153] Becher (2007), 55 (56); Alfen/Fischer in Weber et al. (2006), 19–20; Stöcker (2010), 96; Roquette/Butcher in Roquette and Otto (2005), F. Public–Private Partnership III para 18.

[154] Yescombe (2007), 62–64.

[155] Fourie and Burger (2001), 147 (150).

[156] Corbacho and Schwartz (2008), 85 (93); To date, several countries operate PPP task forces, in Stöcker (2010), 96; PPP task forces, in: Roquette/Butcher in Roquette and Otto (2005), F. Public–Private Partnership III paras 27–29.

[157] Hard to distinguish between renegotiation and adjustment clauses, however adjustment clauses frequently deal with change of law and renegotiation clauses with modification of the contractual framework, in: Griebel (2008), 33–34; Metje (2008), 126–127; These clause must respect the principle of *pacta sunt servanda*, in: Berger (2003), 65 (71) and (92–93).

[158] Establishing a independent consortium, in: Koman (2004).

[159] Christiansen (2008), 143 (148); Diederichs (2009), 547 (548); cf. DIHK and bfai (2006), 12.

[160] EIB (2004); Corbacho and Schwartz (2008), 85 (93); Bonus-penalty contract, in: Werner and Fiedler (2007), 91 (98).

[161] Knütel in Siebel et al. (2008), para 1314.

[162] Deutscher Bundestag (2005), 10; Corbacho and Schwartz (2008), 85 (93); Transparency and concessions, in: Karl (2003), 37 (48–49); Koman (2004).

violating any restrictive national regulations.[163] Transparency would also help to prevent any kind of corruption, e.g. if an annual transparency report were issued. To avoid a "guarantee culture", the use of guarantees by the state should be restricted to events that the private operator is not able to control, handle or anticipate itself.[164] Governments should make sure that all guarantees have a 'quantitative ceiling' and appear in the annual budget report.[165] This would involve all costs (even if they are only predicted) being capped and published. In addition, there could be a contractual agreement like '(...) helping on the downside / profit of the upside (...)', related to financial support by the state to avoid misuse by the private sector,[166] or the government could claim guarantee fees during a warranty case.[167]

A PPP may create new jobs as most private operators recruit their staff within the host state.[168] With the training of staff and the building of a new project, there is also a transfer of know-how.[169] Lastly, in cases of a PPP, the government can also obtain an asset without paying too much.[170]

1.3.8 Assessment of the Benefits and Drawbacks

In most cases where a PPP has failed or has not lived up to the expectations of the parties involved and critical mistakes have been made, such as incorrect risk allocation, insufficient transparency, etc., it is important to understand that the drawbacks of PPPs mostly stem from the behavior of the parties. Consequently, it is wrong to condemn PPPs straightaway. Undesirable party behavior can occur in every form of cooperation, which does not mean that the respective cooperation is problematic. Due to the complexity of the Desertec Concept, all forms of cooperation could face similar problems, which mean that the argument against the complexity of PPPs is not a convincing one.

Overall, all the positives and negatives add up to a list of prerequisites for a successful PPP. This encompasses: political will, public acceptance, an adequate legal framework, a feasibility study (including a PSC), adequate allocation of risks, projects of a suitable size for the country, clear output specifications, flexibility due to the long-term contract, adequate public-sector institutional capacity to handle the PPP programs as a whole and to deal with individual projects, supervision possibilities, and a transparent process. Morocco could also demand that an investor

---

[163] European Commission (2004b), 3.

[164] Hemming (2006), 10; Hemming (2008), 235 (241).

[165] Hemming (2006), 11–12.

[166] Estache et al. (2007), 19.

[167] Hemming (2006), 12.

[168] Metje (2008), 47.

[169] DIHK and bfai (2006), 12.

[170] Irwin (2008), 105 (106).

commits to the OECD Guidelines for Multinational Enterprises, which were updated in 2011.[171] An investor would be considered a multinational enterprise as it is a possible TNC and has a great variety of shareholders. This guideline sets certain standards concerning the behavior of multinational enterprises operating or interacting with different states.

1.4 Setup of PPPs

International PPPs are classified according to features of performance like Build Operate Transfer (BOT), Build Operate Own (BOO), Build Operate Own Transfer (BOOT), Design Build Finance Operate (DBFO) and Design Build Operate Transfer (DBOT).[172] PPPs can be broken down into two basic categories of cooperation, namely contractual (solely based on a contractual partnership between the public and private sector) and institutional (foundation of a new entity to facilitate the cooperation between both sectors) PPPs, whereby the latter can be again divided into a cooperation model and a joint venture.[173] The entity at the core of the institutional PPP is often called a special purpose vehicle (SPV).[174] Its main task is to supply operating material, also known as feedstock contracts (e.g. water, oil), operate, maintain and sell the product.[175]

1.4.1 Cooperation Model and Joint Venture

The main difference between the cooperation model and the joint venture is in the setup of the SPV. The state must hold 51 % of the SPV's shares within the cooperation model.[176] Within a joint venture, this requirement does not exist (mostly 51 to 49 % in favor of the private sector).[177] The advantage of a joint

---

[171] OECD (2011).

[172] Yescombe (2007), 7–8; Bertelsmann Stiftung et al. (2011), 18; cf. Liu and Cheah (2009), 331 (331); Palmer (2009), 13.

[173] European Commission (2004b), 8–9 and 18; cf. Seidel/Mertens in Dauses (2010), H. IV. Öffentliches Auftragswesen, para 206; cf. Emmerich-Fritsche (2010), slides 11 and 16.

[174] Corbacho and Schwartz (2008), 85 (86); Yescombe (2007), 108–109; Concerning SPV, in: Nickel and Kopf (2004); Kistner and Price (1999), 5; Dorner (1994), 75 (83); cf. HyunChan et al. (1997), 93 (93).

[175] Schoeltzke and Claus (2007), 133 (136) and (139).

[176] Stöcker (2010), 101; At least 50 %, in: Knütel in Siebel et al. (2008), para 1282.

[177] Ehricke (2007), 67 (71); Allen (2001), 12; Yescombe (2007), 321.

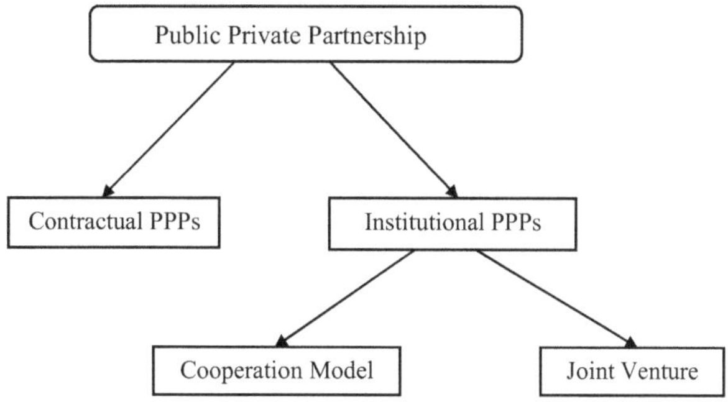

Fig. 1.1 The basic structure of PPPs

venture is that the public sector has better access to information and can thus control the SPV more effectively.[178] Figure 1.1 illustrates the basic setup of PPPs.

1.4.2 Different Types of Contracts Within Contractual and Institutional PPPs

There are numerous types of cooperation within the wide range of PPPs.[179] Besides those mentioned above, there are further PPP forms, such as Design Build Operate (DBO), Lease Own Operate (LOO) or Lease Develop Operate (LDO) and Concession.[180] Instead of using them to distinguish PPPs from one another, they represent the different forms of PPP cooperation and can thus be combined.[181] The most relevant agreements are BOT, BTO (Build Transfer Operate), DBFO (Design-Build-Finance-Operate), BOO, BOOT, PFI (Private Finance Initiative) and Concessions.

The core elements of a BOT agreement combine the responsibilities for design and construction with the operation and maintenance for a certain period of time.[182] At the end of a BOT agreement, there is an obligation to transfer the rights to the operation.[183] Nonetheless, at the end of the contract period, there is also the possibility of returning the task of operation back to the SPV.[184] A BTO agreement

---

[178] Yescombe (2007), 321.

[179] Kumaraswamy and Zhang (2001), 195 (196); Kullack in Heiermann et al. (2008), 1487 para 29; Roquette/Butcher in Roquette and Otto (2005), F. Public–Private Partnership III para 1.

[180] Yescombe (2007), 7–8; Malone (2005), 420 (421); Hathorn (2008), 245 (247).

[181] Yescombe (2007), 8.

[182] Werner and Fiedler (2007), 91 (94); Palmer (2009), 13; Yescombe (2007), 8; Stöcker (2010), 104; Knütel in Siebel et al. (2008), para 1273; Jasper (2001), 51; Bertelsmann Stiftung et al. (2011), 15; Coining part of the BOT, respectively PPP, in: Hunter (2007), 165 (165); Bremer (2005), 23; cf. Bauer (1998), 89 (91).

[183] Palmer (2009), 13; Yescombe (2007), 8; Bremer (2005), 23.

[184] Werner and Fiedler (2007), 91 (94).

is quite similar to a BOT agreement. The difference between both forms of cooperation lies in the fact that a BTO agreement requires transfer of the asset after the completion of construction.[185] After the transfer, a contract between the public authority and the private operator deals with the question of the operation of the asset.[186] The main difference between a BOO agreement and a BOT agreement lies in the fact that a BOO agreement does not include the requirement for the private operator to transfer the company back to the state at the end of the contract.[187] Both forms also differ in cases of a BOOT agreement in which both the property rights and the operation rights are transferred to the state.[188]

A PPP often takes the form of a Design-Build-Finance-Operate (DBFO) scheme, where the government specifies the project and the private company designs, builds, finances and operates it.[189] The private sector does not own the facility; there is just a contract between the two parties, giving the private sector the right to operate the asset.[190] Furthermore, there is a PFI agreement. A PFI agreement is more comparable to a BOT or BOO, where specific performance features shape the appearance of the PPP.[191] PFI and PPP agreements have a lot in common, but a PFI's scope is narrow, making it a special form of a PPP.

Finally, concessions are often the core element of a BOT project,[192] which again underlines that concessions are a part of PPPs.[193] Nowadays, joint ventures are frequently combined with concessions.[194] During the concession period, the concessionaire enjoys privileges over other private operators because the concessionaire is the only one with state permission to fulfill the respective task.[195] For this reason, the object of concession must fall within the exclusive rights of the government.[196] The private operator receives the right to charge the public directly.[197] In return for being able to charge the public (instead of the state), the

[185] Werner and Fiedler (2007), 91 (94); Yescombe (2007), 8; Similar process, but different name (DBO—Design/Build/Operate), in: Palmer (2009), 13; Tettinger (2005), 1 (3); Tettinger (1996), 764 (765); Stöcker (2010), 105; Noltensmeier (2009), 30.

[186] Tettinger (2005), 1 (3); Tettinger (1996), 764 (765); Stöcker (2010), 105; Noltensmeier (2009), 30; Yescombe (2007), 8.

[187] Palmer (2009), 13.

[188] Palmer (2009), 13.

[189] Specification by government is essential: Davies and Eustice (2011), 25; Hemming (2006), 3; Corbacho and Schwartz (2008), 85 (86); Hathorn (2008), 245 (246).

[190] Yescombe (2007), 8; Palmer (2009), 13; cf. Werner and Fiedler (2007), 91 (95).

[191] Metje (2008), 16; Stöcker (2010), 106.

[192] Herdegen (1994), 41 (41); Dorner (1994), 75 (81).

[193] Dorner (1994), 75 (80); Yescombe (2007), 5.

[194] Yescombe (2007), 321.

[195] Metje (2008), 7–8.

[196] Herdegen (2003), 13 (15).

[197] Schede/Pohlmann in Weber et al. (2006), 143; Werner and Fiedler (2007), 91 (99); Jonas and Paulsen (2007), 13 (24); Stöcker (2010), 105; Bertelsmann Stiftung et al. (2011), 16; Roquette/Butcher in Roquette and Otto (2005), F. Public–Private Partnership III para 6; Noltensmeier (2009), 28–29; Matthey (2003), 99 (101).

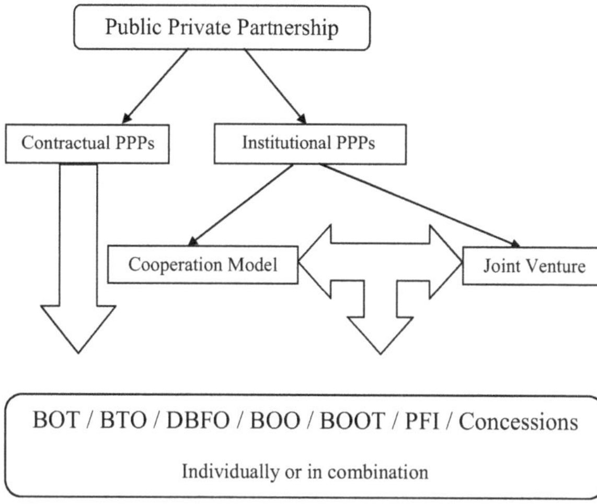

Fig. 1.2 The complete setup of PPPs

private operator must pay a concession fee to the government.[198] At the end of the concession period, there needs to be a clear and detailed agreement on the transfer and whether there is any state payment for the residual value.[199] In cases of a non-competitive award of the concession, only 8 % of all agreements were subject to renegotiation.[200] For a better illustration of PPPs compare Fig. 1.2.

1.5 Best-Suited PPP for the Implementation of the Desertec Concept

The public sector lacks money, is unwilling to bear all the ensuing risks and has limited experience with efficient large scale projects, so most forms of cooperation are unsuitable, leaving only PPPs to implement the project. Since there is no additional data available, this assessment is based on studies that have been published to date.

1.5.1 Contract Partner

In Morocco, l'Office National de l'Electricité (ONE) is in charge as a public monopoly dealing with the generation, transportation and distribution of energy.[201]

[198] Stöcker (2010), 105–106; Koman (2004).

[199] Roquette/Butcher in Roquette and Otto (2005), F. Public–Private Partnership III para 77; Mostly performance of a tender to find the new operator, in: Roquette/Butcher in Roquette and Otto (2005), F. Public–Private Partnership III para 79.

[200] Andres and Guasch (2008), 197 (215).

[201] European Commission (2004a), 20; There is an obligation for ONE to buy the produced energy, in: European Commission (2004a), 20.

Although the Moroccan *loi 13-09 relative aux energies renouvelables* (loi 13-09)[202] is exempt from the ONE monopoly as it allows energy production by third parties,[203] according to Art. 2 of loi 13-09, ONE must guarantee the production of energy (including renewable energy). If the energy production is above 50 MW, it is necessary to conclude a concession with ONE.[204] In cases of private concessions, ONE gives a purchase guarantee.[205] ONE always distributes and sells all of the energy produced,[206] which is also illustrated by the fact that despite loi 13-09, the Moroccan grid is still run as a monopoly.[207]

1.5.2 Institutional PPP and Joint Venture

Only an institutional PPP combined with a joint venture would be a suitable solution to realize the Desertec Project. The foundation of a joint SPV would strengthen the commitment of both parties and a detailed list of control rights (also necessary within a contractual PPP) would shift to the foundation of the SPV. It also has a beneficial effect on the public's perception, as they feel that there is state involvement through the SPV and that the state can actively protect their rights. Studies also revealed that an institutional PPP (e.g. as a joint venture) may have a longer negotiation period, but this allows it to be embedded in the local population.[208] An institutional PPP also affects the creation of jobs. By choosing an institutional PPP, the likelihood that new jobs would be created is three times higher than if other PPP forms were used.[209]

1.5.2.1 Allocation of Shares

Based on the allocation of the shares, there are direct effects on the nationality of the SPV and on the question of dispute resolution. Investment arbitration in particular depends on the nationality of the investor according to Art. 25 of the ICSID Convention as it must differ from that of the host state. On an international level, the theory of incorporation normally prevails and the principle of the main seat of the business (siège social) can be taken alternatively.[210] The theory of

[202] The main idea behind loi 13-09 is to attract new investors and enable BTO projects for renewable energies, in: Sablé and Lauriol (2010).

[203] Le Bihan (2010).

[204] WTO (2009), 97 para 87.

[205] WTO (2009), 97 para 88.

[206] Jaouad (2010).

[207] Jaouad (2010).

[208] Komendantova (2010).

[209] Komendantova (2010).

[210] ICSID [2001] ARB/00/5, 41 para 107; Muchlinski (2009), 341 (348); Metje (2008), 120.

incorporation implies that the place of incorporation or the registered office is essential for the nationality.[211] The main seat of business theory encompasses the assessment of the place of central administration or 'effective seat.'[212] According to the ICJ, the nationality does not depend on the origin of the majority of shareholders.[213] The ICJ underlines this explicitly by mentioning that the effective link theory, which is accepted by the ICJ,[214] cannot be applied to TNCs.[215] Therefore, the ICJ usually checks the nationality according to the incorporation theory.

Within ICSID jurisdiction, the place of incorporation or place of the seat have been accepted as well as applicable methods to determine the investor's nationality.[216] However, Art. 25(2)(b) alternative 2 of the ICSID Convention is an exception to the general rule because it takes the control theory into account,[217] without imposing any specific criteria or tests (e.g. control or place of incorporation) on the contract parties.[218] The idea of the control theory is to lift the 'corporate veil' and assess the controlling shareholders and their nationality.[219] If the place of incorporation or place of the seat theories do not deliver a satisfying result, the only possibility is to check for foreign control, according to Art. 25(2)(b) alternative 2 of the ICSID Convention.[220] The foreign control is checked on the date of submission to the ICSID, as the ICSID Convention does not require a continuous ownership.[221] ICSID tribunals are often unwilling to assess "further layers" beyond

---

[211] Dolzer and Schreuer (2008), 49; Muchlinski (2009), 341 (348); Herdegen (2008), § 15 para 4.

[212] Herdegen (2008), § 15 para 4; Muchlinski (2009), 341 (349).

[213] Pursuant to the gaining of legal independence of the company, there can be no protection by public law because of the control of foreign shareholders, in: ICJ [1970] Barcelona Traction, Light and Power Company, Limited (second phase)—Judgment, 3 (6) para 3, (7) para 8 and (50–51) para 102.

[214] ICJ [1955] Nottebohm Case (second phase)—Judgment, [1955], 4 (22).

[215] Applying diplomatic protection, but not using the effective link theory, in: ICJ [2007] Case concerning Ahmadou Sadio Diallo, Preliminary Objections—General List No. 103, 28 para 64; ICJ [1970] Barcelona Traction, Light and Power Company, Limited (second phase)—Judgement, 3 (42) paras 70–71; This decision was subject to numerous separate opinions, in: ICJ [1970] Barcelona Traction, Light and Power Company, Limited (second phase)—Separate opinion Gerald Fitzmaurice, 64 (79–84) paras 26–34; ICJ [1970] Barcelona Traction, Light and Power Company, Limited (second phase)—Separate opinion Philip C. Jessup, 161 (182–190) paras 38–49; ICJ [1970] Barcelona Traction, Light and Power Company, Limited (second phase)—Separate opinion André Gros, 267 (280–283) paras 22–27.

[216] ICSID [2004] ARB/02/18, 205 (220) para 40; ICSID [2001] ARB/00/5, 41 para 108.

[217] Hobe (2002), 249 (251); Dolzer and Schreuer (2008), 52–53.

[218] ICSID [2010] ARB/07/27, 42–43 paras 156–157; ICSID Convention not defining the terms "nationality" and "foreign control", in: Reed et al. (2004), 17.

[219] Muchlinski (2009), 341 (350).

[220] There is no consistency concerning what "control" actually means, in: Griebel (2008), 66; cf. Knütel in Siebel et al. (2008), para 1271; Opposite opinion that the state will always try to avoid a investor majority, in: Hunter (2007), 165 (167).

[221] ICSID [2010] ARB/07/24—Award, 27–28 para 95; ICSID [2005] ARB/97/3—Decision, 17 para 61; ICJ [1998] Questions of Interpretation and Application of the 1971 Montreal Convention arising from the Aerial Incident at Lockerbie, Judgment—Preliminary Objections, 115 (129) paras 37–38.

the obvious nationality of an investor.[222] Nevertheless, ICSID tribunals look for 'ultimate' or 'real' control of the investor, e.g. if the investor is a Limited, on a second "layer", to enable ICSID jurisdiction.[223] Mr. Weil, who had an opposing opinion, pointed out that he believes that '(...) the origin of the capital is relevant (...) and even decisive.'[224] In situations where control is lacking, ICSID tribunals frequently conclude that they do not have jurisdiction.[225] Within recent ICSID case law, tribunals started to check the question of control of the investor in greater depth and examined the investor's shareholders and the percentage distribution of the shares.[226]

The existing differences on the international scale can also be found on a national level. Based on the applicable jurisdiction it differs if the incorporation theory or the main seat theory is applied.[227] The first is preferred by common law jurisdictions and the later by civil law jurisdiction.[228] Some legal authors suggested that 'certain issues of private international law have been subsumed into an overarching transnational legal regime referred to as the lex mercatoria'.[229]

> However, this is not an uncontroversial concept and the fact remains that significant differences continue to exist between national laws on key substantive and procedural issues pertaining to the international operations of corporations.[230]

It is important to note that even a non-controlling shareholder is an investor if the Bilateral Investment Treaty (BIT) stipulates this.[231] Art. 1(3)(b) of the German–Moroccan Investment Treaty 2004 (G/M-BIT)[232] defines the term investor as every

---

[222] "(A) (...) cut-off point exists beyond which claims would not be permissible as they would have only a remote connection to the affected company (...)", in: ICSID [2004] ARB/01/3—Decision, 21–22 para 52; Prawoko (2005), 143 (148).

[223] ICSID *Noble Energy, Inc. and MachalaPower Cia. Ltda. v The Republic of Ecuador and Consejo Nacional de Electricidad* [2008] ARB/05/12, 28–29 paras 81–83; ICSID *Société Ouest Africaine des Bétons Industriels v Senegal* [1984] ARB/82/1, [1984] ICSID Review, 217 (224–227) paras 34–46.

[224] ICSID *Tokios Tokelės v Ukraine* [2004] ARB/02/18—Dissenting Opinion (Weil), [2004] ICSID Review, 245 (253) para 20.

[225] Prawoko (2005), 143 (147); Nevertheless ICSID case law reveals that there is broad understanding of foreign control, in: Tietje (2005), 47 (54); Overall looser interpretation of investor, in: UNCTAD (2009), 4.

[226] ICSID [2013] ARB/09/8—Award, 36–37 para 136; Extensive check of nationality of the investor, in: ICSID [2013] ARB/10/11 and ARB/10/18—Decision on Jurisdiction, 55–60 paras 184–208; Burden of proof concerning nationality of investor, in: ICSID [2013] ARB/08/9—Decision on Jurisdiction and Admissibility, 99–100 paras 308–312; ICSID [2012] ARB/10/12—Award, 54 para 237.

[227] Muchlinski (2014), para 24.

[228] Muchlinski (2014), para 24.

[229] Muchlinski (2014), para 24.

[230] Muchlinski (2014), para 24.

[231] cf. ICSID [2011] ARB/07/17—Award, 33–34 paras 137–140; ICSID [2010] ARB/02/16—Annulment, 15 para 87; ICSID [2006] ARB/03/15, 42 para 138.

[232] Vertrag zwischen der Bundesrepublik Deutschland und dem Königreich Marokko über die gegenseitige Förderung und den gegenseitigen Schutz von Kapitalanlagen (Traité entre la

company founded according to Moroccan or German company law that has its main seat in the territory of either Morocco or Germany and performs an investment in the other country. This wide understanding of the term investor allows the operation of a German company in Morocco, as well as a Moroccan company founded by German shareholders, to operate in Morocco. In addition, Art. 1(1)(b) of the G/M-BIT stipulates that shares and all other kinds of involvement, including minority shares, are an investment.

Parties should not just agree that the investing company is 'another national' because of foreign control, nor should they ignore this issue completely. Taking the ICJ and ICSID tribunals into account, at the moment is sufficient if the new SPV has its office outside of Morocco and that an investor has at least 50 % +1 share to achieve legal certainty concerning the nationality of the SPV.

1.5.2.2 Moroccan S.A.R.L. or German GmbH

The question of whether a German GmbH or a Moroccan S.A.R.L. is appropriate for this cooperation (also known as an SPV) arises as it is within the interests of the host state to have some influence over the SPV. If there are no divergent terms stipulated in the company agreement, s 47(1) of the German GmbHG regulates that ordinary majority applies. According to s 47(2) of the GmbHG, every Euro of a share grants one vote granting further rights according to s 50(1), 51a(1) and (2), 51b of the GmbHG. Finally, a change to the Articles of Association always requires a 75 % majority according to s 53(2) of the GmbHG.

Prior to an examination of Moroccan company law, it is worth mentioning that there is no requirement to found a company according to Moroccan law.[233] Regulations concerning the establishment, operation and termination of a *société à responsabilité limitée* (S.A.R.L.) are within Loi n° 5–96 *sur la société en nom collectif, la société en commanditesimple, la société en commandite par actions, la société à responsabilité limitée et la société en participation* (loi no. 5–96). According to *loi n° 17–95 relative* aux *sociétés anonyms* (loi no. 17–95), a *société anonyme* (S.A.) is not an option because it is '(. . .) intended for companies of an average size and/or of a family character, where the personality of the contributor is generally more important than the capital contributed.'[234] One benefit of establishing an S.A.R.L. is a "blocking minority", which requires one fourth of the shares, according to Art. 67 of loi no. 5–96 or Art. 71 of loi no. 5–96. Further rights are granted in Art. 75 of loi no. 5–96 and Art. 80 and 82 of loi no. 5–96.

Overall, there is no real difference between a Moroccan S.A.R.L. and a German GmbH. Although there might be more social acceptance of an S.A.R.L. within

République fédérale d'Allemagne et le Royaume du Maroc relatif à l'encouragement et à la protection mutuels des investissements), BGBl 2004 Teil II, 333–340.

[233] El Moukhtari (2010).

[234] Quinn (2009), 6.

Morocco, the abovementioned issue of investor nationality makes a GmbH more preferable. This could avoid questions of effective control of the SPV or even an investor and its shareholders. The risk that an ICSID tribunal could assess these questions in depth is low, but at the same time it is unique project, which might lead to a more detailed review. Finally, the GmbH and S.A.R.L. offer the same rights to all shareholders. Since the investor is raising the necessary funds to invest in a project like the Desertec Concept and will bear most of the risks, they should receive at least 50 %+1 share of the new SPV. If Morocco desires a greater influence, an investor is unlikely to accept these conditions, which would endanger the complete project.

1.5.3 PPP Setup for Desertec

Only two PPPs agreements are suitable to implement the Desertec project. These are BOT/BOOT agreements and concessions. As illustrated above, BOT/BOOT agreements have a lot in common and are quite frequently applied in emerging countries.[235] In cases of energy production, a combination of BOOT models often includes a concession at its core or a global service contract.[236] Morocco also has some BOTs working within its territory.[237] The advantage of BOT/BOOT agreements is that the investor is the owner of the project during the entire contracting period and the host state receives the project at the end of this period. Thus, the concerns and interests of both contracting sides are met. As the choice of agreement—BOT or BOOT—is up to the contracting partner. As BOT and BOOT agreements are both very similar, the next section of this thesis can be applied to both forms of agreements. However, in order to provide a better understanding, only BOT agreements will be discussed...

BOTs are frequently used in combination with a concession in the case of power plants.[238] In Morocco, investors and public parties already have some experience of cooperation in the field of energy, as in the case of *Jorf Lasfar Power Plant Concession* highlights.[239] A concession for the implementation of the Desertec Concept could encompass all of the relevant parts of a BOT. The "build" aspect could be part of the construction contract. The "operate" aspect could be handled by

---

[235] Badawi (2003), 1.

[236] Kistner and Price (1999), 5.

[237] Kistner and Price (1999), 3.

[238] Hunter (2007), 165 (166); Herdegen (2003), 13 (15); Badawi (2003), 3; Sometimes it is even pointed out that concession contracts are repeatedly used for big scale projects, in: Karl (2003), 37 (54); ICSID [2004] ARB/02/5—Decision, 8 para 18 and 11 para 32; ICSID [2001] ARB/98/8—Award, 2 para 3; ICSID [2008] ARB/06/13—Award, 38 para 57 and 39 para 59; Bremer (2005), 60.

[239] La Forge and Nouel (2008), slide 2.

a service concession, which could also include regulations for the "transfer" of the asset.

Finally, it is important that the land on which the plant is being built belongs to the state. The ECJ recently decided a case concerning a concession which was connected to the right of property usage. If the public sector wants to transfer the right of usufruct of a building, it must have that legal right.[240] This is not possible if the right of usufruct is solely ingrained in the right of property.[241] If the owner of the property obeys all the applicable laws, he can use the property in any way he desires.[242] Thus, as long as the private party has the right of usufruct and personal property, the public authority cannot grant a concession of usage.[243]

1.5.4 Assessment of PPP Cooperation

Looking at the contractual structure, the best solution would be an institutional joint-venture PPP with a BOT/concession combination at its core. For a successful PPP, the concession period must be at least 30 years long and, if possible, extend up to 40 years. This would enable an investor to get a return on all its debts and to make a profit. However, the average duration of a concession in Morocco is 20 years.[244] Art. 13 of Loi 13-09 fixes the possible project duration at 25 years, with a possibility of a one-time renewal of 25 years. In practice, these automatically renewable "concessions" already exist.[245] To make the concession even more interesting, it could also include the exclusive right of performance and treat the investor as a preferred candidate for comparable projects in the future.[246] The fact that concessions normally involve a change of consumer (which are normally directly charged) does not exclude its application. With respect to Morocco, it is highly likely that ONE will purchase the energy, which must be negotiated.

Feed-in tariffs have also proven to be highly effective by implementing new energy technologies.[247] Nonetheless, there are no feed-in tariffs in Morocco.[248] Feed-in tariffs are similar to a power purchase agreement (PPA) as they regulate the price of the energy. The nature of a PPA involves a split payment by the private

[240] ECJ [2010] Case C-451/08 preliminary ruling, para 72.

[241] ECJ [2010] Case C-451/08 preliminary ruling, para 73.

[242] ECJ [2010] Case C-451/08 preliminary ruling, para 74.

[243] ECJ [2010] Case C-451/08 preliminary ruling, para 74.

[244] Jaouad (2010).

[245] ICSID [2010] ARB/05/18 and ARB/07/15—Award, 23 para 77.

[246] Already practiced in reality, in: ICSID [2010] ARB/05/18 and ARB/07/15—Award, 24–26 para 79; It can lead to problems, if a comparable product comes up and is favored by the state, in: ICSID [2001] ARB/98/8—Award, 1 paras 1–2.

[247] RAL (2010), 54.

[248] Jaouad (2010); RAL (2010), 54.

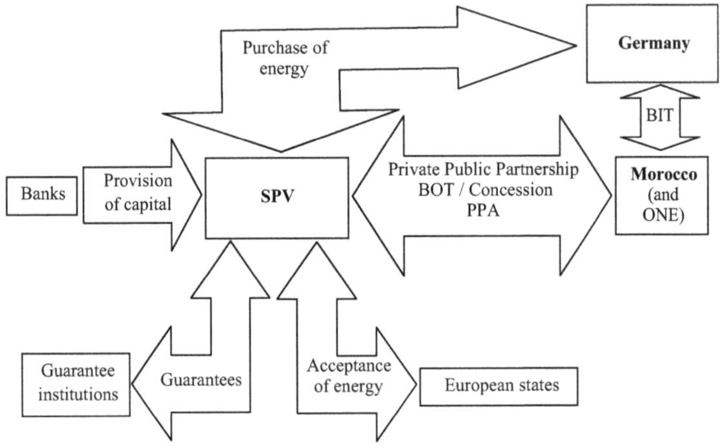

Fig. 1.3 Setup of the cooperation

operator, whereas an Availability Charge ['(...) for making their power station available (...)'] and a Usage Charge ['(...) for the cost of generating power (...)'] accrue.[249] A key element of a PPA lies in the fact that the private operator has the demand risk because the purchaser pays irrespective of whether there is a public need for the energy or not.[250] Morocco does have some experience with PPAs.[251] The difference between a PPA and a feed-in tariff is that the PPA does not regulate the price independently of the demand. It is therefore essential to negotiate an acceptable time frame for both parties and for the regulation to be comparable to a PPA. Finally, it is essential that the SPV does not sell its energy directly to the end consumer as ONE has the exclusive right of sale. Thus, the SPV must negotiate a contract of acceptance of performance (e.g. a PPA between the SPV and ONE). The complete contractual setup is highlighted in Fig. 1.3.

1.5.5 Funding the Desertec Concept

There are several ways to fund this project. Although the public sector will be reluctant or unable to provide extensive funds for the Desertec Concept, they may provide some support. The kingdom of Saudi Arabia and the UAE have mobilized around $800 million dollars to realize solar energy projects within Morocco.[252]

---

[249] RAL (2010), 6.

[250] RAL (2010), 6.

[251] RAL (2010), 24.

[252] Moroccan Investment Development Agency (2014).

The German Federal Ministry for Economic Cooperation and Development (BMZ) has a special budget for supporting PPPs that promote development.[253] As a prerequisite for BMZ funding, this involves cooperation between the private operator and development organizations.[254] Furthermore, there should be Export Credit Agency coverage (ECA coverage).[255] ECA, such as the German Euler Hermes Kreditversicherung-AG (Hermesdeckung), are public export credit insurance companies which promote exports by securing loans.[256] Furthermore, the Clean Technology Fund will guarantee $750 million dollars for its own plans to install a solar power plant in the MENA region. So far, $150 million dollars have been allocated to renewable energy projects in Morocco.[257]

Finally, there is also the possibility of using the Clean Development Mechanism (CDM) that is based on Art. 12 of the Kyoto Protocol. The idea of the CDM is that an investor from an Annex I country (e.g. Germany) will (co-) finance a project in a non-annex I country (e.g. Morocco), and therefore gain certified emission reduction units (CER).[258] Developing countries would benefit accordingly from the assistance, achieve sustainable development and comply with their own emission reduction commitments.[259] In this way, CDM is trying to encourage '(...) environmentally friendly investments (...).'[260] The Kyoto Protocol demands an "additionality", which means that the project will help to reduce emissions. This would not have happened without the CDM.[261]

1.5.6 Guarantees for FDIs

Besides guarantees from the host state to secure the project, there are also other means of protection, such as international guarantee institutions like the Multilateral Investment Guarantee Agency (MIGA).[262] Most importantly, MIGA guarantees are in addition to national guarantees and thus of great relevance vis-à-vis FDIs.[263]

[253] DIHK and bfai (2006), 12.

[254] E.g. GTZ, DEG, SEQUA GmbH, KfW, in: DIHK and bfai (2006), 6 and 12.

[255] Schoeltzke and Claus (2007), 133 (137).

[256] LandesBank Berlin (2014); Importance of insurance and list of other German insurance companies, in: Metje (2008), 63; Concerning Hermes Kreditversicherungs AG and Treuarbeit AG, in: Herdegen (2003), 13 (28).

[257] Climate Investment Fund (2014).

[258] Winter (2009), 289 (290); Byod et al. (2007), 7; Böhringer (2009), 25 (35–36).

[259] Huang and Barker (2009), 2; The CERs can be used to meet Kyoto standards of annex I countries, in: Hepburn (2007), 375 (379).

[260] Huang and Barker (2009), 3.

[261] Byod et al. (2007), 5; Winter (2009), 289 (289).

[262] MIGA is one of the five World Bank Group institutions, in: MIGA (2013a), 5.

[263] Accpeted world wide, in: Lowenfeld (2008), 588–591; Krajewski (2009), 180.

According to Art. 2 of the Convention Establishing the Multilateral Investment Guarantee Agency (MIGA Convention), the objective is 'to encourage the flow of investments for productive purposes among member countries, and to develop member countries in particular, thus supplementing the activities of the (. . .) (World Bank) (. . .).' Therefore, Art. 2(a) of the MIGA Convention stipulates that the agency issues '(. . .) guarantees (. . .) against non-commercial risks in respect of investments in a member country which flow from other member countries (. . .).' These non-commercial risks are currency transfer, expropriation and similar measures, breach of contract, war and civil disturbance, according to Art. 11(a)(i to iv) of the MIGA Convention. Under Art. 13(a)(ii) of the MIGA Convention, any judicial person can receive the Agency's guarantee, as long as '(. . .) (it is) incorporated and has its principal place of business in a member or the majority of its capital is owned by a member or members or nationals thereof (. . .).' Art. 12 of the MIGA Convention regulates eligible investments, which also includes FDIs. Art. 12(d) of the MIGA Convention mentions that certain investment requirements for the MIGA Convention will be applicable. These requirements include 'the economic soundness of the investment and its contribution to the development of the host country' (Art. 12(d)(i) of the MIGA Convention), the 'compliance of the investment with the host country's laws and regulations' (Art. 12(d)(ii) of the MIGA Convention), the 'consistency of the investment with the declared development objectives and priorities of the host country' (Art. 12(d)(iii) of the MIGA Convention) and, finally, 'the investment conditions in the host country, including the availability of fair and equitable treatment and legal protection for the investment' (Art. 12(d)(iv) of the MIGA Convention).

Morocco and Germany are contracting states to the MIGA Convention.[264] According to Art. 13(a)(ii) of the MIGA Convention, the abovementioned SPV would be an investor. The scope and application of Art. 12(d)(ii) and (iii) of the MIGA Convention will be discussed later on in this thesis.

1.5.7 The Desertec Concept and Risks

Like every large scale project, the Desertec Concept faces a lot of different risks. This assessment will only cover severe risks, which are political instability, state accountability and transparency.[265]

[264] MIGA (2013b).

[265] Joffé (2000), 33 (45).

1.5.7.1 The Legal, Political and Economic Situation in Morocco

Germany and Morocco have a good relationship and the market potential for renewable energies in Morocco is good as Mohammad VI's goal is to increase renewable energy.[266] Most of the Moroccan market is liberalized, with only a few products under state influence (e.g. petrol).[267] Morocco already exports electricity to Spain and wants to increase that amount.[268] Morocco has also signed a memorandum of understanding on integration of the Maghreb electricity market and a possible integration into the EU internal market.[269] However, regulations and bureaucracy continue to be obstacles in Morocco.[270]

Nevertheless, there are drawbacks due to the country's political stability. Within the MENA region, there are a large number of young people entering the job market at the same time, which has led to employment problems.[271] The majority of Moroccans live in an unstable and insecure situation. Along with a high rate of illiteracy (43 % over the age of ten), the unemployment rates are high. Statistics show that in 2005 the urban unemployment rate was 18.4 % and that there was 32.7 % unemployment among young people.[272]

In this context, the Desertec Concept is a double-edged sword: on the one hand, it could help to establish new jobs and thus ease unemployment; on the other hand, it would be risky to build a large-scale project in an area with significant social problems. The situation in Morocco is slightly different compared to other MENA states as King Mohammed VI started to modernize the country prior to the revolutions. Among other things, this involved liberalization. As a result, tensions within the population are not as heated as in other MENA states. In addition, the Maghreb countries face difficult economic challenges, e.g. lack of exports and low economic growth rates.[273] The fact that there is almost no trade within the Maghreb region illustrates that the region is very poorly integrated on a regional scale.[274]

Overall, there are doubts within the MENA region. For example, Algerian officials have expressed concerns about the use of the term "partnerships" as they fear that there will be a concentration on Europe.[275] Similar concerns have been voiced in Morocco as it is sometimes said that a PPP cannot always be understood

---

[266] DIHK and bfai (2006), 45 and 46.

[267] European Commission (2004a), 14.

[268] European Commission (2004a), 20.

[269] European Commission (2004a), 20; European Commission (2010a), 4 and 27.

[270] European Commission (2004a), 15.

[271] DIHK and bfai (2006), 30.

[272] European Neighbourhood and Partnership Instrument (2007), 14–15; Illiteracy rate is the highest in all Arab countries, in: Alonsoa and Reya (2007), 571 (575); High unemployment, in: Brenton et al. (2006), 3.

[273] Brenton et al. (2006), 1 and 3.

[274] Brenton et al. (2006), 4.

[275] Lowe (2010).

as a partnership.[276] There seems to be reluctance in relation to incorporating an international investor into already existing solar thermal projects as Morocco wants to realize its own solar projects.[277]

1.5.7.2 Terrorism

Since the terrorist attacks of 9/11, there has been increased violence in the name of *jihadism*.[278] Among other things, one "branch" of the new wave of extremists promotes violence against Islamic countries that '(. . .) are regarded as "apostate", such as Morocco.'[279] The terrorist attacks in Casablanca in 2003 revealed that Morocco has also been subjected to *jihadism*, and the fact that all 14 terrorists stemmed from Casablanca's worst slums (*Sidi Moumen*) underlines the danger.[280] Surviving terrorists confirmed that they committed the terrorist attacks to weaken the Moroccan government.[281] This is also illustrated by the terror attacks in Marrakech in 2011, which killed 18 people. There are indications that this attack was closely related to dissatisfaction with reform processes in Morocco.[282] Furthermore, newly formed or renamed terrorist groups like the *Groupe Salafiste pour la Prédication et le Combat* (GSPC/Algerian Salafist Group for Preaching and Combat) (since 2007, the *Qaedat al-Jihad fi Bilad al-Maghrib al-Islami—al-Qaeda* in the Islamic Maghreb) operate in North Africa.[283] The GSPC was also responsible for several suicide bombings in Morocco after the 2003 bombings, which resulted in minor damage.[284]

In addition, certain socio-economic factors may increase terrorism in Morocco.[285] Such factors include the high unemployment rate and the growing deterioration of living conditions in the slums.[286] Amongst the youth of Morocco, there is also an increasing acceptance of al-Qaeda. Forty-four percent of young

276 Boutayeb (2008), slide 13.

277 Jaouad (2010).

278 Alonsoa and Reya (2007), 571 (571).

279 In addition, Osama Bin Laden identified Morocco as a terrorist target in 2003, in: Alonsoa and Reya (2007), 571 (571–572).

280 Alonsoa and Reya (2007), 571 (572).

281 In addition, terrorists struck allegedly enemies of Moslems (Jews) and place hazardous to the Islamic beliefs (place where alcohol is severed or casinos), in: Alonsoa and Reya (2007), 571 (576–577).

282 Zeit Online (2011).

283 Steinberg and Werenfels (2007), 407 (407).

284 Steinberg and Werenfels (2007), 407 (411).

285 This also took place in accordance with King Hassan II, among other things to prevent spread of revolutionist ideas, in: Alonsoa and Reya (2007), 571 (572–573).

286 Alonsoa and Reya (2007), 571 (575).

Moroccans do not regard al-Qaeda as a terrorist organization.[287] More young Moroccans are joining terrorist groups in neighboring countries like Algeria.[288]

Despite the abovementioned problems, it needs to be recognized that there is a widespread rejection of terrorist violence within Morocco.[289] Morocco has become a fertile ground for terrorism, but terrorists are reluctant to act internally.[290] As a consequence of the Casablanca bombing, Morocco has increased the protection of possible terrorist targets.[291] New forms of political treatment and antiterrorist legislation have been passed to fight terrorism.[292]

1.5.7.3 Moroccan Legal Framework

Moroccan law is very Western orientated,[293] but the king still holds a lot of power.[294] An examination of the Moroccan legal framework reveals that it shares a number of similarities with French law.[295] This is mostly due to Morocco's status as a former French colony. Over the last few years, the Moroccan legal system has been modernized.[296] The handling of investment issues and (international) arbitration in particular are now subject to new regulations.[297] However, there is still considerable discussion about the role and importance of *Shari'a* law within Morocco.[298] The fact that the king is also overseeing the Ministry of Justice as the 'Commander of the Faithful' (considered an important religious duty) highlights the extent to which law and religion are still intertwined.[299] Like the Japanese, Arab parties regard the contract as "open-ended" and thus '(...) rely

[287] Alonsoa and Reya (2007), 571 (577).

[288] Steinberg and Werenfels (2007), 407 (411).

[289] Alonsoa and Reya (2007), 571 (578).

[290] Furthermore, Moroccans actively participated in the Spanish train bombings in 2004, in: Alonsoa and Reya (2007), 571 (579) and (580–581).

[291] Alonsoa and Reya (2007), 571 (584).

[292] These new regulations are closely connected to allegations of mistreatment and torture, in: Alonsoa and Reya (2007), 571 (584); However the whole regency of King Hassan II is closely associated with dictatorship, violent repression and torture, in: Kausch (2009), 165 (165); King Hassan II role, in: Sater (2009), 181 (183); There is also the fear that these new strict regulations lead to a better organization of terroristic groups, in: Alonsoa and Reya (2007), 571 (584–585); New political treatment, in: Alonsoa and Reya (2007), 571 (585–586).

[293] Kutty (2006), 565 (595); Darwazeh and El-Kosheri (2008), 203 (203).

[294] Sater (2009), 181 (181–182); Joffé (2009), 151 (151–152).

[295] Concerning arbitration laws, in: Hammoud and Houerbi (2008), 231 (235).

[296] WTO (2009), 12 para 60; European Neighbourhood and Partnership Instrument (2007), 13; For example the establishment of new commercial courts after a big corruption scandal to secure legal protection of the private sector, in: Sater (2009), 181 (184).

[297] WTO (2009), 12 para 61.

[298] Sater (2009), 181 (182).

[299] King as religious "controller", in: Sater (2009), 181 (187).

more on oral commitments and permanent accommodations (. . .)' than the written contract.[300] Questions also remain about the existence and recognition of the rule of law.[301] Good governance and the rule of law are important prerequisites for a successful investment.[302] There are plans to unify PPP regulations, but so far no implementation has taken place.[303]

There is the general issue of foreign property rights, which have improved over the past decades but still face problems in the MENA region.[304] Due to the close relationship with the French legal system, a short analysis of French regulations on PPPs might be helpful. France has laws for a delegated administration called *Loi sur la Gestion Déléguée*, which, for example, stipulates the fair allocation of risks within a PPP project.[305] The *Loi sur la Gestion Déléguée* does not expressively forbid state purchase of the product.[306]

As mentioned above, loi 13-09 regulates all matters involving renewable energies and is therefore highly relevant for the implementation of the Desertec Concept. Art. 1(1) of loi 13-09 stipulates that solar energy is subject to this law. All renewable power plants that produce more than 2 MW need authorization as per Art. 3 of loi 13-09. Authorizations are not transferable to anyone else (Art. 15 of loi 13-09). Art. 5 of loi 13-09 stipulates the possibility to connect different plants to the local grid of Morocco. Any change in the project requires authorization by the respective administration (Art. 17 of loi 13-09). The primary market for any project will be Morocco (Art. 25–26 of loi 13-09). Nevertheless, there is the possibility of exporting energy, as regulated in Art. 27–30 of loi 13-09.

MENA countries lack experience with new types of sponsors, business models, markets and technology. Such inexperience creates risks for an investor.[307] However, there are positive movements within Morocco and its legal framework. Issues concerning international arbitration will be assessed later on in this thesis.

1.6 Conclusion

This introduction illustrates that it is possible to implement the Desertec Concept through a PPP. Such a PPP must include the foundation of an SPV, including the state of Morocco as a minority shareholder, and should also be set up as a

300 Berger (2009), 217 (229–230); cf. Kutty (2006), 565 (609–611).

301 Sater (2009), 181 (185–188).

302 Joffé (2000), 33 (44).

303 Also point out that there should be no infringement of local rights and laws, in: Boutayeb (2008), slide 12.

304 Joffé (2000), 33 (40–41).

305 Talbi (2008), slide 1.

306 La Forge and Nouel (2008), slide 4.

307 RAL (2010), 50.

BOT/concession agreement. Besides the PPP, a PPA should be concluded. The price for the produced energy is likely to be higher than the average Moroccan energy costs, so this new industry needs help at the beginning of its life. Furthermore, subsidies might be another way of helping an SPV to start up. All these possible tools need to be in accordance with the law. The concession period is closely connected to the charged price. Although the figures are not complete, the concession period should be at least 30 years. Within this time, the SPV should be able generate enough money to pay off all its debts and generate a profit. Due to the small amount of information available, this is just a prediction, especially given that there is a lack of detailed information on the intended energy supply for the MENA region.

The Moroccan economy and currency are healthy at the moment, which makes it unsurprising that Dii chose Morocco as initial potential partner. There is a great chance that any investor would have chosen Morocco too. Although Morocco is a kingdom, the political system is mostly free. Stability is an issue that requires further observation. The continuing revolutions and even civil wars within the MENA area might be the deciding factor when it comes to the actual implementation of the Desertec Concept. The lack of a Moroccan PPP guideline is a drawback, but, as Moroccan authors have pointed out, there is an awareness of important PPP regulations. Due to the close relations with the French legal system, French regulations might help to solve problems or deliver ideas.

Corruption is still a major problem in Morocco. Combined with the large-scale project and the great amount of money connected with it, it is important to prevent any form of corruption. Suspicion of corruption can lead to a decline in public acceptance. Indeed, some North African authors have already mentioned the tensions that currently exist. To overcome Moroccan or African reservations, the foundation of a "Moroccan" SPV is another step to demonstrate an investor's desire for cooperation.

Overall, there are some risks surrounding the PPP, but they are not extremely high or different from the risks faced by other PPP projects. The abovementioned risks, mistakes, concerns, suggested solutions and positive examples should function as a checklist or catalog for the implementation of the Desertec Concept. There are no unified rules or requirements for PPPs. It is essential to take all positive PPPs into account, especially in the case of African PPPs. All in all, the Desertec Concept has the potential to be extremely successful.

References

Ahadzi M, Bowles G (2004) Public–private partnerships and contract negotiations: an empirical study. Constr Manag Econ 22:967–978

Allen G (2001) The private finance initiative (PFI). Available via research paper 01/117, House of Commons Library. http://www.parliament.uk/documents/commons/lib/research/rp2001/rp01-117.pdf. Accessed 30 Apr 2014

Alonsoa R, Reya MG (2007) The evolution of Jihadist terrorism in Morocco. Terrorism Polit Violence 19:571–592

Andres LA, Guasch LJ (2008) Negotiating and renegotiating PPPs and concessions. In: Schwartz G, Corbacho A, Funke K (eds) Public investment and public–private partnerships. New York, pp 197–227

Badawi I (2003) Globalization of the B.O.T. system and its taxation problems. Available via review of business. http://www.allbusiness.com/accounting/550780-1.html. Accessed 21 Apr 2010

Bauer H (1998) Verwaltungsrechtliche und verwaltungswissenschaftliche Aspekte der Gestaltung von Kooperationsverträgen bei Public Private Partnership. DÖV 89–97

Baur J, Blask H (2002) Regelungszuständigkeit der EG im Bereich Energie. Energiewirtschaftliche Tagesfragen, 636–641

Becher G (2007) Risikobewertung im PPP-Wirtschaftslichtkeitsvergleich aus der Sicht des privaten Partners. In: Nicklisch F (ed) Betreibermodelle – BOT/PPP-Vorhaben im In- und Ausland. München, pp 55–61

Berger K-P (2003) Äquivalenzstörungen, Neuverhandlungsklauseln und Vertragsanpassung bei internationalen Konzessionsverträgen sowie Probleme der Streitschlichtung. In: Baur JF, Hobe S (eds) Rechtsprobleme von Auslandsinvestitionen. Baden-Baden, pp 65–97

Berger K-P (2009) The international arbitrator's dilemma: transnational versus home jurisdiction. Arbitr Int 25:217–238

Bertelsmann Stiftung, Clifford Chance Pünder and Initiative D21 (2011) Prozessleitfaden Public Private Partnership. http://www.bertelsmann-stiftung.de/bst/de/media/xcms_bst_dms_18439_18440_2.pdf. Accessed 30 Apr 2014

Biobay.de (2009) Initiative für Solar-Strom aus Afrika gegründet. http://www.biobay.de/news/desertec-initiative-fuer-solar-strom-aus-afrika. Accessed 15 Oct 2010

Blanchard P (1998) Contractual structures which guarantee the internal profitability of independent power production projects in developing countries. Int Bus Law 4(5):417–476

Böhringer A-M (2009) Die Bedeutung der Klimaschutzkonvention und des Kyoto-Protokolls für die Energiewirtschaft. In: Ruppel N (ed) Europäisches und Internationales Energiewirtschaftsrecht. Stuttgart, pp 25–59

Boutayeb N-E (2008) PPP et services publics locaux au Maroc. Available via presentation at public private partnerships symposium Rabat (Morocco) within the OECD framework of the good governance for development (GfD) in Arab countries initiative. http://www.oecd.org/dataoecd/31/44/41751561.ppt. Accessed 01 Oct 2010

Bremer BG (2005) Public private partnership. Köln

Brenton P, Baroncelli E, Malouche M (2006) Trade and investment integration of the Maghreb, Middle East and North Africa. The World Bank working paper series no. 44

Bundesumweltministerium (2005) Studie: Potenziale für erneuerbare Energien im Mittelmeerraum. http://www.bmu.de/erneuerbare/energien/doc/35439.php. Accessed 25 June 2010

Bungenberg M (2009) Europäische Wirtschaftsverfassung zwischen Freiheit und Regulierung am Beispiel des Umweltschutzes. In Fastenrath U, Nowak C (eds) Der Lisabonner Reformvertrag. Berlin, pp 205–217

Byod E, Hultman NE, Roberts T, Corbera E, Ebeling J (2007) The clean development mechanism: an assessment of current practice and future approaches for policy. Available via Tyndall – working paper 114. http://www.tyndall.ac.uk/sites/default/files/wp114.pdf. Accessed 01 May 2014

Christiansen H (2008) The OECD principles for private sector participation in infrastructure. In: Schwartz G, Corbacho A, Funke K (eds) Public investment and public–private partnerships. New York, pp 143–161

Climate Investment Fund (2014) Morocco. https://www.climateinvestmentfunds.org/cifnet/?q=country/morocco. Accessed 31 Mar 2014

Corbacho A, Schwartz G (2008) PPPs and fiscal risks: should governments worry? In: Schwartz G, Corbacho A, Funke K (eds) Public investment and public–private partnerships. New York, pp 85–104
Corner D (2006) The United Kingdom Private Finance Initiative: the challenge of allocating risk. OECD J Budg 5:37–55
Darwazeh N, El-Kosheri AS (2008) Arbitration in the Arab World: an interview with Professor Ahmed Sadek Wl-Kosheri. J Int Arbitr 25:203–210
Dauses MA (ed) (2010) EU-Wirtschaftsrecht, 26th supp. München
Davies P, Eustice K (2011) Delivering the PPP promise – a review of PPP issues and activity. Available via PwC. http://www.pwc.com/en_GX/gx/government-infrastructure/pdf/promisereport.pdf. Accessed 30 Apr 2014
Davis K (2005) PPPs and infrastructure investment. Aust Econ Rev 38:439–444
Desertec Foundation (2009a) Aufgaben & Ziele. http://www.desertec.org/de/stiftung/aufgaben-ziele/. Accessed 25 June 2010
Desertec Foundation (2009b) Von der Vision zur Realität. http://www.desertec.org/de/konzept/status-quo/. Accessed 24 June 2010
Desertec Foundation (2009c) Zusammenfassung. http://www.desertec.org/de/konzept/zusammenfassung/. Accessed 25 June 2010
Desertec Foundation (2013) Pressemitteilung. http://www.desertec.org/de/presse/pressemitteilungen/130701-die-desertec-stiftung-verlaesst-das-industriekonsortium-Dii/. Accessed 24 Mar 2014
Desertec Foundation (2014a) Netzwerke. http://www.desertec.org/de/organisation/netzwerk//. Accessed 24 Mar 2014
Desertec Foundation (2014b) Red paper, 3rd edn. http://www.desertec.org/fileadmin/downloads/desertec-foundation_redpaper_3rd-edition_deutsch.pdf. Accessed 24 Apr 2014
Desertec Industries (2010a) About us. http://www.dii-eumena.com/home/about-us.html. Accessed 29 June 2010
Desertec Industries (2010b) Our mission. http://www.dii-eumena.com/home/our-missioparahtml. Accessed 29 June 2010
Desertec Industries (2014) Partners. http://www.Dii-eumena.com/about-us/partners.html. Accessed 24 Mar 2014
Desertec-UK (2010) The Desertec CONCEPT and Desertec-UK. http://www.trec-uk.org.uk/. Accessed 24 Apr 2014
Deutsch M, Hobohm J (2010) Strom aus der Wüste – Ja, aber aus welcher Quelle. Energiewirtschaftliche Tagesfragen 54–59
Deutsches Zentrum für Luft- und Raumfahrt (2010) DESERTEC: Solarstrom aus der Wüste. http://www.dlr.de/desktopdefault.aspx/tabid-5885/9548_read-18787/. Accessed 03 Oct 2010
Deutsche Industrie- und Handelskammer in Marokko (2008) Marktprofil Marokko. http://www.een-bayern.de/een/inhalte/Laender/Anhaenge/Marktprofil-Marokko.pdf. Accessed 17 July 2010
Deutsche Physikalische Gesellschaft e.V. (2005) Klimaschutz und Energieversorgung in Deutschland 1990–2020. http://www.dpg-physik.de/static/info/klimastudie_2005.pdf. Accessed 29 Apr 2014
Deutscher Bundestag (2005) BTDrucks. 15/5668
Deutscher Bundestag (2009) BTDrucks. 16/12363
Diederichs J (2009) Wirtschaftlichkeitsuntersuchungen bei PPP-Projekten. NZBau 547–552
DIHK and bfai (2006) Mit Public Private Partnerships nach Nahost und Nordafrika. Berlin
Dolzer R, Schreuer C (2008) Principles of International Investment Law. Oxford
Dorner G (1994) BOT-Projekte in der nationalen und internationalen Praxis – Wechselwirkung durch Vertragsvielfalt und Mehrparteienverbund. In: Nicklisch F (ed) Rechtsfragen privatfinanzierter Projekte. Heidelberg, pp 75–89
Düren M (2008) Die Zukunft der Energieversorgung. Spiegel der Forschung 6–10

Ehricke U (2007) Equity-Joint Venture-Verträge als Instrument zur Sicherung der Versorgung mit ausländischen Energieträgern. In: Leible S, Lippert M, Walter C (eds) Die Sicherung der Energieversorgung auf globalisierten Märkten. Tübingen, pp 67–80

EIB (2004) The EIB's role in public–private partnerships (PPPs). http://www.eib.org/attachments/thematic/eib_ppp_eparapdf. Accessed 20 Sept 2010

El Moukhtari G (2010) Assistante du Directeur Général AHK Marokko. Email received 08 Oct 2010

Emmerich-Fritsche A (2010) Privatisierung. Available via Vorlesung Öffentliches Wirtschaftsrecht. http://www.rph1.jura.uni-erlangeparade/material/oeffwirt/oeffwirt12.pdf. Accessed 01 Oct 2010

Estache A, Juan E, Trujillo L (2007) Public–private partnerships in transport. Available via policy research working paper 4436 – The World Bank. http://www-wds.worldbank.org/servlet/WDSContentServer/WDSP/IB/2007/12/12/000158349_20071212085739/Rendered/PDF/wps4436.pdf. Accessed 30 Apr 2014

European Commission (2004a) European neighbourhood policy – country report Morocco – commission staff working paper SEC(2004) 569. http://ec.europa.eu/world/enp/pdf/country/morocco_enp_country_report_2004_eparapdf. Accessed 07 Sept 2010

European Commission (2004b) Green paper on public-private partnerships and community law on public contracts and concessions – COM(2004) 327 final

Europäische Kommission (2006) Grünbuch: Eine europäische Strategie für nachhaltige, wettbewerbsfähige und sichere Energie. KOM(2006) 105 endgültig, Brüssel

European Commission (2010a) ENP action plan Morocco. http://ec.europa.eu/world/enp/pdf/action_plans/morocco_enp_ap_final_eparapdf. Accessed 04 Oct 2010

European Commission (2010b) Morocco: EU Commission approves €135 million in grants for 2010. http://www.europa-eu-uparaorg/articles/en/article_10011_eparahtm. Accessed 08 Sept 2010

European Commission (2010c) NIF – projects in the southern region. http://ec.europa.eu/europeaid/where/neighbourhood/regional-cooperation/irc/investment_projects_south_en.htm. Accessed 24 Apr 2014

European Neighbourhood and Partnership Instrument (2007) Morocco – strategy paper 2007 – 2013. http://eeas.europa.eu/enp/pdf/pdf/country/enpi_csp_morocco_en.pdf. Accessed 24 Apr 2014

FAZ.Net (2014a) Strom aus der Wüste. http://www.faz.net/s/RubC9401175958F4DE28E143E68888825F6/Doc~EC2C0E78A39654406A48652CCC01AF374~ATpl~Ecommon~Scontent.html. Accessed 24 Apr 2014

FAZ.Net (2014b) Auch Eon steigt aus Desertec aus. http://www.faz.net/aktuell/wirtschaft/auch-eon-steigt-aus-desertec-aus-12891203.html. Accessed 14 Apr 2014

FAZ.Net (2014c) Auch Bilfinger verlässt Desertec. http://www.faz.net/aktuell/wirtschaft/wirtschaftspolitik/auch-bilfinger-verlaesst-wuestenstromprojekt-desertec-12894051.html. Accessed 14 Apr 2014

FAZ.Net (2014d) Der Traum vom Wüstenstrom ist gescheitert. http://www.faz.net/aktuell/wirtschaft/wirtschaftspolitik/wuestenstrom-projekt-desertec-ist-gescheitert-13207437.html. Accessed 14 Nov 2014

Fourie FCN, Burger P (2001) Fiscal implications of public–private partnerships (PPPs). S Afr J Econ 69:147–168

German Aerospace Center (DLR), Institute of Technical Thermodynamics, Section Systems Analysis and Technology Assessment (2005) Concentrating solar power for the Mediterranean region. http://www.dlr.de/tt/Portaldata/41/Resources/dokumente/institut/system/projects/MED-CSP_Full_report_final.pdf. Accessed 24 Apr 2014

German Aerospace Center (DLR), Institute of Technical Thermodynamics, Section Systems Analysis and Technology Assessment (2006) Trans-Mediterranean interconnection for

concentrating solar power. http://www.dlr.de/tt/Portaldata/41/Resources/dokumente/institut/system/projects/TRANS-CSP_Full_Report_Final.pdf. Accessed 24 Apr 2014
German Aerospace Center (DLR), Institute of Technical Thermodynamics, Section Systems Analysis and Technology Assessment (2007) Concentrating solar power for seawater desalination. http://www.dlr.de/tt/Portaldata/41/Resources/dokumente/institut/system/projects/aqua-csp/AQUA-CSP-Full-Report-Final.pdf. Accessed 24 Apr 2014
Griebel J (2008) Internationales Investitionsrecht. München
Hammoud L, Houerbi S (2008) ICC arbitration in the Arab World. J Int Arbitr 25:231–240
Happ R (2002) Investitionsschiedsverfahren unter dem Energiechartavertrag: AES Summit Generation/Ungarn. RdE 39–44
Harries H (1996) Sicherungsstrukturen der Projektfinanzierung. In: Nicklisch F (ed) Partnerschaftliche Infrastrukturprojekte. Heidelberg, pp 19–27
Hathorn M (2008) IPSASB: service concession arrangements. In: Schwartz G, Corbacho A, Funke K (eds) Public investment and public–private partnerships. New York, pp 245–255
Heiermann W, Riedl R, Rusam M, Kuffer J (eds) (2008) Handkommentar zur VOB, 11th edn. Wiesbaden
Hemming R (2006) Public–private partnerships. Available via realizing the potential for profitable investment in Africa seminar. https://www.imf.org/external/np/seminars/eng/2006/rppia/pdf/hemmiparapdf. Accessed 20 Sept 2010
Hemming R (2008) PPPs: some accounting and reporting issues. In: Schwartz G, Corbacho A, Funke K (eds) Public investment and public–private partnerships. New York, pp 235–244
Hepburn C (2007) Carbon trading: a review of the Kyoto mechanism. Annu Rev Environ Resour 32:375–393
Herdegen M (1994) Der Konzessionsvertrag aus öffentlich-rechtlicher Sicht: das Beispiel des Kanaltunnelprojekts. In: Nicklisch F (ed) Rechtsfragen privatfinanzierter Projekte. Heidelberg, pp 41–51
Herdegen M (2003) Rechtsprobleme des internationalen Konzessionswesens. In: Baur JF, Hobe S (eds) Rechtsprobleme von Auslandsinvestitionen. Baden-Baden, pp 13–35
Herdegen M (2008) Internationales Wirtschaftsrecht, 7th edn. München
Hesse J (1994) Fragen der Projektfinanzierung aus der Sicht des Anlagenerrichters. In: Nicklisch F (ed) Rechtsfragen privatfinanzierter Projekte. Heidelberg, pp 31–39
Heun-Rehn S, Dratwa F (2010) In varietate concordia – Strategie und Ziele der "neuen" EU-Energie und Klimapolitik. In: Dratwa F, Ebers M, Pohl A, Spiegel B, Strauch G (eds) Energiewirtschaft in Europa. Heidelberg, pp 101–125
Hobe S (2002) Enteignung zu Zwecken des Umweltschutzes: Zur ICSID Compania del Desarollo de Santa Elena, S.A. und Republik von Costa Rica (ILM 39 [2000], 1317). IPRax, pp 249–251
Höfler H (2007) Risikoallokation als Erfolgsfaktor im PPP-Projekt. In: Nicklisch F (ed) Betreibermodelle – BOT/PPP-Vorhaben im In- und Ausland. München, pp 35–54
Huang Y, Barker T (2009) The clean development mechanism and sustainable development: a panel data analysis, Tyndall Centre for Climate Change Research. Available via Tyndall – working paper 130. http://www.tyndall.ac.uk/sites/default/files/wp130.pdf. Accessed 01 May 2014
Hunter R (2007) Investitionsschutz und Streitbeilegung bei BOT-Projekten im Ausland. In Nicklisch F (ed) Betreibermodelle – BOT/PPP-Vorhaben im In- und Ausland. München, pp 165–174
HyunChan C, Masahiko K, Tsuneaki Y (1997) Build Operate Transfer (BOT) project financing in Asian developing countries, pp 93–104. http://library.jsce.or.jp/jsce/open/00524/1997/05-0093.pdf. Accessed 30 Apr 2014
Irwin T (2008) Controlling spending commitments in PPPs. In: Schwartz G, Corbacho A, Funke K (eds) Public investment and public–private partnerships. New York, pp 105–117
Jasper J-E (2001) Privatisierung und EG-Vergaberecht. Osnabrück
Joffé G (2000) Foreign investment and the rule of law. Mediterr Polit 5:33–49
Joffé G (2009) Morocco's reform process: wider implications. Mediterr Polit 14:151–164

Jonas I, Paulsen H (2007) PPP: Begriffe, Modelle, Erfolgsvoraussetzungen, Risiken. In: Nicklisch F (ed) Betreibermodelle – BOT/PPP-Vorhaben im In- und Ausland. München, pp 13–33
Karl J (2003) Aktuelle Probleme des Investitions- und Enteignungsschutzes am Beispiel des Internationalen Konzessionsvertrages. In: Baur JF, Hobe S (eds) Rechtsprobleme von Auslandsinvestitionen. Baden-Baden, pp 37–61
Kausch K (2009) The European Union and political reform in Morocco. Mediterr Polit 14:165–179
Kistner R, Price HW (1999) Financing solar thermal power plants. Prepared for the proceedings of the ASME renewable and advanced energy systems for the 21st century conference. http://www.osti.gov/bridge/servlets/purl/3532-YwOlGi/webviewable/3532.pdf. Accessed 30 Apr 2014
Klinski S (2005) Zur Vereinbarkeit des EEG mit dem Elektrizitätsbinnenmarkt – Neubewertung unter Berücksichtigung der Richtlinien 2003/54/EG und 2001/77/EG. ZNER 207–215
Knies G (2008) Das Desertec Konzept. http://www.lohas.de/component/option,com_docman/task,doc_download/gid,225/Itemid,102/. Accessed 25 June 2010
Koman A (2004) Das Grünbuch der Kommission über Public Private Partnerships: ein weiterer Schritt auf dem Weg zu größerer Rechtssicherheit. ZfBR, via juris
Krajewski M (2009) Wirtschaftsvölkerrecht, 2nd edn. Heidelberg
Kumaraswamy M, Zhang XQ (2001) Governmental role in BOT-led infrastructure development. Int J Proj Manag 19:195–205
Kutty F (2006) The Shari'a factor in International Commercial Arbitration. Loyola Los Angel Int Comp Law Rev 28:565–624
La Forge H, Nouel GL (2008) Methods for comparing PPPs and other procurement approaches – The Moroccan experience. Available via presentation at public private partnerships symposium Rabat (Morocco) within the OECD framework of the good governance for development (GfD) in Arab countries initiative. http://www.oecd.org/dataoecd/31/26/41751478.ppt. Accessed 01 May 2014
LandesBank Berlin (2014) Bestellerkredit. http://www.lbb.de/landesbank/de/87_IN/020_aussenhandelsfinanzierung/030_bestellerkredit.html. Accessed 31 Mar 2014
Le Bihan W (2010) Le nouveau cadre législatif du secteur des énergies renouvelables au Maroc. Available via CMS. https://www.cms-bfl.com/Flash-info-Maroc--Le-nouveau-cadre-legislatif-du-secteur-des-energies-renouvelables-au-Maroc-11-08-2010. Accessed 18 Nov 2010
Lehnert W, Vollprecht J (2009) Neue Impulse von Europa: Die Erneuerbare-Energien-Richtlinie der EU. ZUR 307–317
Lewiner C (2010) How to respond to an increasing energy demand without increasing the carbon dioxide emissions? In: Dratwa F, Ebers M, Pohl A, Spiegel B, Strauch G (eds) Energiewirtschaft in Europa. Heidelberg, pp 41–46
Liu J, Cheah CYJ (2009) Real option application in PPP/PFI project negotiation. Constr Manag Econ 27:331–342
Lowe C (2010) EU sees solar power imported from Sahara in 5 years. Available via Reuters. http://www.reuters.com/article/2010/06/20/us-energy-eu-renewables-interview-idUSTRE65J1ZO20100620. Accessed 01 May 2014
Lowenfeld AF (2008) International economic law, 2nd edn. Oxford
Maghraoui D (2009) Introduction: interpreting reform in Morocco. Mediterr Polit 14:143–149
Makinda S, Okumu, FW (2008) The African Union. London
Malone N (2005) The evolution of private financing of government infrastructure in Australia—2005 and beyond. Aust Econ Rev 38:420–430
Matthey P (2003) Die Bedeutung für das Konzessionswesen. In: Baur JF, Hobe S (eds) Rechtsprobleme von Auslandsinvestitionen. Baden-Baden, pp 99–111
Metje TM (2008) Der Investitionsschutz im internationalen Anlagenbau. Tübingen
MIGA (2013a) Annual report 2013. http://www.miga.org/resources/index.cfm?aid=3554. Accessed 31 Mar 2014

MIGA (2013b) MIGA member countries. http://www.miga.org/about/index_sv.cfm?stid=1695. Accessed 31 Mar 2014
Moroccan Investment Development Agency (2014) Investment opportunities. Available via invest in Morocco. http://www.invest.gov.ma/?Id=24&lang=en&RefCat=2&Ref=145. Accessed 31 Mar 2014
Morris C (2009) Sunrise over Morocco. Available via pv Magazin. http://www.petiteplanete.org/wp-content/uploads/2009/10/04022_marokko.pdf. Accessed 01 May 2014
Muchie M, Habib A, Padayachee V (2006) African integration and civil society: the case of the African Union. Transformation 61:3–24
Muchlinski P (2009) The diplomatic protection of foreign investors: a tale of judicial caution. In: Binder C, Kriebaum U, Reinisch A, Wittich S (eds) International investment law for the 21st century – essays in honour of Christoph Schreuer. Oxford, pp 341–362
Muchlinski P (2014) Corporations in international law. Available via Max Planck encyclopedia of public international law. http://opil.ouplaw.com/view/10.1093/law:epil/9780199231690/law-9780199231690-e1513#law-9780199231690-e1513-div2-8. Accessed 07 May 2014
Nickel T, Kopf H (2004) Public Private Partnerships: Ein Ausweg aus der Finanzkrise der öffentlichen Hand? ZfBR (found via juris)
Nicklisch F (2007) Betreibermodelle – ein Königsweg. In: Nicklisch F (ed) Betreibermodelle – BOT/PPP-Vorhaben im In- und Ausland. München, pp 3–10
Noltensmeier S (2009) Public Private Partnership und Korruption. Berlin
n-tv.de (2009) Öko-Strom aus der Wüste. http://www.n-tv.de/wirtschaft/Oeko-Strom-aus-der-Wueste-article367279.html. Accessed 24 Apr 2014
n-tv.de (2011) Marokko will neue Verfassung. http://www.n-tv.de/politik/Marokko-will-neue-Verfassung-article3718866.html. Accessed 24 Apr 2014
OECD (2008) Public–private partnerships. Paris
OECD (2011) Guidelines for multinational enterprises. http://www.oecd.org/daf/inv/mne/48004323.pdf. Accessed 14 Apr 2014
OECD (2014) Country risk classifications of the participants to the arrangement on officially supported export credits. http://www.oecd.org/tad/xcred/cre-crc-current-english.pdf. Accessed 25 Mar 2014
ZEIT Online (2011) Tote durch Terroranschlag in Café in Marrakesch. http://www.zeit.de/gesellschaft/zeitgeschehen/2011-04/marokko-explosion-marrakesch-cafe. Accessed 01 May 2014
Pache E, Bielitz J (2006) Das Verhältnis der EG zu den völkerrechtlichen Verträgen ihrer Mitgliedstaaten. EuR 41:316–339
Pahl T (1994) Die Entwicklung von Projektfinanzierungsstrukturen aus der Sicht des Financiers. In: Nicklisch F (ed) Rechtsfragen privatfinanzierter Projekte. Heidelberg, pp 23–30
Palmer G (2009) Public–private partnerships. Available via literature review – draft. http://www.dpwg-lgd.org/cms/upload/pdf/PublicPrivatePartnership__Lit__Review.doc. Accessed 01 Oct 2010
Pfaffhausen P (2008) Gestaltung von public private partnerships. In: Bauer H, Büchner C, Brosius-Gersdorf F (eds) Verwaltungskooperation. Potsdam, pp 95–106
Pfeiffer J (2010) Zwei-Wege-Strategie für Europas Energiesicherheit. In: Dratwa F, Ebers M, Pohl A, Spiegel B, Strauch G (eds) Energiewirtschaft in Europa. Heidelberg, pp 131–140
Prawoko P (2005) An analysis of ICSID case law on the state of nationality of foreign investors. Griffin's View 6:143–158
Quinn S (2009) Moroccan company law reform and manufacturing firms' access to manufacturing firms' access to bank credit: a before/after panel evaluation. http://www.csae.ox.ac.uk/conferences/2009-EdiA/papers/103-Quinparapdf. Accessed 15 Oct 2010
RAL (2010) Identification mission for the Mediterranean solar plan. http://ec.europa.eu/energy/international/international_cooperation/doc/2010_01_solar_plan_report.pdf. Accessed 21 Dec 2010
Reed L, Paulsson J, Blackaby N (2004) Guide to ICSID arbitration. The Hague

Renner A, Jasper J (2010) Wettbewerb und Sicherheit in der Energieversorgung – worauf es wirklich ankommt. In: Dratwa F, Ebers M, Pohl A, Spiegel B, Strauch G (eds) Energiewirtschaft in Europa. Heidelberg, pp 91–99
Ringel C, Bitsch C (2009) Die Neuordnung des Rechts der Erneuerbaren Energien in Europa. NVwZ 807–811
Roquette AJ (2007) Die Beendigung von PPP-Verträgen – Erfahrungen in Großbritannien. In: Nicklisch F (ed) Betreibermodelle – BOT/PPP-Vorhaben im In- und Ausland. München, pp 147–164
Roquette AJ, Otto A (eds) (2005) Vertragsbuch. München
Roth F (2007) Die Beendigung von Betreiberverträgen. In: Nicklisch F (ed) Betreibermodelle – BOT/PPP-Vorhaben im In- und Ausland. München, pp 107–116
Ruchay D (1996) Partnerschaftliche Entsorgungsprojekte. In: Nicklisch F (ed) Partnerschaftliche Infrastrukturprojekte. Heidelberg, pp 69–87
Ruchser M (2010) Opinion: Desertec or Mediterranean solar plan – whose sun is shining brighter? Available via DW-World.de. http://www.dw-world.de/dw/article/0,,5861222,00.html. Accessed 24 Apr 2014
Sablé L, Lauriol T (2010) Energies renouvelables: le Maroc veut parier sur le solaire. Available via Le Cerrcle/Les Echos. http://lecercle.lesechos.fr/economie-societe/energies-environnement/221130471/energies-renouvelables-le-maroc-veut-parier-sur-le. Accessed 30 Apr 2104
Sater JN (2009) Reforming the rule of law in Morocco: multiple meanings and problematic realities. Mediterr Polit 14:181–193
Schäfer D (2008) Solarthermie. Spiegel der Forschung 11–15
Schmidt S (2008) Die EU als Retterin der AU? GIGA Focus Afrika, vol 5. Hamburg
Schöbener B, Herbst J, Perkams M (2010) Internationales Wirtschaftsrecht. Heidelberg
Schoeltzke W, Claus JC (2007) BOT-Projekte in Emerging Markets. In: Nicklisch F (ed) Betreibermodelle – BOT/PPP-Vorhaben im In- und Ausland. München, pp 133–145
Schüssler F (2008) Energy partnership Africa – Europe: concentrated solar power between technical realization and ethic responsibility. Erdkunde 62:221–230
Siebel UR, Röver J-H, Knütel C (eds) (2008) Rechtshandbuch Projektfinanzierung und PPP, 2nd edn. München
Soest C v (2008) Wacklige "Bausteine" und schwache "Mauer": Die Afrikanische Union hat ein Integrationsproblem. GIGA Focus Afrika 4
Steinberg G, Werenfels I (2007) Between the 'near' and the 'far' enemy: Al-Qaeda in the Islamic Maghreb. Mediterr Polit 12:407–413
Stigler H, Bachhiesl U (2010) Künftige Herausforderungen für die europäische Energiewirtschaft. In: Dratwa F, Ebers M, Pohl A, Spiegel B, Strauch G (eds) Energiewirtschaft in Europa. Heidelberg, pp 183–202
Stöcker N (2010) Entwicklung des Verwaltungskooperationsvertrages unter Berücksichtigung des Vergaberechts. Frankfurt am Main
Straßburg W (1996) Projektfinanzierung und Risikomanagement aus der Sicht eines Energieversorgungsunternehmens. In: Nicklisch F (ed) Partnerschaftliche Infrastrukturprojekte. Heidelberg, pp 29–47
Talbi A (2008) PPP: l'expérience marocaine. Available via presentation at public private partnerships symposium Rabat (Morocco) within the OECD framework of the good governance for development (GfD) in Arab countries initiative. http://www.oecd.org/dataoecd/31/23/41751518.ppt. Accessed 01 May 2014
Tettinger PJ (1996) Die rechtliche Ausgestaltung von Public Private Partnership. DÖV 764–770
Tettinger PJ (2005) Public Private Partnership, Möglichkeiten und Grenzen – ein Sachstandsbericht. NWVBl 1–10
The Club of Rome (2014) Club of Rome. http://www.clubofrome.org/?p=375. Accessed 24 Apr 2014
Tietje C (2005) Die Beilegung internationaler Investitionsstreitigkeiten. In: Marauhn T (ed) Streitbeilegung in den internationalen Wirtschaftsbeziehungen. Tübingen, pp 47–62

UNCTAD (2009) Latest developments in investor-State dispute settlement – IIA monitor no. 1 (2009). http://www.unctad.org/en/docs/webdiaeia20096_eparapdf. Accessed 5 Sept 2010
Verveniotis G (1996) Sicherung des "Monopol"-Betriebs und des Kapital und Zinsrückflusses u.a. gegenüber staatlichen Hoheitsentscheidungen des Gastlandes. In: Nicklisch F (ed) Partnerschaftliche Infrastrukturprojekte, Heidelberg, pp 115–127
Wagner UN (1996) BOT-Projekte – Chance oder Risiko? In: Nicklisch F (ed) Partnerschaftliche Infrastrukturprojekte. Heidelberg, pp 3–18
Wahl K-D (1994) Das Project Implementation Agreement aus der Sicht des Anlangenerrichters/ Investors. In: Nicklisch F (ed) Rechtsfragen privatfinanzierter Projekte. Heidelberg, pp 15–22
Wang SQ, Tiong LK (2000) Case study of government initiatives for PRC's BOT power plant project. Int J Proj Manag 18:69–78
Weber M, Schäfer M, Hausmann FL (eds) (2006) Public private partnership. München
Weh M (2008) Transnationale Unternehmen im System kollektiver Sicherheit. Regensburg
Werenfels I, Westphal K (2010) Solarstrom aus Nordafrika. SWP Studie 1–40
Werner J, Fiedler A (2007) Betrieb, Betriebs-Erträge und Beendigung des Projekts. In: Nicklisch F (ed) Betreibermodelle – BOT/PPP-Vorhaben im In- und Ausland. München, pp 91–105
Wetzel D (2013) Chinesen steigen groß bei Wüstenstrom-Projekt ein. Available via Die Welt. http://www.welt.de/wirtschaft/article122871903/Chinesen-steigen-gross-bei-Wuestenstrom-Projekt-ein.html. Accessed 24 Mar 2014
Winker P, Preußner C (2008) Solarkraftwerke in Nordafrika und Energiepartnerschaft mit Europa. Spiegel der Forschung 34–38
Winter G (2009) Das Klima ist keine Ware. ZUR 289–297
Wippel S (2005) The Agadir agreement and open regionalism. EuroMeSCopaper 45:5–28
Wiwen-Nilsson T (1996) The contractual arrangements of a BOT-project from the government perspective – the Arlanda Link example. In Nicklisch F (ed) Partnerschaftliche Infrastrukturprojekte. Heidelberg, pp 91–105
Wolfs M, Woodroffe S (2002) Structuring and financing international BOO/BOT desalination projects. Desalination 142:101–106
WTO (2009) Trade policy review – report by the kingdom of Morocco WT/TPR/G/217
Yescombe ER (2007) Public–private partnerships. Burlington

Cases

ECJ

Case C-451/08 [2010] *Helmut Müller GmbH v Bundesanstalt für Immobilienaufgaben* preliminary ruling

International Court of Justice

Belgium v Spain [1970] Barcelona Traction, Light and Power Company, Limited (second phase) – Judgment, [1970] ICJ Reports, 3
Belgium v Spain [1970] Barcelona Traction, Light and Power Company, Limited (second phase) – Separate opinion Gerald Fitzmaurice, [1970] ICJ Reports, 64
Belgium v Spain [1970] Barcelona Traction, Light and Power Company, Limited (second phase) – Separate opinion Philip C. Jessup, [1970] ICJ Reports, 161
Belgium v Spain [1970] Barcelona Traction, Light and Power Company, Limited (second phase) – Separate opinion André Gros, [1970] ICJ Reports, 267

Libyan Arab Jamahiriya v United States of America [1998] Questions of Interpretation and Application of the 1971 Montreal Convention arising from the Aerial Incident at Lockerbie, Judgment – Preliminary Objections, [1998] ICJ Reports, 115
Liechtenstein v Guatemala [1955] Nottebohm Case (second phase) – Judgment, [1955] ICJ Reports, 4
Republic of Guinea v Democratic Republic of the Congo [2007] Case concerning Ahmadou Sadio Diallo, Preliminary Objections – General List No. 103, 17

ICSID Tribunals

Aguaytia Energy LLC v Republic of Peru [2008] ARB/06/13 – Award
Ambiente Ufficio S.p.A. and others v Argentine Republic [2013] ARB/08/9 – Decision on Jurisdiction and Admissibility
Autopista Concesionada de Venezuela, C.A. v Bolivarian Republic of Venezuela [2001] ARB/00/5
Compañiá de Aguas del Aconquija S.A. and Vivendi Universal S.A. v Argentine Republic [2005] ARB/97/3 – Decision
El Paso Energy International Company v The Argentine Republic [2006] ARB/03/15
Enron Corporation and Ponderosa Assets, L.P. v Argentine Republic [2004] ARB/01/3 – Decision
Gustav F W Hamester GmbH & Co KG v Republic of Ghana [2010] ARB/07/24 – Award
Impregilo S.p.A. v Argentine Republic [2011] ARB/07/17 – Award
Ioannis Kardossopoulos and Ron Fuchs v The Republic of Georgia [2010] ARB/05/18 and ARB/07/15 – Award
KT Asia Investment Group B.V. v Republic of Kazakhstan [2013] ARB/09/8 – Award
Mobil Corporation and others v Bolivarian Republic of Venezuela [2010] ARB/07/27
Niko Resources (Bangladesh) Ltd. v Bangladesh Petroleum Exploration & Production Company Limited ("Bapex") and Bangladesh Oil Gas and Mineral Corporation [2013] ARB/10/11 and ARB/10/18 – Decision on Jurisdiction
PSEG Global Inc., The North American Coal Corporation, and Konya Ilgın Elektrik Üretim ve Ticaret Limited Şirketi v Republic of Turkey [2004] ARB/02/5 – Decision
Sempra Energy International v The Argentine Republic [2010] ARB/02/16 – Annulment
Standard Chartered Bank v United Republic of Tanzania [2012] ARB/10/12 – Award
Tanzania Electric Supply Company Limited v Independent Power Tanzania Limited [2001] ARB/98/8 – Award
Tokios Tokelės v Ukraine [2004] ARB/02/18, [2004] ICSID Review, 205

Conference Speeches

Jaouad OC (2010) Renewable energies: roadmap for Morocco. Presented at the SEPA 2010 (Desert Sun, Yours, Mine, Ours?) conference, University of Giessen, Giessen November 2010
Komendantova N (2010) Investment into large-scale deployment of renewable energy sources in North Africa. Presented at the SEPA 2010 (Desert Sun, Yours, Mine, Ours?) conference, University of Giessen, Giessen, November 2010

Chapter 2
Arbitration

This chapter deals with the question of international arbitration. This includes an assessment of the best suited arbitration institution for the Desertec Concept. International arbitration is frequently necessary because investors distrust the judicial independence of the host state's courts,[1] and its judicial system as a whole.[2] This is especially valid for states in the Mediterranean Sea because most of them apply the Shari'a.[3] The judicial system is often not independent and can be highly corrupt.[4] Different understandings concerning legal terms and interpretation within Islamic countries concerns foreign investors.[5] Especially in cases of a large scale project arbitration is often recommended.[6] To find a good solution for the Desertec Concept, it is necessary to review the different arbitration forms. This includes an overview of arbitration institutions and arbitration regulations of institutional and ad hoc arbitration.

As an example from the Islamic perspective, Turkish law views PPPs pertaining to the service of general interest as contracts subject to public law.

---

[1] Vannieuwenhuyse (2009), 115 (116); Not desirable for an investor, in: Dolzer and Schreuer (2008), 214; Walter (2006), 815 (816); Horn (2008), 587 (592); Sattar (2010), 51 (51); Moses (2008), 220; Reed et al. (2004), 8; Hobe and Müller (2009), 65 (65); Ozumba (2009), 4; Griebel (2008), 119.

[2] UNCTAD (2010b), 14; Sattar (2010), 51 (51); Also the interference in arbitral awards or the narrow interpretation of modern arbitration laws cause serious problems, in: Sattar (2010), 51 (57); All these relate to the close connection or state-controlled position of the other partner, in: Vannieuwenhuyse (2009), 115 (116); Dolzer and Schreuer (2008), 214.

[3] Vogl (2010), 32 (33); Kausch (2009), 165 (168); Euro-Mediterranean Association for Cooperation and Development e.V. (2010), 2.

[4] Kausch (2009), 165 (168); Sayed (2008), 289 (290).

[5] Turkish law views PPPs pertaining to the service of general interest as contracts subject to public law, contrary to the common understanding that PPPs are part of the civil law, in: Ulusoy and Oguz (2007), 5021 (5025).

[6] cf Redfern et al. (2004), 50 para 1-106; cf Metje (2008), 159.

R. Gömmel, *Investing into North African Solar Power*,
DOI 10.1007/978-3-319-15756-6_2

2.1 International Arbitration

Despite the fact that an increasing number of national courts are dealing with issues of international law and resolving these cases, the amount of arbitration institutions and arbitrators is constantly growing.[7] This is also due to the general acceptance of international arbitration as a method of (international) dispute settlement.[8] Over the past few decades, there has been an increase in arbitration usage instead of litigation in local courts.[9] Authors frequently view international courts and tribunals as being advantageous as they are impartial "organs" with legal expertise which have experience of solving legal disputes.[10]

2.1.1 International Courts

In the case of the Desertec Concept, the only possible international court would be the International Court of Justice (ICJ). The ICJ is the principal judicial organ of the UN (Art. 92 Charter of the United Nations (UN Charter)) and one of the principal organs of the UN (Art. 7(1) of the UN Charter). All members of the UN are *ipso facto* parties to the Statute of the International Court of Justice (ICJ-Statute), pursuant to Art. 93(1) of the UN Charter. All decisions of the ICJ are binding for UN member states, as Art. 94(1) of the UN Charter highlights. Nevertheless, Art. 95 of the UN Charter underlines that the ICJ is not the only institution which settles international disputes.

According to Art. 36(1), 38 of the ICJ Statute, the ICJ has jurisdiction concerning all matters the parties refer to it and decides in accordance with international law. The ICJ is not limited to certain international law claims or in its *ratione materiae*.[11] The jurisdiction depends on the consent of the parties as Art. 36(2) of the ICJ Statute implies, but the parties are not obliged to consent to ICJ jurisdiction.[12] Art. 34(1) of the ICJ Statute regulates that only states can be party to the ICJ. Therefore, the ICJ is not an option in cases where natural or judicial persons are involved. As an investor like Dii is generally a private company, the ICJ is not applicable to the Desertec Concept.

---

[7] Wood (2007), 575 (575).

[8] Redfern et al. (2004), 1 para 1-01.

[9] Loncle (2005), 3 (3).

[10] Ulfstein et al. (2007), 3 (8).

[11] Ebner (2005), 107 and 109.

[12] Ebner (2005), 107.

2.1.2 Differences Between Commercial and Investment Arbitration

International commercial and investment arbitration have some general features in common. Both are forms of international arbitration.

> International arbitration is a private form of dispute resolution with numerous goals, namely to protect investors' rights in a neutral forum, to quicken the time of litigation and lower expenses, and to offer procedural flexibility allowing businesses to continue operation.[13]

Besides investor's distrust concerning judicial independence the above mentioned corruption is frequently an issue for investors. Two cases in Pakistan, where Pakistan among other things objected to *lis pendens* within an ICSID proceeding, also illustrate the misuse of judicial power, as courts purposely interfered with international arbitration for the benefit of Pakistan.[14] However, in most cases involving PPPs, the host state wants to use its own judicial system for the settlement of conflicts.[15]

International investment and commercial arbitration have a similar historical background. The most important similarity relates to the general acceptance of international arbitration.[16] As a result of the US–UK Jay Treaty of 1794, numerous mixed commissions were set up to deal with 'problems of injury to aliens'.[17] The Jay Treaty stipulated the establishment of mixed commissions to resolve economic disputes related to the American independence.[18] This was the starting point for both forms of arbitration to be an accepted tool of legal dispute resolution. Nevertheless, international commercial arbitration has developed differently than international investment arbitration. Arbitration between two private or judicial parties is an old and long established method of dispute resolution. On the other hand, the possibility to sue the state in front of an arbitration court is a right which has only emerged recently.

Consequently, there is a need to distinguish between international commercial arbitration and international investment arbitration. Investment arbitration almost exclusively deals with investor-state dispute settlement (ISDS). Most international commercial arbitration institutions were not capable of dealing with the question of a state's exercise of public authority concerning foreign nationals in the ISDS.[19] The main idea of international investment arbitration was to close this gap, by "transforming" features of international commercial arbitration into the

---

[13] Shookmann (2010), 361 (361); cf Kouris (2005), 127 (127).

[14] Sattar (2010), 51 (60–61); E.g. Pakistan, in: Shookmann (2010), 361 (368).

[15] Hunter (2007), 165 (166).

[16] Tietje (2010), 5 (6).

[17] Shaw (2008), 823; Tietje (2005), 47 (47); cf Tietje (2010), 5 (6).

[18] Tietje (2005), 47 (47); Tietje (2010), 5 (6).

[19] Harten and Loughlin (2006), 121 (126).

governmental realm.[20] The major difference between international investment arbitration and international commercial arbitration is that "investment" claims '(...) are based on international law established by investment treaties (treaty claims) (e.g. BITs).'[21] Thus, international investment arbitration encompasses a kind of review of governmental acts, public interests and policies.[22] Commercial arbitration, on the other hand, deals with issues of contractual behaviors.[23] International investment arbitration results from a sovereign act, whereas international commercial arbitration is constituted by a private act.[24] To find the best arbitration institution for the Desertec Concept, both forms of arbitration require a separate assessment.

Both forms of international arbitration share an important aspect. By accepting arbitration, the state freely limits its own sovereignty.[25] Thus, the state accepts international arbitration as the valid legal method to solve the dispute in question.

2.1.2.1 History of Investment Protection and Arbitration

Investment protection has been a part of international law for a long time and first appeared in a major way at the foundation of the Hanse in the thirteenth to fourteenth century.[26] The next big step followed in 1796, when the US and France negotiated a Treaty in Friendship, Commerce and Navigation, which dealt with the issue of '(...) protection of alien property by the rules of international law.'[27] Combined with the 1794 Jay Treaty, international investment protection was on its way to becoming one of the most important tools of international law. Today investment arbitration is an important part of modern international law.[28]

2.1.2.2 International Investment and the Treatment of Aliens

Since investment protection is a part of international investment, it mainly deals with the treatment of aliens.[29] This was also highlighted by Judge Read, who

[20] Harten and Loughlin (2006), 121 (126).

[21] Also a state can be a party in international investment and commercial arbitration, in: Horn (2008), 587 (591); Moss (2009), 782 (784); Within investment arbitration, the state gives a binding commitment to all future cases, unlike in commercial arbitration, where it is a case to case decision, in: Harten and Loughlin (2006), 121 (142).

[22] International law is above domestic regulations, in: Moss (2009), 782 (784); Harten and Loughlin (2006), 121 (146); Horn (2008), 587 (591).

[23] Moss (2009), 782 (784–785).

[24] Harten and Loughlin (2006), 121 (140).

[25] Ruchay (1996), 69 (85).

[26] Tietje (2010), 5 (5).

[27] Dolzer and Schreuer (2008), 11.

[28] Sornarajah (2010), 77–79.

[29] Whereby expropriation was of main concern, in: Metje (2008), 73; Tietje (2010), 5 (5–6); Verdross and Simma (2010), 801–803 paras 1212–1213; cf Herrmann et al. (2007), § 17 para 789; Shaw (2008), 823; Michaelis, Salomon in Hilf and Oeter (2010), § 15 para 7.

pointed out that at the moment when a foreigner is allowed to enter the host state, this state has certain rights and responsibilities related to the home state of the foreigner.[30] The state has no obligation to accept all possible investments and thus constrain its own sovereignty.[31] Once the state allows the investment, it is subject to minimum standards of investment protection.[32] Consequently, it is necessary to limit state sovereignty in order to establish an investor-friendly environment.[33]

The treatment of aliens also helped to develop international customary rules for the protection of alien property.[34] The rules on the treatment of aliens do not forbid the expropriation of foreigners, as long as the expropriation happens because of public needs or interests.[35] The treatment of aliens has always been an important issue of international law and disputed because of different approaches (e.g. supported by either developed or developing countries).[36] There are two approaches towards the level of protection, which are the principle of national treatment and the international minimum standard.[37] The first one is mainly supported by developing countries and stipulates that all aliens are to be treated like nationals, whereas the second approach (championed by developed countries) encompasses a minimum standard for foreigners, even if it is above the national standard.[38] In 1868, the Argentine jurist Carlos Calvo introduced the idea (Calvo Doctrine) that '(. . .) the international rule should (. . .) be understood as allowing the host state to reduce the protection of alien property when also reducing the guarantees for property held by nationals.'[39] In 1938, US Secretary of State Cordell Hull mentioned in a dispute between the US and Mexico that 'international law allowed expropriation of foreign property, but required (. . .) prompt, adequate and effective compensation (. . .)' (Hull Formula).[40] This dispute took place between the nineteenth and twentieth century and the sum of all approaches created the international minimum standard.[41] With the fall of the Soviet Union in the 1990s, the

---

[30] ICJ [1955] Nottebohm Case (second phase)—Dissenting Opinion Judge Read, 34 (46–47).

[31] Maier (2010), 95 (104); Dolzer and Schreuer (2008), 7; Harten (2007), 83; cf Zampetti and Sauvé (2007), 211 (220).

[32] Harten (2007), 87; Dolzer and Schreuer (2008), 7.

[33] Dolzer and Schreuer (2008), 9.

[34] Herrmann et al. (2007), § 17 para 789; cf Michaelis, Salomon in Hilf and Oeter (2010), § 15 para 7.

[35] Schweisfurth (2006), 591 para 58.

[36] Shaw (2008), 823.

[37] Shaw (2008), 824.

[38] Shaw (2008), 824.

[39] Lowenfeld (2008), 473; This doctrine was a reaction towards the so-called gunboat diplomacy of some states, in: Dolzer and Schreuer (2008), 12

[40] Dolzer and Schreuer (2008), 13; Lowenfeld (2008), 475–476.

[41] Schöbener et al. (2010), 228, Kap. 4 §15 para 25; Krajewski (2009), 172; Dolzer and Schreuer (2008), 13.

dispute between Hull and Calvo was over, and a clear set of minimum standards was agreed to as customary international law.[42]

Overall, this illustrates that investment protection has always been a dichotomy between sovereignty (e.g. over natural resources) and investment protection (state responsibility for injury to aliens).[43] Furthermore, the development and ratification of human rights has led to a minimum standard within international law.[44]

2.1.2.2.1 International Investment Arbitration and International Law

The basic principle of international law was that states as only subject of international law were capable to have rights and duties assigned to it.[45] This was modified after the appearance of international organizations which possess a functional international personality.[46] Nowadays international organizations have a partial international legal personality limited to the previous determined purpose of the organization and a particular (relative) international legal personality towards its members or towards other subjects, which have an own international legal personality and recognize the organization (e.g. based on a bilateral contract).[47] As a next step, international human rights conventions, e.g. the European Convention on Human Rights, provided natural persons with the rights and duties of international law.[48] However, down to the present day, this remains an exception within international law, and natural or legal persons still lack an international legal personality within international law.

It is not surprising that international law plays an important role in investment law and arbitration.[49] In cases of expropriation, it defines standards of protection; in cases of BITs, it helps to interpret disputed regulations.[50] Before the utilization of

[42] Dolzer and Schreuer (2008), 15–16; Article 38(1)(b) ICJ Statute explicitly mentions customary international law as a source of international law. Customary international encompasses two elements, (1) state practice (substantial, uniform practice of a substantial amount of states) and (2) *opinio juris* (the practicing state must recognize (or belief) that its activity is legally binding), in: Aust (2007), 11; Shaw (2008), 72–93; Birnie et al. (2009), 22–25; Herdegen (2010), § 16 para 1.

[43] Krajewski (2009), 170; Garcia-Bolivar (2010), 2; cf Metje (2008), 73–74; Balancing the "playing field", in: Nmehielle (2001), 21 (28).

[44] Shaw (2008), 825–826; cf Dolzer and Schreuer (2008), 13; The relationship of human rights obligations and investment agreements is subject of an increasing amount of research, whereby investment treaties still do not refer to human rights in: UNCTAD (2009) Selected Recent Developments in IIA Arbitration and Human Rights—IIA MONITOR No. 2, 1–2; E.g. the status of the foreigner and includes rights like diplomatic protection or fair and equitable treatment, in: ICJ [1955] Nottebohm Case (second phase)—Dissenting Opinion Judge Read, 34 (47).

[45] Muchlinski (2009), 341 (342).

[46] Muchlinski (2009), 341 (342).

[47] Hofmann (2012), 2.

[48] Tietje (2003), 5 (16); Tietje (2005), 47 (61); Muchlinski (2009), 341 (342).

[49] Hobér (2008), 545 (545).

[50] Hobér (2008), 545 (545).

BITs, private investors were only able to fall back on diplomatic protection, which is part of customary international law.[51] The main rules and prerequisites for diplomatic protection are laid down in the 2006 ILC Draft articles on Diplomatic Protection (ILC Diplomatic Protection). The International Law Commission (ILC) was founded in 1947 by the UN General Assembly. There is no obligation under international law for the state to extend diplomatic protection to nationals.[52] Furthermore, the utilization of diplomatic protection "transforms" a claim of an investor into a claim of the state.[53] The utilization of diplomatic protection is difficult and thus not desirable for investors as there is no guarantee that the home state will act.[54] This is also underlined by the fact that the home state of the investor does not have any obligation (except a moral one) to pay reparation to the respective investor.[55] Furthermore, the ICJ pointed out that diplomatic protection will only be exercised after all possible local solutions have been exhausted beforehand.[56]

2.1.2.2.2 The ICSID as a New Tool of Investment Protection

The most recent development within international investment protection took place in 1965. Within international investment law, the foundation of the ICSID provided a new dispute settlement mechanism, where investors are able to directly litigate against the host state.[57] The establishment of the ICSID was also due to the increasing number of investment disputes and the need for an institutional arbitration.[58] Therewith the international customary law concerning international investment, dating back to the Hanse, was transferred into treaty based law, by stipulating certain requirements within the ICSID Convention.

The ICSID is an autonomous international institution under the Convention on the Settlement of Investment Disputes between States and Nationals of Other States (ICSID Convention, also known as the Washington Convention).[59] The Executive

---

[51] ICJ [2007] Case concerning Ahmadou Sadio Diallo, Preliminary Objections—General List No. 103, 17 para 39; Muchlinski (2009), 341 (342–343); Vannieuwenhuyse (2009), 115 (118–119); Dolzer and Schreuer (2008), 211.

[52] Vannieuwenhuyse (2009), 115 (116); Muchlinski (2009), 341 (343).

[53] Muchlinski (2009), 341 (343).

[54] Tietje (2005), 47 (50); cf Dolzer and Schreuer (2008), 211; Tietje (2003), 5 (7–8); cf Metje (2008), 167.

[55] Muchlinski (2009), 341 (343).

[56] ICJ [2007] Case concerning Ahmadou Sadio Diallo, Preliminary Objections—General List No. 103, 18 para 44.

[57] This also includes the exclusion of diplomatic protection, in: Vannieuwenhuyse (2009), 115 (119); Egonu (2007), 479 (480); ICSID as the main actor in international investment disputes, in: Hauschka (2005), 1550 (1553).

[58] Schöbener and Markert (2006), 65 (67); Arbitration wave in Africa and Asia, in: Lowenfeld (2008), 536.

[59] ICSID (2014) About ICSID; cf Lörcher (2005), 11 (11).

Directors of the World Bank formulated the ICSID Convention, which was opened for signature in 1965.[60] Thus, the ICSID is a part of the so-called World Bank Group, however it is an own legal entity of international law.[61] According to its Preamble, the goal of the ICSID Convention is to promote foreign investment. Nevertheless, the ICSID is not a court as it only delivers a forum to investment arbitration.[62] The ICSID Convention delivers a comprehensive procedure for investment arbitration and consequently excludes the application of national arbitration regulations.

Today, ICSID tribunals deal with more than half of all investment disputes worldwide.[63] The amount of new cases has been constantly growing since the mid-1990s.[64] Although the headquarters are in Washington D.C., the ICSID is an international arbitration institution.[65] One of the main purposes of the ICSID Convention is the balance between the parties, because an investor (private) faces a state.[66] Most importantly, the ICSID Convention '(. . .) imposes an international law obligation upon host states to comply (. . .).'[67]

2.1.2.2.3 The Emergence of BITs

As mentioned above, the main difference between international investment arbitration and international commercial arbitration is that international investment arbitration deals with treaty claims. Unlike contract claims, treaty claims have an international law background. Most treaty claims within the jurisdiction of the ICSID tribunals relate to BITs. BITs are investment contracts between states. Their main purpose is to legally coordinate investment of nationals of state A in the territory of state B and vice versa. Today, the basis for investment protection is a BIT,[68] which emerged in the mid-1950s.[69] However, the roots of bilateral investment treaties date back to the eighteenth century.[70] Today, BITs are the backbone

[60] ICSID (2014) About ICSID.

[61] Tietje (2003), 5 (8).

[62] Krajewski (2009), 179; Schöbener et al. (2010), 313–314, Kap. 4 § 19 para 377.

[63] Leading investment forum worldwide, in: Vannieuwenhuyse (2009), 115 (116); Griebel (2008), 115; cf Dolzer and Schreuer (2008), 222.

[64] Reinisch (2008), 107 (110).

[65] McIlwrath and Savage (2010), 51 para 1–113.

[66] Alexandrov (2009), 322 (323).

[67] Reinisch (2008), 107 (113); Alexandrov (2009), 322 (329).

[68] Tietje (2010), 5 (6); Mostly due to the complexity of foreign direct investments (FDIs) and portfolio investments, in: Tietje (2003), 5 (7); cf Krajewski (2009), 171; cf Schöbener et al. (2010), 48, Kap. 1 §4 para 159.

[69] Tschanz and Viñuales (2009), 729 (731); Karl (2003), 37 (38).

[70] Metje (2008), 75.

of investment protection.[71] For example, BITs address the question of the definition of investment and thus securing investor protection rights.[72] Interestingly, the wording of most BITs has not changed since their appearance 50 years ago.[73] Through BITs, the state deliberately renounces certain parts of its sovereignty to be able to attract foreign investments.[74] Consequently, the BIT binds both parties.[75]

The fact that an individual can sue the state because of the ICSID Convention and the BIT is almost unique. Therefore, the ICSID and BITs go beyond the classical case of international law, where states sue each other.[76] This means that the ICSID Convention and BITs are intertwined with international law.[77] Although some mention that a BIT favors investors, there are also a lot of examples of where the state has prevailed.[78]

2.1.2.2.4 The Relationship Between ICSID Convention and General International Law

Art. 27(1) of the ICSID Convention expressively prohibits the home state of the investors to use diplomatic protection if the host state has agreed to ICSID jurisdiction. Vice versa, section V No. 33 of the Report of the Executive Directors of the International Bank for Reconstruction and Development on the Convention on the Settlement of Investment Disputes between States and Nationals of Other States (Report of the Executive Directors) stipulates that the investor cannot ask his home country to exercise diplomatic protection. No. 32 Report of the Executive Directors offers an interpretation guide of Art. 27 of the ICSID Convention in section V. It points out that if both parties have not agreed to other remedies, the intention of the parties to have recourse arbitration excludes any other remedy.

The only exception to this rule is when a contracting state does not comply with an ICSID award (Art. 27(1) of the ICSID Convention). One ICSID tribunal recently concluded that '(...) (t)he customary international law minimum standard of

[71] Clapham (2009), 437 (437); Harten and Loughlin (2006), 121 (123); Metje (2008), 76–77; Semler (2003), 97 (98); Zampetti and Sauvé (2007), 211 (214–215); cf Lowenfeld (2008), 554–555; cf Dolzer and Schreuer (2008), 21; Krajewski (2009), 173; cf Schöbener et al. (2010), 223, Kap. 4 §13 para 7; cf Vannieuwenhuyse (2009), 115 (127).

[72] Zampetti and Sauvé (2007), 211 (216).

[73] Karl (2003), 37 (38).

[74] Dolzer and Schreuer (2008), 23.

[75] Desbordes and Vicard (2009), 372 (375).

[76] Alexandrov (2009), 322 (322); Harten and Loughlin (2006), 121 (129).

[77] ICSID [2009] ARB/06/5—Award, 30–31 para 78; Nevertheless customary international law is still important in cases of investment arbitration, in: Maier (2010), 95 (104).

[78] Maier (2010), 95 (101–102); He also points out that a too strict investor protection would be contrary to a functional international investment regime, in: Maier (2010), 95 (106); State sovereignty is not protected enough by ICSID in: Mortenson (2010), 257 (312); ICSID favors investor, in: Rosenberg (2010), 8 (9).

treatment is just that, a minimum standard.'[79] It pointed out that '(i)t is meant to serve as a floor, an absolute bottom, below which conduct is not accepted by the international community' and that '(. . .) (it) is not meant to vary from state to state or investor to investor.'[80] Nevertheless, the sole arbitrator in one ICSID case mentioned that the state's responsibility concerning the minimum standard is in proportion to its resources.[81]

The ICSID and the closely related development of investment arbitration have helped to move away from the old "gun-boat" strategies by creating a legal framework which enables all investors to file a claim in front of a neutral forum.[82] In earlier times, states used "gun-boat" strategies to protect the rights of their investors abroad, using their economic and political power.[83]

The fact that a natural or legal person can file a claim against a state is unique within international law. Combined with the obligation to comply, the ICSID Convention is an effective international investment protection tool. There is also the possibility to conduct international commercial arbitration against a state, but unlike the ICSID Convention, there are no comprehensive legal regimes within commercial arbitration. The ICSID Convention also includes certain compliance mechanisms which do not exist within international commercial arbitration. However, this will be discussed in a later section of this thesis dealing with the best choice of investment arbitration institutions.

2.1.3 International Commercial Arbitration

It is frequently highlighted that international commercial arbitration is one of the best ways to solve international commercial disputes.[84] International arbitration is defined as:

> (. . .) [A] process by which parties agree to the binding resolution of their disputes by adjudicators, known as arbitrators, who are selected by the parties either directly or indirectly via a mechanism chosen by the parties.[85]

Thus, the key aspects of international arbitrations are that the arbitration is according to a party agreement, the dispute is resolved by selected arbitrators of the parties, and the decision is binding.[86] Procedural flexibility, speed, costs and

---

[79] ICSID [2009] In accordance with the UNCITRAL Arbitration Rules, 263–264 para 615.

[80] ICSID [2009] In accordance with the UNCITRAL Arbitration Rules, 263–264 para 615.

[81] In the case, the arbitrator mentioned that the minimum standard of due diligence is a modified objective standard, in: ICSID [2009] ARB/07/21—Award, 20–21 para 81;

[82] UNCTAD (2010b), 13; BITs allow direct claims against the state, in: Maier (2010), 95 (101).

[83] UNCTAD (2010b), 1.

[84] Buchanan (1988), 511 (512); Partasides and Fullelove (2010), 1 (1); Kutty (2006), 565 (570–571).

[85] McIlwrath and Savage (2010), 5 para I-015.

[86] SCC (2014b); McIlwrath and Savage (2010), 5 para I-015; Partasides and Fullelove (2010), 1 (1); Buchanan (1988), 511 (512); Kutty (2006), 565 (570).

confidentiality are also typical characteristics of international commercial arbitration.[87] The best definition of international commercial arbitration can be found in Art. 1(3) 1985 of the UNCITRAL Model Law on International Commercial Arbitration (amended in 2006) (UNCITRAL Model Law), which states that:

> [A] the parties to an arbitration agreement have, at the time of the conclusion of that agreement, their places of business in different States; or (b) one of the following places is situated outside the State in which the parties have their places of business: (i) the place of arbitration if determined in, or pursuant to, the arbitration agreement; (ii) any place where a substantial part of the obligations of the commercial relationship is to be performed or the place with which the subject matter of the dispute is most closely connected; or (c) the parties have expressly agreed that the subject matter of the arbitration agreement relates to more than one country.

No comparable definition exists for the term commercial. It is generally defined in a broad way, including: '(...) all matters arising from all relationships of a commercial nature, whether contractual or not.'[88]

2.1.3.1 International Commercial Arbitration and Domestic Courts

The roots of international arbitration go back to the national level.[89] To create an effective legal framework for international arbitration, states adopted several conventions e.g. the 1958 (New York) United Nations Convention on the Recognition and Enforcement of Arbitral Awards (NY Convention) and the 1985 (amended 2006) UNCITRAL Model Law.[90] Consequently, a link exists between international arbitration and national law, as this is the legal basis.

International arbitration does not operate in a vacuum, which is illustrated by the fact that several legal frameworks, e.g. the arbitration institution rules, the applicable arbitration law (domestic and/or international law) and the municipal law of enforcement are relevant in a case of effective arbitration.[91] The benefits of national courts are related to their power and thereupon '(...) to command parties to appear before them (...) to compel not only parties but third-party entities to turn over information relevant to the resolution of the dispute (...) and the power (...) to enforce a judgment (...).'[92]

---

[87] Kutty (2006), 565 (569–570); Buchanan (1988), 511 (512); Partasides and Fullelove (2010), 1 (1–2); SCC (2014b); Ayad (2009), 93 (93); Importance of confidentiality, in: Egonu (2007), 479 (479); Importance of confidentiality, in: Tietje (2005), 47 (60); Importance of confidentiality, in: Fiebiger (2010), 269 (270); Importance of confidentiality, in: Tietje (2003), 5 (15).

[88] Redfern et al. (2004), 18 para 1–32.

[89] Kutty (2006), 565 (571); Buchanan (1988), 511 (512); Partasides and Fullelove (2010), 1 (4).

[90] Partasides and Fullelove (2010), 1 (4); cf Buchanan (1988), 511 (512).

[91] Brower and Sharpe (2003), 643 (648); Kutty (2006), 565 (571); McIlwrath and Savage (2010), 23–24 para 1-031; cf Buchanan (1988), 511 (512).

[92] Mentioned reasons and additional benefits, in: Wood (2007), 575 (578).

Other authors point out that national courts are not the key player in arbitration, but serve a vital part as an'(...) executive partner to provide effectiveness to arbitral proceedings.'[93] This is closely connected to the fact that arbitration is "only" an alternative dispute resolution, which needs the support of national courts.[94] ICC practice also indicates that there is no desire to decouple an international arbitration completely from national courts and their judicial tools.[95] The same applies for international investment arbitration. If the country in question adopted Art. 2 of the NY Convention and interfered with or sabotaged an arbitration award by a national court, this act would be a violation of the state's obligation, according to the Convention.[96]

2.1.3.2 Characteristics of International Commercial Arbitration

The biggest difference between international commercial and investment arbitration rests in the fact that parties differ. In the case of investment arbitration the parties are the investor (private) and the host state (public), whereas in commercial arbitration, there are mostly non-state actors participating. One ICSID tribunal underlined this requirement as it dismissed a case, because the investment was between two private parties.[97] Within international investment arbitration one party must be the host state. There are two different types within international commercial arbitration, namely ad hoc and institutional arbitration. Institutional arbitration is:

> (...) [W]here the parties choose to refer their disputes to arbitration conducted under the rules of a particular institution (...) (, which) (...) will assist in matters such as initiation of the arbitration, the formation and appointment of the tribunal and general administration of the arbitration.[98]

Ad hoc arbitration on the other hand is a form of arbitration performed by the parties without an institution.[99] International commercial arbitration is often chosen to avoid the publicity stemming from a dispute.[100] Unlike in investment disputes,

---

[93] Sattar (2010), 51 (52) and (73); Nariman (2004), 123 (129); Delaume (1983), 784 (785); Schöbener and Markert (2006), 65 (106); McIlwrath and Savage (2010), 27 para 1-037 and 30 paras 1-044 and 1-045; Nmehielle (2001), 21 (36).

[94] Judges and arbitrators work together in the field of international law, in: Nariman (2004), 123 (125); Redfern et al. (2004), 1 para 1-01; Sattar (2010), 51 (52); cf Nmehielle (2001), 21 (36).

[95] Sattar (2010), 51 (53) and (54).

[96] Kovacs (2008), 421 (424); Sattar (2010), 51 (68).

[97] ICSID [2013] ARB/11/18—Award, 30 para 144.

[98] Kutty (2006), 565 (572–573); Redfern et al. (2004), 47 para 1–99; Partasides and Fullelove (2010), 1 (2).

[99] Redfern et al. (2004), 47 para 1–99; Kutty (2006), 565 (571); Partasides and Fullelove (2010), 1 (2).

[100] Moss (2009), 782 (792).

the public interest is often low, because mostly privates are involved in commercial arbitration.[101] Some authors highlighted the benefits of privacy and confidentiality as “two sides of the same coin.”[102]

In the following section, the most important arbitration institutions will be compared. There are several arbitration institutions within the MENA region, such as the Institut Euro Mediterraneen d’Arbitrage (IEMA), the Centre Euro-Mediterraneen de Mediation et d’Arbitrage (CEMA), the Cairo Regional Centre for International Commercial Arbitration (CRCICA), Arab Centre for Commercial Arbitration (ACCA) and the Camera Arbitrale di Milano (CAM). Furthermore, there are also the “traditional” commercial arbitration institutions. These encompass the London Court of International Arbitration (LCIA), the International Chamber of Commerce (ICC) and the Stockholm Chamber of Commerce (SCC).[103] Finally, two important documents remain, which have shaped the existing international commercial arbitration tremendously. These are the UNCITRAL Model Law on International Commercial Arbitration (UNCITRAL Model Law), 1976 (amended 2010) the UNCITRAL Arbitration Rules (UNCITRAL Rules), and the 1985 (amended 2006) UNCITRAL Model Law in International Commercial Arbitration (UNCITRAL CAR).

2.1.3.3 The UNCITRAL Model Law and Arbitration Rules

For more than 40 years, the UNCITRAL was the UN’s main legal framework in the field of international trade.[104] Through conventions, model laws and legal guidelines, the UNCITRAL tries to harmonize, modernize and enhance international trade.[105] The UNCITRAL has released several important documents, such as the 1958 NY Convention, the UNCITRAL Rules, and the UNCITRAL CAR. These three documents are the most relevant legal texts if it comes down to international commercial arbitration. The UNCITRAL CAR is a proposal for national legislators[106] and has been adopted in at least 60 different states.[107] The UNCITRAL Rules on the other hand are an arbitration framework which the parties can choose.[108]

---

[101] Moss (2009), 782 (793).

[102] Kouris (2005), 127 (128); Confidentiality is the most advantageous feature of international arbitration, in: Trakman (2002), 1 (1).

[103] Wood (2007), 575 (579); cf Metje (2008), 159; ICC, in: Berger (2003), 65 (85); ICC, in: Lörcher (2001), 275 (276–277).

[104] UNCITRAL (2014a); Hughes (2010), 149 (149); Concerning the UNCITRAL Rules, in: Kutty (2006), 565 (576).

[105] Hughes (2010), 149 (149); UNCITRAL (2014a); cf Ahdab and Stackpool-Moore (2008), 275 (276).

[106] UNCITRAL (2014b); Dolzer and Schreuer (2008), 228; Hughes (2010), 149 (149–150); Proposal for modernization, in: Brower and Sharpe (2003), 643 (650).

[107] Hughes (2010), 149 (150); cf, smaller number of states, in: Harten (2007), 52.

[108] UNCITRAL (2014b).

Up to the present day, there has been application of the UNCITRAL Rules in international commercial arbitration and international investment arbitration.[109] For utilization of UNCITRAL Rules, both parties must agree on a respective arbitration clause in the contract.[110] There is a reluctance to define the term commercial, so it is left to the parties' discretion.[111] The UNCITRAL Rules only offer arbitration rules for a particular case, but they do not include an arbitration "body" such as the one the ICSID offers with special tribunals.[112] This legal framework can be used in cases of ad hoc arbitration or within existing institutions like ICSID or the LCIA.[113] If it comes down to ad hoc arbitration, the UNCITRAL Rules are the most popular legal framework.[114]

2.2 International Commercial Arbitration in Practice

The arbitration agreement is the cornerstone of international arbitration.[115] International law, and in most cases national law, require a written arbitration agreement.[116] A clear formulation of the arbitration agreement is important to avoid the question of tribunal jurisdiction or award enforcement.[117] The seat of arbitration determines which court supervises and supports the process, and what nationality the award has.[118] The seat of an arbitration institution can be different from the seat of the actual arbitration.[119] It is vital that the seat of arbitration favors international arbitration.[120] It is beneficial to conduct an arbitration agreement prior to a dispute and not afterwards.[121] Lastly, arbitration relies on the consent of both parties. [122]

[109] Horn (2008), 587 (588).

[110] Private autonomy of the parties, in: UNCITRAL (1981), 75 (78) para 17 accessed 07 January 2011; Great freedom for parties, in: Hanotiau and Caprasse (2008), 721 (721); Horn (2008), 587 (588).

[111] UNCITRAL (1981), 75 (80) para 31.

[112] Dolzer and Schreuer (2008), 227–228; Hughes (2010), 149 (150).

[113] Kutty (2006), 565 (571–572); Dolzer and Schreuer (2008), 228; Hughes (2010), 149 (150).

[114] Especially after the UNCITRAL Rules where used in the 1980s during the Iran–US Claims, in: Hughes (2010), 149 (150); McIlwrath and Savage (2010), 63 para 1–139.

[115] Hauschka (2005), 1550 (1552); Partasides and Fullelove (2010), 1 (3).

[116] Partasides and Fullelove (2010), 1 (3).

[117] McIlwrath and Savage (2010), 11–12 para 1-003; cf Partasides and Fullelove (2010), 1 (3).

[118] Especially the nationality of the award can be a deciding factor in the case of enforcement, in: McIlwrath and Savage (2010), 21 para 1-025; It is advised that the law governing the contract should be equal to the law of the seat of arbitration, in: Partasides and Fullelove (2010), 1 (4).

[119] McIlwrath and Savage (2010), 21–22 para 1-026.

[120] Partasides and Fullelove (2010), 1 (4).

[121] Partasides and Fullelove (2010), 1 (3).

[122] Wood (2007), 575 (586); Reason for arbitration, e.g. confidentiality, in: Harten and Loughlin (2006), 121 (140).

Consent to arbitration can be given in advance or after the dispute has arisen, whereas the latter limits arbitration to the respective dispute.[123]

2.2.1 *International Commercial Arbitration Institutions*

Which form of international commercial arbitration is better suited for the Desertec Concept? As mentioned above, there is institutional and ad hoc arbitration. Besides helping to establish a tribunal, institutional arbitration offers a kind of quality control during the process.[124] Some claim that institutional arbitration offers a greater acceptance of awards and the costs are more predictable.[125] It is also beneficial that institutional arbitration offers a book or rules for the participants, and there is a variety of well-trained arbitrators and institutions to review the award before transfer.[126] On the other hand, institutional arbitration can be expensive and tedious.[127] Nevertheless, there is also some support for ad hoc arbitration. Ad hoc arbitration can be fitted to each case individually and therefore is very flexible.[128] The claim that ad hoc arbitration offers more flexibility has not been proven through closer examination.[129] Ad hoc arbitration completely depends on the cooperation of the parties, so it can be delayed purposely by them.[130]

As mentioned above, the Desertec Concept is a large scale project. However, the literature is also divided on the issue on the best type of arbitration for large scale projects. Some authors in the field of international arbitration claim that ad hoc arbitration is best,[131] whereas others view institutional arbitration as being more practical.[132] To find a good solution for the Desertec Concept, it is necessary to review the different arbitration forms. This includes an overview of arbitration institutions and arbitration regulations of institutional and ad hoc arbitration.

---

[123] Harten (2007), 61.

[124] McIlwrath and Savage (2010), 39 para 1-070.

[125] Metje (2008), 158–159; McIlwrath and Savage (2010), 65–66 paras 1-145 and 1-146.

[126] Redfern et al. (2004), 48–49 para 1-102.

[127] Redfern et al. (2004), 49 para 1-103.

[128] Redfern et al. (2004), 49–50 paras 1-104 and 1-105.

[129] McIlwrath and Savage (2010), 63 para 1–141.

[130] Redfern et al. (2004), 50–51 para 1–107.

[131] Redfern et al. (2004), 50 para 1–106.

[132] In cases of ad hoc arbitration the parties must regulate the entire procedure, in: Euro-Mediterranean Association for Cooperation and Development e.V. (2010), 1; Metje (2008), 159.

2.2.1.1 International Center of Commerce

The 1923 International Court of Arbitration (ICA) is the oldest and most influential arbitration institution of the ICC in Paris and operates globally.[133] In 1919, after the end of World War I, businessmen from the US, the UK, Italy, Belgium and France founded the ICC to promote world peace,[134] but waited until 1923 to establish the ICA.[135] In 1946, the ICC was granted the highest level of consultative status within the UN.[136] The ICA is only an administrative body and its main purpose is to appoint the arbitrator(s).[137] It only has an administrative function and is not involved in arbitration award decisions.[138] Its focus is for example on the appointment or replacement of arbitrators.[139] All cases under the ICC Rules of Arbitration amended in 2012 (ICC Rules) are decided by a specially constituted ICA tribunal.[140]

Furthermore, the ICC encompasses the ICC national committees and the ICC case managers (part of the Secretariat of the ICC Court),[141] which are also involved in the arbitration process.[142] The Secretary General is appointed according to Art. 9 of the ICC Constitution (amended 2012) and has mostly administrative tasks. The Executive Board assesses the performance of the Secretary General (Art. 9(1)(b) ICC Constitution). Art. 2(2)(a), (b) and (c) and 3 of ICC Constitution include further details on the national committees. According to Art. 5(1) of the ICC Constitution, there is an annual meeting of the World Council, which is the supreme authority of the ICC. Art. 5(3) of the ICC Constitution clarifies that the World Council has only administrative functions. Furthermore, Art. 6 of the ICC Constitution mentions the Executive Board, which is in charge of implementing strategy, policies and programs of the ICC and oversees the financial affairs. There are also ICC Commissions (Art. 10 of the ICC Constitution), which prepare policy statements, recommendations and technical documents for the ICC. ICC Arbitration Rules are binding to parties that have agreed to comply with the award.[143] Awards

---

[133] Reiner, Jahnel in Schütze (2011), 26; McIlwrath and Savage (2010), 39 para 1-072; Kutty (2006), 565 (573); Metje (2008), 170; cf Dolzer and Schreuer (2008), 227; ICC is voice of the business world, in: Hodges et al. (2010), 55 (55) and (58); Concerning global operation, in: McIlwrath and Savage (2010), 39 para 1-072.

[134] ICC (2014); These renowned businessmen shared the idea that most countries are connected through trade and commerce and increasing this cooperation would lead to an avoidance of war, in: Hodges et al. (2010), 55 (55).

[135] Ergan (2012), 83 (84).

[136] ICC (2014).

[137] Dolzer and Schreuer (2008), 227; cf Metje (2008), 170.

[138] Hodges et al. (2010), 55 (55–56); McIlwrath and Savage (2010), 39–40 para 1-073.

[139] McIlwrath and Savage (2010), 39–40 para 1-073.

[140] Reiner, Jahnel in Schütze (2011), 26; Metje (2008), 170.

[141] Hodges et al. (2010), 55 (57).

[142] McIlwrath and Savage (2010), 40 para 1-074. 1-075.

[143] Pengelley (2009), 859 (864).

of international arbitration chambers like the ICC are subject to national arbitration laws.[144]

There is no other international arbitration institution with the same amount of experience in the field of international commercial arbitration.[145] The world's top arbitration experts guarantee a neutral and well-organized arbitration process.[146] Finally, there are several cases which have been negotiated under the ICC within the MENA region.[147]

2.2.1.2 London Court of International Arbitration

Like the ICC, the LCIA is an international institution,[148] and it offers a service which is as good as that of the ICC.[149] The LCIA has existed since 1986, and succeeded the 1891 London Chamber of Arbitration.[150] The LCIA deals with questions of commercial disputes as well as investment disputes.[151] It consists of a "three-tier structure", encompassing the Company, the Arbitration Court and the Secretariat.[152] The LCIA is a non-profit company limited by guarantee (incorporated in England and Wales).[153] The Company is managed by the Board of Directors, which does not interfere in the dispute resolution.[154] The Arbitration Court is the authority which appoints the arbitrators for the respective tribunal.[155] The Secretariat is responsible for the daily administration work of the LCIA.[156]

The LCIA and its benefits are quite comparable to the other arbitration institutions.[157] The LCIA is also cost-effective for high value arbitrations, offers special provisions on the security of costs, and special provisions on confidentiality.[158] Around 6 % of all cases at the LCIA stem from Africa.[159] One of the drawbacks for

---

[144] Semler (2003), 97 (99).

[145] Hodges et al. (2010), 55 (71).

[146] One of the best known arbitration institutions, in: Kutty (2006), 565 (572); Hodges et al. (2010), 55 (71–72).

[147] Aljazy (2000), 1 (2–4).

[148] One of the best known arbitration institutions, in: Kutty (2006), 565 (572); McIlwrath and Savage (2010), 47 para 1-099.

[149] Metje (2008), 172; Bellhouse et al. (2010), 101 (101).

[150] Dolzer and Schreuer (2008), 227; Bellhouse et al. (2010), 101 (101); cf Kutty (2006), 565 (574); cf McIlwrath and Savage (2010), 47 para 1-099.

[151] Dolzer and Schreuer (2008), 227.

[152] Bellhouse et al. (2010), 101 (103); LCIA (2014a).

[153] LCIA (2014a); Bellhouse et al. (2010), 101 (103).

[154] Bellhouse et al. (2010), 101 (103); LCIA (2014b).

[155] LCIA (2014b); Bellhouse et al. (2010), 101 (103).

[156] LCIA (2014b); Bellhouse et al. (2010), 101 (103).

[157] Bellhouse et al. (2010), 101 (112).

[158] Bellhouse et al. (2010), 101 (112).

[159] Bellhouse et al. (2010), 101 (102).

some parties is that the LCIA is not as involved in arbitration proceedings as other institutions.[160] In addition, the LCIA Arbitration Rules amended in 1998 (LCIA Rules) include only few provisions on multi-party arbitrations.[161] Unlike the ICC Court, the Arbitration Court does not scrutinize the arbitration award.[162] The LCIA does not differ a lot to the ICC, since most arbitrators are the same.[163] Finally, there is the question of the focus of the LCIA. Some claim that the LCIA is the proper choice if the applicable law is Anglo-American and future settlements should be in Anglo-American parts of the world[164]; although this is disputed.[165]

2.2.1.3 Stockholm Chamber of Commerce

Unlike the ICSID, ICC and the LCIA, the 1917 Stockholm Chamber of Commerce (SCC) is a regional arbitration institution which can arbitrate international cases.[166] Nevertheless, the SCC is a neutral, competent and high-quality arbitration institution.[167] It also attracts cases from all over the world.[168] The SCC Arbitration Rules amended in 2010 (SCC Rules) are similar to the UNCITRAL Rules.[169] The Arbitration Institute of the SCC is an independent entity within the Stockholm Chamber of Commerce, as stipulated in Art. 1 Appendix I of the SCC Rules. Art. 2 Appendix I of the SCC Rules mentions that the SCC does not decide the disputes itself, but assists. The SCC encompasses arbitrational dispute resolution within the framework of the SCC Institute administration (Art. 1 Appendix I of the SCC Rules). The daily management is handled by the Secretariat.[170] According to Art. 8 Appendix I of the SCC Rules, the Board can also assign functions to the Secretariat. The Board is the decision-making component within the SCC Institute and decides on questions of e.g. arbitrator nomination or prima facie decisions on jurisdiction, as stipulated in Art. 6 Appendix I of the SCC Rules.

---

[160] Kutty (2006), 565 (574); Bellhouse et al. (2010), 101 (112).

[161] Bellhouse et al. (2010), 101 (112).

[162] Bellhouse et al. (2010), 101 (103).

[163] McIlwrath and Savage (2010), 47–48 para 1–100.

[164] Bellhouse et al. (2010), 101 (102); Metje (2008), 172; 30 % of all cases stem from UK and US, in: McIlwrath and Savage (2010), 47 para 1-099.

[165] Bellhouse et al. (2010), 101 (112).

[166] Metje (2008), 173–174; Concerning establishment, in: SCC (2014a); McIlwrath and Savage (2010), 53 para 1–121.

[167] McIlwrath and Savage (2010), 53 para 1–121.

[168] SCC (2014a); Bagner and Dandenell (2010), 123 (123); McIlwrath and Savage (2010), 54 para 1–123.

[169] Hobér and Mckechnie (2007), 261 (261 et seq.); Metje (2008), 174.

[170] SCC (2014a); Bagner and Dandenell (2010), 123 (124).

SCC arbitration costs are relatively low and its process is fast.[171] The Swedish courts are also very arbitration-friendly and thus assist SCC arbitration.[172] Finally, through the amendments in 2010, the SCC Rules were modernized to better fit the contemporary arbitration needs. However, the SCC has not play an important role within the MENA region so far.

2.2.1.4 Cairo Regional Centre for International Commercial Arbitration

In 1978, the African–Asian Legal Consultative Committee (AALCC) decided to establish the CRCICA, as an institution promoted and established by developing state and as an alternative to the Western arbitration centers.[173] Nevertheless, the CRCICA is also a regional and not an international arbitration institution.[174] At the beginning, the CRCICA was on a 3-year "period of probation", but become a permanent institution in 1984.[175] The CRCICA is an independent, non-profit international organization.[176] It encompasses the Board of Trustees of the CRCICA, the High Legal Committees of the CRCICA, and the Board of Consultants of the CRCICA.[177]

Today, it is the most important arbitration institution on the African continent.[178] In addition, it is the best known arbitration institution in the Middle East.[179] The arbitration is according to the 2011 CRCICA Arbitration Rules (CRCICA Rules). It enjoys '(. . .) privileges and immunities of independent organizations in Egypt.'[180] This is also mentioned in Art. VIII of the Headquarters' Agreement for the Cairo Regional Centre for International Commercial Arbitration (Headquarters' Agreement). Art. IX of the Head Quarter's Agreement specifies that the purpose of granting the CRCICA immunity from state influence is to guarantee that the CRCICA functions successfully.

---

[171] cf Bagner and Dandenell (2010), 123 (137).

[172] Bagner and Dandenell (2010), 123 (137).

[173] Brower and Sharpe (2003), 643 (653); Euro-Mediterranean Association for Cooperation and Development e.V. (2010), 5; Vogl (2010), 32 (34).

[174] Mentioned under regional institutions, in: McIlwrath and Savage (2010), 53 et seq.

[175] Vogl (2010), 32 (34); Euro-Mediterranean Association for Cooperation and Development e.V. (2010), 5.

[176] CRCICA (2014b).

[177] CRCICA (2014a).

[178] Darwazeh and El-Kosheri (2008), 203 (206); Euro-Mediterranean Association for Cooperation and Development e.V. (2010), 5; cf in: Brower and Sharpe (2003), 643 (654); McIlwrath and Savage (2010), 57 para 1–129.

[179] McIlwrath and Savage (2010), 54–55 para 1–125; cf Brower and Sharpe (2003), 643 (653–654); Focus is on African-Asian commercial and investment disputes, in: Euro-Mediterranean Association for Cooperation and Development e.V. (2010), 5.

[180] CRCICA (2014b).

Its main goal is to support the progress of economic development within Africa and the Middle East.[181] The arbitration is mostly according to UNCITRAL Rules, with the influence of Egyptian culture and arbitration traditions.[182] Nevertheless, well-known arbitrators from all over the world belong to the list of arbitrators.[183] By cooperating with the CAM, the CRCICA also focuses on the MENA region.[184] The CRCICA also established the Alexandria Centre for International Arbitration together with the Alexandria Businessmen Association.[185] Later on, the Mediation & ADR Centre was established, which offers alternative dispute resolutions (ADR) in addition to arbitration.[186] However, European parties do not tend to use the CRCICA in cases which are not connected to Africa and the Middle East. Even in these cases, Western parties seem to rely more on the ICC for arbitration.

2.2.1.5 Arab Centre for Commercial Arbitration

In 1987, 14 Arab states concluded the Amman Arab Convention on Commercial Arbitration (AACCA).[187] This convention established the Arab Centre for Commercial Arbitration (ACCA) in Rabat (Morocco) (Art. 4 of the AACCA). The AACCA is modeled on the ICSID Convention.[188] In general, the AACCA applies to '(. . .) natural or juristic persons of any nationality, linked by commercial transactions with one of the contracting States or one of its nationals, or which have their main headquarters in one of these States' according to Art. 2 of the AACCA. Art. 22 of the AACCA mentions that the parties can agree on the seat of arbitration, but it must be approved by the arbitration tribunal. One of the requirements of the ACCA is that all arbitrations are conducted in Arabic, according to Art. 23(1) of the AACCA. There are doubts that any investor is willing to accept this concerning arbitration. Knowledge of Arabic is quite rare in comparison to knowledge of French and English. Furthermore, most international arbitration cases are conducted in English, French or Spanish. This also leads to a certain legal equality, due to its frequent practice and usage of legal procedures and terms. Consequently, the ACCA is not an alternative in the case of the Desertec Concept.

---

[181] CRCICA (2014b); Wider use of arbitration in the Islamic region, in: Brower and Sharpe (2003), 643 (654).

[182] Vogl (2010), 32 (34).

[183] Euro-Mediterranean Association for Cooperation and Development e.V. (2010), 5.

[184] Euro-Mediterranean Association for Cooperation and Development e.V. (2010), 6.

[185] Vogl (2010), 32 (34); Euro-Mediterranean Association for Cooperation and Development e.V. (2010), 6.

[186] Euro-Mediterranean Association for Cooperation and Development e.V. (2010), 6; Vogl (2010), 32 (34); CRCICA (2014b).

[187] Ouerfelli (2008), 241 (243–244); Brower and Sharpe (2003), 643 (654).

[188] Redfern et al. (2004), 642 para 10–60; Brower and Sharpe (2003), 643 (654).

2.2.1.6 Camera Arbitrale di Milano

The CAM was established in 1985,[189] and is the most important arbitration institution in the Mediterranean area.[190] It belongs to the Milano Chamber of Commerce and offers different types of arbitration and ADR.[191] Yet, it is an autonomous entity attached to the Chamber of Commerce of Milan.[192] The CAM is a national arbitration institution, and has a more "national caseload" than the SCC.[193] Nevertheless, the CAM belongs to the most well-known European arbitration institutions which can handle international arbitration cases.[194]

The Preamble of the Arbitration Rules of the Milan Chamber of Arbitration amended in 2010 (CAM Rules) stipulates the exact tasks of the CAM, which include the appointment of the arbitrators and the administration of the arbitral proceedings. Furthermore, the Preamble of the CAM Rules mentions the Arbitral Council, which has general competence '(...) over all matter relating to the administration of arbitral proceedings and issues all orders relating thereto, without prejudice to the Secretariat's functions (...).' Finally, the Preamble of the CAM Rules deals with the Secretariat. Its tasks are administrational (e.g. signing the Arbitral Council's orders or forwarding the award to the parties). The head of the Secretariat is the Secretary General.

The CAM established a new program called *Projet Méditerranéen*, which should help to make commercial arbitration more efficient and to support the EU Barcelona Process.[195] The project includes the setup of an arbitration network with different important arbitration institutions within the MENA region to exchange 'best practice.'[196] Up to the present day the CRCICA and the chamber of commerce of Beirut, Syria, Algeria and Istanbul have joined the above-mentioned project.[197] The ultimate goal is to accomplish a unified area of commercial arbitration.[198] Furthermore, a new institute (*Institut de l'Arbitrage et de la Médiation en*

[189] Cicogna (2010), 347 (349).

[190] Vogl (2010), 32 (32) and (34); Euro-Mediterranean Association for Cooperation and Development e.V. (2010), 4.

[191] Euro-Mediterranean Association for Cooperation and Development e.V. (2010), 4; Vogl (2010), 32 (34).

[192] CAM (2014a); Cicogna (2010), 347 (349).

[193] Cicogna (2010), 347 (349); McIlwrath and Savage (2010), 53 para 1–121.

[194] McIlwrath and Savage (2010), 55–56 para 1–127; cf Cicogna (2010), 347 (349).

[195] Vogl (2010), 32 (34); Euro-Mediterranean Association for Cooperation and Development e.V. (2010), 4.

[196] Euro-Mediterranean Association for Cooperation and Development e.V. (2010), 4–5; Vogl (2010), 32 (34).

[197] Euro-Mediterranean Association for Cooperation and Development e.V. (2010), 4–5; Except Istanbul mentions all different institutions, in: Vogl (2010), 32 (34).

[198] Vogl (2010), 32 (34); Euro-Mediterranean Association for Cooperation and Development e.V. (2010), 5.

Méditerranée) is going to be established to coordinate the network, advanced training courses and the ADR.[199]

By publishing all its statistical data, the CAM tries to be as transparent as possible.[200] In 2008, 27 % of all cases in front of a CAM tribunal were considered to be international arbitration.[201] Furthermore, the arbitration rules of the CAM are easier to understand than the rules of the CRCICA.[202] Overall, the CAM is an interesting alternative within the Mediterranean region. However, they do not deal with large scale projects as often as the ICC does. Thus, there may be concerns due to the CAM's lack of experience.

2.2.2 *Arbitration Proceedings*

After introducing the different institutions, the following section deals with a comparison of the different arbitration rules. Due to the historical and structural comparison of the different institutions, only the ACCA fails to meet the requirements. Therefore, analyzing all arbitration rules might help to find the best arbitration proceeding for the Desertec Concept. The comparison includes all institutional arbitration frameworks and the UNCITRAL Rules as the prominent example of ad hoc arbitration.

2.2.2.1 Jurisdiction of Arbitration

There are articles dealing with the jurisdiction of each arbitration institution (Art. 1 and 9(1) of the SCC Rules/Art. 6 of the ICC Rules/Art. 23 of the LCIA Rules/Art. 23 of the CRCICA Rules/Art. 1(1) and 11 of the CAM Rules/Art. 23(1) of the UNCITRAL Rules) The ICC Rules regulate the question of validity of the arbitration agreement in a detailed way, whereby the Court decides about it (Art. 6(9) of the ICC Rules). In addition the latest reform of the ICC Rules established an emergency arbitrator (Art. 29 ICC Rules and Appendix V to the ICC Rules with jurisdiction according to Art. 6 Art. Appendix V to the ICC Rules). The emergency arbitrator procedure is for particularly urgent interim measures. 23(1) and (2) of the CRCICA Rules even says that the tribunal can check the validity of the contract of which the arbitration agreement is part of. Similar to the CRCICA, the LCIA Rules stipulate that the tribunal decides upon its own jurisdiction and the validity of the arbitration agreement, Art. 23.1 of the LCIA Rules. However, the LCIA tribunal

[199] CAM (2014b); Vogl (2010), 32 (34); Euro-Mediterranean Association for Cooperation and Development e.V. (2010), 5; cf Cicogna (2010), 347 (349).

[200] Cicogna (2010), 347 (350).

[201] Cicogna (2010), 347 (350).

[202] Vogl (2010), 32 (37).

does not have the competence to check the whole contract. The SCC Rules do not really stipulate regulations concerning jurisdiction, except that the Board may decide about the tribunal jurisdiction (Art. 9(i) of the SCC Rules). In the case of the CAM, the Arbitral Council decides about the admissibility of the arbitration (Art. 11(1) of the CAM Rules). The decision of the Council is not binding for the tribunal, according to Art. 11(2) of the CAM Rules. According to Art. 23(1) of the UNCITRAL Rules, the arbitral tribunal rules on its own jurisdiction, and the validity of the arbitration agreement and the respective contract.

2.2.2.2 Arbitrator

It is important that the arbitrators are aware of the law of the country where the award is going to be enforced.[203] Otherwise, the award could be just a "waste of paper".[204] This includes a well balanced choice of arbitrators, as there are stereotypes that Western arbitrators only reflect their cultural background.[205] However, there are often very few arbitrators from Arab countries. This also reflects ICC arbitration. Within ICC practice, only a small amount of chosen arbitrators stem from the Middle East or Africa.[206]

All the different arbitration rules include special provisions on the appointment of the arbitrators. One general principle that all rules stipulate that the arbitrator must be impartial and independent (Art. 14(1) of the SCC Rules/Art. 11(1) of the ICC Rules and Art. 2(4) Appendix V to the ICC Rules/Art. 5.2 of the LCIA Rules/ Art. 8(4) of the CRCICA Rules/Art. 1(1) and 6–7 (Code of Ethics of Arbitrators) of the CAM Rules/Art. 8 and 6(7) of the UNCITRAL Rules).

The number of arbitrators is also an important, because parties might be confident with 1, 3 or more arbitrators. In some arbitration rules, there are three arbitrators (Art. 12 of the SCC Rules/Art. 10(1) of the CRCICA Rules/Art. 7 (1) of the UNCITRAL Rules). However, there are also few arbitration rules which use just one arbitrator, unless this has been agreed to by the parties (Art. 13(1) and (2) of the ICC Rules and Art. 2(1) Appendix V to the ICC Rules/Art. 5.4 of the LCIA Rules/Art. 13(2) of the CAM Rules).

Furthermore, there is the possibility to challenge an arbitrator under all the arbitration rules (Art. 15 of the SCC Rules/Art. 14(4) of the ICC Rules and Art. 3 Appendix V to the ICC Rules/Art. 10.3 of the LCIA Rules/Art. 11–13 of the CRCICA Rules/Art. 19 of the CAM Rules/Art. 12 of the UNCITRAL Rules). Finally, there are regulations concerning the release or replacement of an arbitrator

[203] Euro-Mediterranean Association for Cooperation and Development e.V. (2010), 1; cf Ahdab and Stackpool-Moore (2008), 275 (278).

[204] Euro-Mediterranean Association for Cooperation and Development e.V. (2010), 1.

[205] Ahdab and Stackpool-Moore (2008), 275 (277).

[206] Although the number is increasing, in: Nariman (2004), 123 (125); Ahdab and Stackpool-Moore (2008), 275 (277–278); Opposite opinion, in: Hammoud and Houerbi (2008), 231 (240).

(Art. 16 and 17 of the SCC Rules/Art. 12 of the ICC Rules/Art. 10 of the LCIA Rules/Art. 13–14 of the CRCICA Rules/Art. 20 of the CAM Rules/Art. 14 of the UNCITRAL Rules).

2.2.2.3 The Arbitration Process

Some arbitration institutions also include a code of conduct for the arbitration (Art. 19 of the SCC Rules/Art. 22 ICC Rules/Art. 14 of the LCIA Rules/Art. 17 of the CRCICA Rules/Art. 17 of the UNCITRAL Rules). Only the CAM Rules do not have a section or article specifically governing how the arbitration is to be conducted.

All arbitration rules stipulate that the language can be chosen by the parties or will be chosen by the tribunal (Art. 21(1) of the SCC Rules/Art. 4(1)(h) and 20 of the ICC Rules and Art. 1(1)(g) and (4) of the Appendix V to the ICC Rules/Art. 5 (1) and (2) of the CAM Rules/Art. 19(1) of the UNCITRAL Rules). Only Art. 19 (1) of the CRCICA Rules regulates that the language is only subject to the party's agreement. Furthermore, Art. 17(1) of the LCIA Rules highlights that the language of arbitration will be according to the language of the arbitration agreement if the parties do not choose differently. If there are more language versions of the arbitration agreement, the tribunal will decide which language is applicable for the process, according to Art. 17.2 of the LCIA Rules.

All the arbitration rules include provisions on the seat of arbitration (Art. 20 of the SCC Rules/Art. 18 of the ICC Rules and Art. 4 of the Appendix V to the ICC Rules/Art. 16 of the LCIA Rules/Art. 18 of the CRCICA Rules/Art. 4 of the CAM Rules/Art. 18 of the UNCITRAL Rules). However, Art. 20(3) of the SCC Rules specifically mentions that no matter where the hearing, meeting or deliberation is held, the arbitration will nonetheless be treated if it were being done at the seat of arbitration.

2.2.2.4 Applicable Law, Amendments and Timetable

All arbitration rules say that the applicable law relates to the choice by the parties (Art. 22(1) of the SCC Rules/Art. 21(1) of the ICC Rules and 1(1)(g) of the Appendix V to the ICC Rules/Art. 22.3 of the LCIA Rules/Art. 35(1) of the CRCICA Rules/Art. 2(1) and 3(1) and (2) of the CAM Rules/Art. 35(1) of the UNCITRAL Rules). If the parties have not chosen any law, the tribunals can determine which law is best for the respective case (Art. 22(1) of the SCC Rules/ Art. 21(1) of the ICC Rules and 1(1)(g) of the Appendix V to the ICC Rules/Art. 22.3 of the LCIA Rules/Art. 35(1) of the CRCICA Rules/Art. 3(3) of the CAM Rules/Art. 35(1) of the UNCITRAL Rules). Only the CAM Rules mention that certain mandatory provisions always apply to arbitration, according to Art. 2(2) of the CAM Rules.

There is also the possibility to file amendments or supplements to the claim (Art. 25 of the SCC Rules/Art. 25(5) of the ICC Rules/Art. 22.1(a) of the LCIA Rules/Art. 22 of the CRCICA Rules/Art. 27 of the CAM Rules/Art. 22 of the UNCITRAL Rules).

Furthermore, some arbitration rules have a special regulation concerning a provisional timetable (Art. 23 of the SCC Rules/Art. 24(2) of the ICC Rules/Art. 17(2) of the UNCITRAL Rules). The advantage of this kind of timetable lies in the fact that the proceeding is better structured, more efficient and foreseeable for the parties. Art. 32(1) of the CAM Rules stipulates that the tribunal '(...) shall file the final award with the Secretariat within six months from its constitution, unless otherwise agreed by the parties in the arbitration agreement.' The CRCICA Rules do not have any provision concerning arbitration duration within its statutes, but Art. 3(2) of the CRCICA Rules offers the possibility that the parties agree on a maximum duration of the arbitration process.

2.2.2.5 Multi-Party Arbitration

As illustrated above, PPP projects include a lot of different actors. Within a PPP, all contracts are individually concluded. As they are unconnected, they can develop differently.[207] Therefore, there is a need for multi-party arbitration.[208] There are certain requirements for multi-party arbitration. Firstly, an examination of all necessary multi-party arbitration requirements is vital. To be able combine several arbitration process into one proceeding, all disputing parties must agree in writing.[209] It is also possible that there is a post-dispute consent after the initiation of proceedings.[210] However, it is hard to assume at what point such a pooling is desired or not.[211] Furthermore, all parties must have the same chance and possibility to influence the arbitration tribunal's setup.[212]

This raises the question of whether there is any benefit in combining related disputes into multi-party arbitration.[213] There are cost benefits in cases of multi-party arbitration.[214] Furthermore, a previous coordination can help to prevent inconsistent or contradictory results.[215] In addition, there must be a solution for

[207] Nicklisch (1994), 53 (54) and (55).

[208] Nicklisch (1994), 53 (53).

[209] Lamm et al. (2009), 54 (55–56); Nicklisch (1994), 53 (57).

[210] Lamm et al. (2009), 54 (64).

[211] Different arbitrational courts exclude a pooling and the fact that different contracts refer to each other is not sufficient enough, in: Nicklisch (1994), 53 (57–58).

[212] This requirement is not problematic, if the respective parties previous agreed upon a multi-party arbitration and its setup, in: Nicklisch (1994), 53 (58); Even the French *Cour de Cassation* pointed out that ability to choose an arbitrator is part of the principle of equality and thus belongs to public policy, in: Nicklisch (1994), 53 (59).

[213] Nicklisch (1994), 53 (57).

[214] Lamm et al. (2009), 54 (69).

[215] Lamm et al. (2009), 54 (70).

the eventuality that not all the parties in a multi-party case are able to select an arbitrator.[216]

However, international multi-party litigations are rather rare. By dividing the process into single phases, where each party was able to state their claims, the result was a delay of the litigation.[217] The biggest problem of multi-party arbitration lies in the fact that most rules of procedure are set up for two-party litigation.[218] Furthermore, the principle of confidentiality is harder to observe, which represents the main benefit of choosing arbitration.[219] Longer proceedings also lead to higher costs and decrease the proceedings' efficiency.[220] Finally, legal unity cannot be guaranteed, because of the different development of contracts.[221]

De facto multi-party arbitrations are often split into the respective two-party constellations and because of less confidentiality, parties frequently reject a 'legal pooling.'[222] In addition, the questions of higher costs, less efficiency and legal unity do not support the application of multi-party arbitration.

With regard to multi-party arbitration, all arbitration regulations include a section concerning procedural details (Art. 13(4) of the SCC Rules/Art. 7 and 8 of the ICC Rules/Art. 8 of the LCIA Rules/Art. 8 of the CRCICA Rules/Art. 15 of the CAM Rules). Finally, there is also an increasing demand for multi-contract arbitration, but only Art. 6(2) and 4(2) of the ICC Rules deal with this issue specifically. Overall, multi-party arbitration is not advisable due to the numerous drawbacks.

2.2.2.6 Evidence, Hearings, Witnesses and Interim Measures

All arbitration rules have special statutes relating to evidence and hearings, whereas most of them are quite similar (Art. 26 and 27 of the SCC Rules/Art. 20(5) and 21 of the ICC Rules/Art. 22.1(f) and 16(2) of the LCIA Rules/Art. 27 and 28 of the CRCICA Rules/Art. 25 and 24 of the CAM Rules/Art. 27 and 28 of the UNCITRAL Rules). The SCC Rules and the ICC Rules are the most detailed arbitration rules concerning issues of evidence. In the Appendix IV to the ICC Rules new regulations concerning the documentary of evidence were introduced as part of a new case management technique.

There are special regulations concerning witnesses and appointed experts. Unlike the LCIA there is no requirement for a witness statement in writing (Art. 28(1) of the SCC Rules/Art. 25(3) of the ICC Rules/Art. 20.3 of the LCIA Rules/

---

[216] Possible solution in Article 1126 NAFTA, in: Lamm et al. (2009), 54 (71–72).

[217] Procedural efficiency, in: Lamm et al. (2009), 54 (69); Nicklisch (1994), 53 (60).

[218] Nicklisch (1994), 53 (60).

[219] Lamm et al. (2009), 54 (72–74); Nicklisch (1994), 53 (61).

[220] Nicklisch (1994), 53 (61–62).

[221] Further references, in: Nicklisch (1994), 53 (62).

[222] Nicklisch (1994), 53 (65–66).

Art. 27(2) of the CRCICA Rules/Art. 26 (1) of the CAM Rules/Art. 27–29 of the UNCITRAL Rules). The question of the tribunal appointing an expert is mostly the same within all arbitration rules (Art. 29 of the SCC Rules/Art. 25(3) of the ICC Rules/Art. 20(1), 20(2) and 20(7) of the LCIA Rules/Art. 27 of the CRCICA Rules/ Art. 26 of the CAM Rules).

With regard to interim measures, almost all tribunals have the same regulations (Art. 32 of the SCC Rules/Art. 28 and 29 of the ICC Rules and Art. 1(1)(e) of the Appendix V to the ICC Rules/Art. 25 of the LCIA Rules/Art. 26 of the CRCICA Rules/Art. 22(2) of the CAM Rules/Art. 26 of the UNCITRAL Rules). In almost all cases, the possibility to conclude an interim measure exists "if it deems appropriate." Based on the fact that the SCC process is a national arbitration, the interim measures are according to Swedish law. Therefore, they can be more effective than international interim tools.

2.2.2.7 Confidentiality, Consolidation and LCIA Specialties

Another important fact of arbitration is confidentiality. All arbitration rules include special regulations concerning the confidentiality of tribunals (Art. 46 of the SCC Rules/Art. 22(3), Art. 6 of the Appendix I of the ICC Rules, Art. 1 of the Appendix II to the ICC Rules/Art. 30 of the LCIA Rules/Art. 40 of the CRCICA Rules/Art. 8 of the CAM Rules). The UNCITRAL Rules addresses the issue of confidentiality in Art. 34(5) and demands that both parties consent. Parties can agree on special confidentiality agreements, e.g. by stipulating it within the arbitration agreement.[223] These agreements are always at risk of being challenged, interpreted differently by one party or violating the law.[224] It is important to analyze the applicable law and its regulations concerning the possibility to conclude confidentiality agreements.[225] Furthermore, confidentiality agreements only apply *inter partes*, which means third parties must sign separate agreements for example.[226] It must be kept in mind that there also exemptions to confidentiality, e.g. agreed waivers, time, geographic limits, or on the ground of public policy.[227]

There is also the possibility of consolidation in most arbitration rules, which means that combinations of different arbitrations involving the same parties and concerning the same legal relationship are possible (Art. 11 of the SCC Rules/Art. 10 of the ICC Rules/Art. 25 of the LCIA Rules/Art. 22(3) of the CAM Rules). Art. 11 of the SCC Rules points out that after consulting the parties, the Board can decide on consolidation. Consolidation is not regulated within the UNCITRAL Rules or the CRCICA Rules.

---

[223] Trakman (2002), 1 (11).

[224] Trakman (2002), 1 (12).

[225] Horn (2008), 587 (594); Trakman (2002), 1 (13).

[226] Trakman (2002), 1 (14–15).

[227] Trakman (2002), 1 (16–17).

Only the LCIA Rules and the ICC Rules include regulations concerning an expedited formation of the tribunal (Art. 29 and Appendix V of the ICC Rules/ Art. 9 of the LCIA Rules). This is only possible in an exceptionally urgent case (Art. 29(1) ICC Rules and Art. 1(1)(e) of the Appendix V of the ICC Rules/Art. 9(1) of the LCIA Rules). It can help to minimize the outage of a project. Especially the updated Appendix V of the ICC Rules encompasses a detailed and comprehensive set of regulations concerning the application of an emergency arbitration. It is noteworthy the ICC tribunal will decided in the form on an order and not an award according to Art. 6(1) of the Appendix V of the ICC Rules. This might cause issues with a later enforcement, which is going to be discusses later. Lastly, only Art. 22(1)(a) of the LCIA Rules allows the direct involvement of a third party, if one of the disputing parties applies, in the case in question.

2.2.2.8 Arbitration Awards and Costs

All arbitration rules have special provisions on the requirements of an award (Art. 36 of the SCC Rules/Art. 30–35 of the ICC Rules/Art. 26 of the LCIA Rules/Art. 33–41 of the CRCICA Rules/Art. 30 of the CAM Rules/Art. 34 of the UNCITRAL Rules).

In cases where three arbitrators are in a tribunal, the decision will be by majority (Art. 35(1) of the SCC Rules/Art. 31(1) of the ICC Rules/Art. 26(3) of the LCIA Rules/Art. 33(1) of the CRCICA Rules/Art. 30(1) of the CAM Rules/Art. 33(1) of the UNCITRAL Rules).

Some even include a special provision on the effect of the award (Art. 40 of the SCC Rules/Art. 34(6) of the ICC Rules/Art. 26(9) of the LCIA Rules/Art. 34 of the CRCICA Rules). It is doubtful that this provision has any influence on the enforcement of the award, since other legal documents (e.g. NY Convention) deal specifically with this issue.

Finally, there is also the question of cost. All arbitration rules have special regulations concerning the calculation and composition of the arbitration costs (Art. 43 of the SCC Rules/Art. 30 and 31 of the ICC Rules/Art. 28 of the LCIA Rules/Art. 38 to 41 of the CRCICA Rules/Art. 35 to 38 of the CAM Rules/Art. 40 of the UNCITRAL Rules).

2.2.3 *Best Suited Arbitration Process*

Overall, almost all arbitration institutions and the UNCITRAL Rules have similar regulations concerning international commercial disputes. There are only a few "special" or exceptional regulations in the different arbitration rules, which make some sets arbitration rules stand out. However, these regulations are not as important as they exclude application of arbitration rules, which do not include them. Thus, it is difficult to recommend a certain set of arbitration rules.

2.2.3.1 Ad Hoc and Institutional Arbitration

First of all, the question of institutional arbitration or ad hoc arbitration is important. There are arguments supporting both these forms of arbitration. With regard to the Desertec Concept, both types have advantages and disadvantages. As mentioned above in the final stage, the Desertec Concept involves a lot of different actors. It would be advisable to unify the different contracts, especially concerning the question of arbitration. Institutional arbitration has several advantages. There is a list of arbitrators in cases of institutional arbitration. All these arbitrators have experience in the field of international arbitration.

Furthermore, cases within institutional arbitration are more predictable than in ad hoc arbitration. Institution rules and case law can make it easier for the parties to foresee the outcome of a case. This is also an example of the "know-how" of arbitration institutions, which the parties can use. Compensation of higher costs can happen through a faster arbitration process because the parties cannot delay institutional arbitration, unlike in the case of ad hoc arbitration, where delay caused by the parties is a big problem. Finally, the fact that institutional arbitration awards are more widely recognized than ad hoc awards underlines that parties to the Desertec Concept should choose institutional arbitration.

2.2.3.2 SCC, LCIA and CRCICA

Secondly, one question remains—which institutional arbitration framework best suits the Desertec Concept. It is important that the chosen arbitration court has some experience with conducting an arbitration process.[228] As mentioned above, European dominance is often understood as neo-colonialism in Africa.[229] Therefore, the selected arbitration institution should be acceptable for MENA countries. As highlighted above, almost all courts have negotiated cases with African or Middle Eastern parties. There are no indicators that any MENA country completely opposes any of the above-mentioned arbitration institutions. Most authors view the ICC, SCC and the LCIA as Western arbitration courts. Since the CAM is a rather new player in the field of international arbitration, it is impossible to categorize them.

First of all, there is the LCIA, which has experience with African parties. Yet, the big problem lies in the fact that the LCIA is not as involved in the arbitration process as other arbitration institutions. This involvement is one of the benefits, which ad hoc arbitration cannot offer, due to the missing of a permanent institution. In the case of the Desertec Concept, which potentially involves various actors, guidance is essential. Since the Desertec Concept is going to provide energy for numerous people, institutional "leadership" is vital to guarantee a fair and fast

---

[228] Euro-Mediterranean Association for Cooperation and Development e.V. (2010), 3.

[229] Euro-Mediterranean Association for Cooperation and Development e.V. (2010), 3.

arbitration process. In addition, although the LCIA deals with international arbitration, its main focus seems to be on international disputes related to the Anglo-American region. Most parties choosing the LCIA stem from a comparable cultural background, which supports the aforementioned assumption. Thus, MENA parties might not feel comfortable dealing with the LCIA.

Secondly, there is the SCC. One of the biggest advantages of the SCC is the Swedish courts, which can help to support arbitration. However, this only applies if Sweden is the chosen place of arbitration. As mentioned above, the SCC's focus is not on Africa, which led in the past to an insignificant amount of cases with African parties being brought in front of the SCC tribunals. There are no doubts concerning the capability of the SCC as an international arbitration institute. However, as with the LCIA, MENA countries might not feel comfortable in front of the SCC. Due to the lack of cases involving parties from the MENA region, the SCC might not have the necessary experience and knowledge concerning contractual, interpretational and trade traditions.

In the case of the CRCICA, it is almost the other way around. Certainly the CRCICA knows a lot about Arab traditions, but it is a rather uncommon arbitration institute for Western parties. Like the SCC, it is a regional arbitration institute with a focus on African and Arab countries. Several documents also underline its independence from the Egyptian government. Western parties might be reluctant to choose the CRCICA due to its limited experience and practice. CRCICA Rules also have the most loopholes compared to the other arbitration rules. The two main problems are the arbitration duration and multi-party arbitration. There is no time limit for the arbitration process in front of the CRCICA. This can result in a long arbitration and therefore affect a lot of potential customers of the energy. Although the CRCICA Rules mention the possibility to conduct multi-party arbitration, there are no detailed regulations. As highlighted above, multi-party arbitration is difficult as it is. CRCICA Rules are also the most difficult arbitration rules of those examined above. Consequently, the CRCICA is not advisable for the Desertec Concept.

2.2.3.3 CAM and ICC

Only two arbitration institutions remain, namely the CAM and the ICC. Unlike all the other arbitration institutions, the CAM is a national institution. However, it also handles international arbitration. The most interesting thing about the CAM is its focus on the Mediterranean Sea and its adjacent states. It is also the only European arbitration institution actively cooperating with African and Arab arbitration institutions. This sharing of knowledge and competencies makes it stand out. The CAM Rules include all major arbitration regulations. However, the CAM does not have a lot of experience with large-scale projects, a factor which is very important considering the size of the Desertec Concept. If Italy is chosen as the seat of arbitration, the CAM might need the help of the Italian courts, which have a reputation of working rather slowly. Despite its lack of experience, the CAM could be the perfect compromise in the case of the Desertec Concept as it combines European and MENA tradition. The focus on the European/MENA region is noticeable as well as most of the Desertec Concept will take place within this region.

On the other hand, there is the ICC. The ICC is the most experienced arbitration institution in the world. The list of arbitrators includes people from all over the world. It is also a well-known arbitration institution within Africa and the Middle East. It is well recognized all over the world. It also has some experience with large-scale projects. A special feature of ICC arbitration is the review of the award before it is transferred to the parties.

The New ICC Rules came into force on 1 January 2012 (updating the Rules from 1998) and will, in general, apply to any ICC arbitration commenced on or after that date. Most changes within the ICC Rules are set out to make the arbitration process faster and cheaper compared to the Rules of 1998. Besides the abovementioned, there are further interesting new regulations within the ICC Rules of 2012. Art. 22 (1) of the ICC Rules now stipulates that the tribunal and the parties shall make every effort to conduct the arbitration in an expeditious and cost-effective manner, taking into account the complexity and value of the dispute. According to Art. 22(2) of the ICC Rules of the new arbitration rules, in order to ensure effective case management, the arbitral tribunal is authorized to adopt any procedural measures it considers appropriate, provided that they are not contrary to any agreement of the parties. Accordingly to Art. 22(3) of the ICC Rules, upon the request of any party, the arbitral tribunal is authorized to make orders concerning the confidentiality of the arbitration proceeding and may take measures to protect trade secrets and confidential information.

Art. 23 of the ICC Rules introduces the "Terms of References", which cannot be found within the LCIA, SCC, CAM and CRCICA Rules. The Terms of References are prepared by the tribunal in charge at the beginning of the arbitration proceeding (Art. 23(1) of the ICC Rules). Combined with Art. 24 of the ICC Rules and Appendix IV of the ICC Rules, the Terms of References are part of the early case management conference, which encompasses all parties and is part of the case management techniques. There is also the possibility of later case management conferences. Furthermore, Art. 9 of the ICC Rules has special regulations concerning multiple contracts. There it is confirmed that claims arising out of or in connection with more than one contract may be made in a single arbitration, irrespective of whether such claims are made under more than one arbitration agreement under the Rules. Art. 27 of the ICC Rules forces the tribunal to notify the Secretariat about its draft concerning an award after the proceedings are closed. Combined with Art. 22(1), this underlines the fact that the goal is to speed up the whole process. In addition, new penalties have also been introduced. According to Art. 37(5) of the ICC Rules, the tribunal may take into account circumstances which it considers relevant, including the extent to which each party has conducted the arbitration in an expeditious and cost-effective manner, to rule on the arbitration costs. If parties purposely hamper the arbitration process, they might face additional costs, even if the ruling is in their favor.

With regard to the Desertec Concept, the actual construction of the project will dictate whether the investor decides to build a decentralized or centralized network. In the case of a decentralized network, the CAM is a good option for arbitration. Due to its focus on the MENA region and its cooperation, it is the perfect compromise for all parties involved. Since not all the solar plants will be built at once, the CAM might have the chance to prove itself. If an investor wants to build a single plant, the ICC would be the more appropriate option. The ICC has the best experience with European-African dispute resolution and handling of large scale projects. Thanks to its new arbitration rules, the ICC's processes are no longer as lengthy and as costly as they were in the past. The new option to penalize party behavior makes it stand out from all other arbitration institutions. Thus, even a decentralized network would be best hosted by the ICC. The CAM offers a good alternative, but based on its lack of experience, it must be seen as the second best choice. Irrespective of the choice of arbitration institution, the complexity and size of the Desertec Concept project may lead to a long arbitration process. This underlines the importance of a well negotiated PPP to avoid conflicts right from the start of the project.

2.2.4 Enforcement of Awards According to the NY Convention

In 1958, the NY Convention replaced the 1927 Geneva Convention on the Execution of Foreign Arbitral Awards (Geneva Convention), after the ICC lobbied for a new, more advanced convention on the enforcement of arbitral awards.[230] One of the reasons was to get rid of the double exequatur (meaning that an award needed to be confirmed in the jurisdiction where it was made before it could be enforced in another jurisdiction) under the Geneva Convention.[231] Until today authors view the Geneva Convention as the starting point of the globalization of arbitration.[232] As the NY Convention only deals with the enforcement of awards, there are questions how Art. 6(1) of the Appendix V of the ICC Rules is going to be handled. In the case of an emergency arbitration the verdict is the form of an order. Since the ICC Rules use both terms (award and order) it is not clear how local courts will deal with this issue. A risk remains that local courts treat both terms differently. This could weaken emergency arbitration regulations of the new ICC Rules.

The new formulation in the NY Convention is in Art. V(1)(e), and highlights that an award must be binding. Although there is no definition of the term "binding", courts consider '(. . .) that an award is binding if there is no way of bringing an

[230] Nariman (2004), 123 (127); Harten and Loughlin (2006), 121 (125); Harten (2007), 50.

[231] Moses (2008), 212–213; BGH finally accepts the doctrine of merger and abandons the double exequatur in Germany, in: Plaßmeier (2010), 82 (82–83).

[232] Nariman (2004), 123 (126).

appeal on the merits.'[233] The most important advancement of the NY Convention lies in the fact that the arbitral award must not be in accordance with the law of the state where it is enforced—it is sufficient that the award complies with the laws of the state where the arbitration took place (Art. I, III and V of the NY Convention). Thus, the respondent state can only apply to review the decision or set the award aside according to the domestic law of the seat of arbitration in the courts of that state where the arbitration took place.[234] The NY Convention recognizes the above-mentioned importance of domestic courts, as several important provisions address them. This underlines the necessity of domestic courts to create a functional and effective international arbitration system. Even though the NY Convention did not live up to the expectations of international lawyers and arbitrators, 137 countries ratified it, and it became one of the most important conventions of international arbitration.[235] Therefore, the NY Convention expanded the scope for the enforcement of arbitral awards.[236] A lot of authors highlight the "pro-enforceability" of awards of the NY Convention.[237] Although the NY Convention does not mention states at all, there is no doubt that it also includes enforcement against states.[238]

Art. I(1), sentence 1 of the NY Convention sets the scope by stipulating that the Convention applies to the recognition and enforcement of arbitral awards in a state other than the state where the arbitration took place. Furthermore, in sentence 2, the Convention mentions that arbitral awards are '(. . .) not considered as domestic awards in the (s)tate where their recognition and enforcement are sought.' The difference between recognition and enforcement is that recognition means that a court regards the award as valid and binding, and enforcement means that the court uses all official means available to implement the award.[239]

There are only two reservation possibilities within the NY Convention. Art. I(3) of the NY Convention mentions that either the principle of reciprocity or a limitation to commercial dispute is possible reservations to the application of the Convention. Art. III of the NY Convention requires states to recognize arbitral awards as binding and to enforce them in accordance with national law and the NY Convention. According to Art. V of the NY Convention, arbitration awards can only be attacked on a limited

[233] In addition in most cases there is no option of appeal against arbitral awards, which makes them automatically binding, in: Moses (2008), 212 and 213.

[234] Harten and Loughlin (2006), 121 (136); cf Delaume (1985), 319 (344); cf Redfern et al. (2004), 4 para 1-05; Nmehielle (2001), 21 (42).

[235] Harten (2007), 51; Starting point for international utilization of arbitration, in: Nariman (2004), 123 (128); Ignorance of local judges to apply the NY Convention, in: Nariman (2004), 123 (132); Wood (2007), 575 (581); cf Lörcher (2001), 275 (277); Concerning the amount of ratifications, in: Saunders and Salomon (2007), 467 (468); NY Convention is a success story, in: Bjorklund (2009), 302 (308).

[236] Definition of enforcement and recognition, in: Ma (2005), 32–33; Moses (2008), 202; Harten (2007), 51; Harten and Loughlin (2006), 121 (125); cf Wood (2007), 575 (581).

[237] Hanotiau and Caprasse (2008), 721 (722); Ozumba (2009), 10; Ma (2005), 24; cf Nariman (2004), 123 (127).

[238] Bjorklund (2009), 302 (308).

[239] Moses (2008), 203.

scale. Both paragraphs of Art. V of the NY Convention mention that the award 'may be refused', which means that the judge is not forced to suspend the international arbitration award.[240] However, national courts cannot review the merits of the case, underlining the principle '(...) that a national court should not interfere with the substance of the arbitration.'[241] This includes that the award is not valid under the applicable law under certain cases: (1)(a), violation of the right of fair hearing (1)(b), missing or wrong interpretation of an arbitration clause (1)(c), unconstitutional establishment of the arbitration tribunal (1)(d), the award is not binding yet or has been suspended (1)(e), the subject matter cannot be settled by arbitration (2)(a), and a violation of public policy (2)(b). The incapacity mentioned in Art. V(1)(a) of the NY Convention can involve issues of sovereign immunity.[242] Furthermore, Art. V(1)(e) of the NY Convention is important, because in some countries certain matters are not arbitrational (e.g. criminal matters, bankruptcy validity of trademarks).[243]

There are two loopholes in the NY Convention, namely in Art. V(1)(e) and V(2)(b). As mentioned above, the award must be in accordance with the law of the seat of arbitration and only a court of the seat is able to review the award and challenge it. Concerning Art. V(1)(e) of the NY Convention, it is solely up to the respective local court to decide on the award's *vacatur* only according to domestic law.[244] Hence, the court can circumvent the strict invalidity requirements of Art. V of the NY Convention. Furthermore, Art. V(2)(b) offers a way to undermine arbitration awards due to the missing definition of public policy. Courts have frequently interpreted public policy in a very narrow way and only a few awards were suspended due to public policy reasons.[245] With regard to the Desertec Concept, the issue of public policy is subject to a separate examination. Finally, states have frequently highlighted the importance of the finality of arbitral awards.[246] Finality means that the arbitration court has '(...) the final word on the facts (...)' and that enforcement of a national court does not affect finality.[247]

---

240 Hanotiau and Caprasse (2008), 721 (722).

241 Hanotiau and Caprasse (2008), 721 (723); A review of all arbitration awards would destroy all advantages of arbitration, except in cases of violation of public policy, in: Hanotiau and Caprasse (2008), 721 (723).

242 Moses (2008), 208–209.

243 Moses (2008), 216.

244 Moses (2008), 214.

245 Hanotiau and Caprasse (2008), 721 (722); Harten (2007), 51; Wood (2007), 575 (581); Moses (2008), 218–219; Ozumba (2009), 5.

246 Clapham (2009), 437 (438); Important feature of international arbitration, in: Partasides and Fullelove (2010), 1 (10).

247 Clapham (2009), 437 (439); There is also a difference between appeal and annulment, whereas annulment supports finality over consistency and correctness, in: Clapham (2009), 437 (439); Distinction can also be found, in: Schreuer (2009), 901–903 paras 8–13; Comparable to Article 52 ICSID Convention, Article 43 UNCITRAL also adopted finality over consistency and correctness, thus Article 43 UNCITRAL does not offer the possibility to appeal and Article 34(2)(b)(ii) UNCITRAL offers the chance that an award is set aside, where it conflicts with public policy of the relevant state, in: UNCITRAL (1985), para 297; In addition states prefer finality in investor-state arbitration; Clapham (2009), 437 (448–450).

Although there are loopholes within the NY Convention, it represents a major step within international arbitration. Acceptance and enforceability of arbitration awards have increased tremendously. A possible investor must be aware of the fact that Morocco has options to circumvent an arbitration award. The NY Convention offers cases where enforcement can be refused, despite the binding character of the award. Morocco should be aware that abusing exceptional provisions could affect their reputation at an international level.

2.3 International Investment Arbitration

After finding the best forum for commercial arbitration, the next step is to find the best arbitration institution for investment arbitration. It is also possible to have ad hoc investment arbitration. However, based on the reasons above, ad hoc investment arbitration cannot be considered.

As mentioned above, investment arbitration deals with the ISDS. Usually the ICSID, ad hoc tribunals according to UNCITRAL arbitration rules or the ICC deal with matters of international investment arbitration.[248] Although courts like the ICC and LCIA also deal with matters of investment arbitration, their main field of action is still commercial arbitration.[249] Over the last few years, there has been an increase in the usage of international commercial arbitration to solve investment disputes.[250] Some parties chose investment arbitration, because it '(. . .) is swifter, cheaper, more flexible, and more familiar for economic operators.'[251]

2.3.1 Alternative Dispute Resolution Instead of Investment Arbitration

The costs of investment arbitration have increased dramatically over the last few years.[252] Furthermore, the long duration of arbitration processes and the issue of a missing appeal option against final verdicts are considered to be the disadvantages of investment arbitration.[253] To avoid these drawbacks the United Nations

[248] Zampetti and Sauvé (2007), 211 (228–229); Tietje (2010), 5 (8–9); Reinisch (2008), 107 (111); Vannieuwenhuyse (2009), 115 (127); Karl (1994), 809 (813).

[249] Dolzer and Schreuer (2008), 225.

[250] Moss (2009), 782 (793).

[251] UNCTAD (2010b), 14; Concerning arbitration in general, in: Ozumba (2009), 4.

[252] UNCTAD (2010b), 16.

[253] Further aspects arguing against investment arbitration benefits, in: UNCTAD (2010b), 16–21.

Conference on Trade and Development (UNCTAD) proposes ADR and dispute prevention policies (DPPs) instead of investment arbitration.[254] Both approaches require consent and cooperation because they are voluntary processes.[255] Both concepts also offer the possibility to find a solution without interpretation of the treaties and assessment of any violation.[256] Although most BITs do not mention ADR explicitly, they do not prohibit them either.[257]

ADR is a very complex term and is often used interchangeably with meditation, which is frequently confused with conciliation.[258] Therefore, utilization of the term ADR must happen in a very restrictive and well-defined way. With regard to the G/M-BIT, Art. 11 does not mention ADR or DPPs, which does not exclude them either. The term "gütlich beileg(en)"[259] might be an indicator that ADR is a possible solution to prevent arbitration. The ICSID Arbitration Rules also include something comparable to ADR. According to Rule 21(2) of the Arbitration Rules, the parties can request a pre-hearing to reach an amicable settlement prior to a process. With regard to the ICC, it is questionable if all ADR possibilities can be applied to an investor-state conflict.[260] The ICC also supports the setup of a "dispute board", which remains in place during the whole contract period and can help to solve issues which come up.[261]

Overall, ADR and DPPs are interesting new approaches to prevent investment arbitration. They are worthwhile supporting, especially in the case of the Desertec Concept but only in a supplementary way. As arbitration institutions or BITs already include forms of ADR, there is no need for them to be part of separate contracts between the host state and the investor.[262]

2.3.2 *Review of the ICSID Convention*

It is necessary to review the ICSID Convention in respect of arbitration regulations. The ICSID Convention encompasses three different types of proceeding, namely the arbitration proceeding, the conciliation proceeding and additional facility

---

[254] UNCTAD (2010b), 22 et seq.; cf Metje (2008), 145–146 and 155–156; The difference between ADR and arbitration rests in the fact that ADR does not require a previous consent and that during ADR process consent is required. DPP on the other hand comes from the host state and the investor's "duty" is to make use of the offered new mechanism, in: UNCTAD (2010b), 23.

[255] UNCTAD (2010b), 23.

[256] UNCTAD (2010b), 24.

[257] UNCTAD (2010b), 46.

[258] McIlwrath and Savage (2010), 174 para 4-005.

[259] To settle amicably.

[260] UNCTAD (2010b), 57–58.

[261] UNCTAD (2010b), 58.

[262] Metje (2008), 157.

proceeding. The first one is a "court-like" proceeding resulting in a binding verdict, and the second one is a proceeding led by a mediator which tries to achieve a consensual abatement of differences.[263] Like the first proceeding, the third one also reaches a binding judgment.[264]

ICSID has a simple structure, consisting of the Administrative Council and the Secretariat. The Administrative Council is the governing body of the ICSID and encompasses functions like the election of the Secretary-General and the adoption of the ICSID budget (Art. 4 to 8 of the ICSID Convention). The Secretariat consists of a Secretary-General, a Deputy Secretary-General and staff (Art. 9 of the ICSID Convention), whereby the Secretary-General is the legal representative organ of ICSID (Art. 11 of the ICSID Convention). The tasks of the Secretariat include providing institutional support for the initiation and assisting with the ICSID proceedings in the constitution arbitral tribunals (Art. 36 and 38 of the ICSID Convention). According to Art. 18 of the ICSID Convention, the Centre has full international legal personality and due to Art. 19 of the ICSID Convention immunity and privileges within the contracting states to fulfill its tasks.

2.3.2.1 Jurisdiction of Arbitration

Art. 25 of the ICSID Convention deals with the jurisdiction of the Centre. The specialty of Art. 25(1) of the ICSID Convention lies in the fact that the jurisdiction is limited to '(...) any legal dispute arising directly out of an investment (...).' According to Art. 41(1) of the ICSID Convention, the respective tribunal must decide on its jurisdiction in a case-by-case analysis.

However, jurisdiction is not as simple as it might seem. With regard to investments, there are a lot of different contracts involved. The most prominent are the BIT and the investment contract between the investor and the host state. The investment contract between an investor and the host state is also known as a state contract.[265] Concerning ICSID jurisdiction, the BIT is relevant as the major tool, which allows ICSID arbitration. If the BIT excludes this possibility, it is difficult to get under the ICSID Convention's umbrella. As mentioned above, breaches of the BIT (treaty claims) can be subject of ICSID tribunals' jurisdiction. BITs give the investor contractual rights against the host state, although there is no direct contractual relationship.[266]

---

[263] Hobe and Müller (2009), 65 (67); Tietje (2003), 5 (10).

[264] Tietje (2003), 5 (10).

[265] Metje (2008), 64; Schöbener et al. (2010), 242, Kap. 4 §16 para 81.

[266] Nariman (2004), 123 (134).

2.3.2.2 Treaty and Contract Claims

The main idea of the ICSID Convention is that contracts '(...) create rights in a domestic legal order whereas treaties create separate rights in the international legal order.'[267] A breach of a contract law is not automatically a breach of international law.[268] The distinction between contract and treaty claims also relates to the distinction between mere commercial acts and state interference (e.g. in the case of a state involvement because of public policy).[269] There is also a definition for both terms, whereby:

> [T]reaty claims are those that are based on an international agreement for breaches of international obligations (...) contained in BITs (...) and (...) contract claims are (...) based on rights that arise under (e.g.) concession agreement(s).[270]

Therefore, a difference between a breach of contractual obligations and a breach of the BIT exists.[271] Treaty and contract claims are different from each other, even if they arise out of the same fact.[272] This does not rule out the possibility that a breach of contract might also be a breach of the BIT.[273]

The tribunals frequently check the "fundamental basis of the claim".[274] However, there is no consistency concerning the scope of assessment, as some tribunals just mention that their case fulfills all requirements.[275] It is not sufficient that one party alleges that there is treaty claim. ICSID tribunals concluded that there are two major concerns to be addressed, which are:

> (...) [T]o ensure that courts and tribunals are not flooded with claims which have no chance of success, or may even be of an abusive nature; and equally to ensure that, in considering issues of jurisdiction, courts and tribunals do not go into the merits of cases without sufficient prior debate.[276]

Hence, the tribunal '(...) has considered whether the facts as alleged by the Claimant in this case, if established, are capable of coming within those provisions of the BIT which have been invoked (...).'[277] The ICJ decided the same question in a comparable. It pointed out that it cannot limit itself by the mere party statement

---

[267] Shookmann (2010), 361 (372); cf ICSID [2002] ARB/97/3—Annulment, 89 (128) para 96.

[268] Herdegen (2003), 13 (17), cf ICSID [2011] ARB/07/17—Award, 40 para 177.

[269] ICSID [2009] ARB/07/12, 34–35 paras 105–108; Further reference concerning commercial and investment contract, in: Nathan (1995), 27 (31).

[270] Rosenberg (2010), 8 (10).

[271] ICSID [2002] ARB/97/3—Annulment, 89 (128) para 96; cf Hauschka (2005), 1550 (1555).

[272] Fiebiger (2010), 269 (269); ICSID [2006] ARB/04/13, 25 para 79; ICSID [2005] ARB/03/29, 40 para 148.

[273] Fiebiger (2010), 269 (269); ICSID [2006] ARB/04/13, 22 para 64, 25 para 80, 84 para 258.

[274] ICSID [2002] ARB/97/3—Annulment, 89 (131) paras 100–101.

[275] ICSID [2003] ARB/01/13, 307 (351–352) para 145.

[276] ICSID [2005] ARB/03/3, 83 para 254.

[277] ICSID [2005] ARB/03/3, 83 para 254.

that the claim does or does not exist.[278] In a Separate Opinion, Judge Higgins emphasized the need to examine whether a dispute exists based on the claimant's allegation in order to find out if ICJ jurisdiction is plausible.[279]

ICSID tribunals have a chance to check its jurisdiction as the tribunal has the possibility at the 'very beginning to look behind the claimant's factual claims.'[280] However, the claimant must provide some evidence for ICSID jurisdiction.[281] This evidence must usually prove that the alleged claim is within the ICSID's jurisdiction.[282] If an investor bases its claim merely on the breach of the investment contract and not the BIT, the dispute settlement provisions of the BIT will not apply.[283] The investor must base its claim on the breach of BIT regulations to gain BIT dispute settlement options.[284] Some authors highlight that the investor should always try to formulate contract claims as treaty claims.[285] Art. 11(1) and (2) of the G/M-BIT stipulates that a disagreement concerning the investment is sufficient for dispute settlement.[286] Furthermore, Art. 11(2) of the G/M-BIT stipulates that ICSID jurisdiction is applicable for any investment dispute.

2.3.2.3 Waivers and Different Dispute Settlement Agreements

There are two common problems related to the issue of ICSID jurisdiction. The first one deals with the issue of explicit waivers. The investor might explicitly waive its right to ICSID jurisdiction, e.g. in a concession contract.[287] This can lead to denial of ICSID jurisdiction.[288] It is within the interest of many governments that the

[278] ICJ [1996] Oil Platforms, Preliminary Objection—Judgment [1996], 803 (809–810) para 16; cf ICJ [1999] Legality of Use of Force, Provisional Measures (Order of 2 June 1999) [1999] ICJ Reports, 481 (490) para 25.

[279] ICJ [1996] Oil Platforms, Separate Opinion of Judge Higgins [1996], 847 (856) para 32.

[280] ICSID [2003] ARB/01/13, 307 (351–352) para 145.

[281] ICSID [2007] ARB/05/07, 24 para 83; ICSID [2006] ARB/04/13, 22–23 paras 69–71; ICSID [2005] ARB/03/29, 52 para 195; ICSID [2003] ARB/01/13, 307 (351–352) para 145; ICSID [1999] ARB/98/4 (2002), 881 (891).

[282] ICSID [2007] ARB/05/07, 24–25 paras 85–86; ICSID [2006] ARB/03/15, 15 paras 42–44; cf ICSID [2004] ARB/02/8—Decision, 73–74 para 180.

[283] Loncle (2005), 3 (9).

[284] Loncle (2005), 3 (9).

[285] Rosenberg (2010), 8 (11).

[286] Article 11(1) G/M-BIT: Meinungsverschiedenheiten in Bezug auf Kapitalanlagen zwischen einem der Vertragsstaaten und einem Investor des anderen Vertragsstaates sollen, soweit möglich, zwischen den Streitparteien gütlich beigelegt werden.

Article 11(2) G/M-BIT: Kann die Meinungsverschiedenheit (...) nicht beigelegt werden, so wird sie auf Verlangen des Investors des anderen Vertragsstaats einem Schiedsverfahren unterworfen.

[287] ICSID [2008] ARB/06/13—Award, 45 paras 69–70.

[288] ICSID [2008] ARB/06/13—Award, 54 paras 96–97.

investor waives its right to ICSID jurisdiction.[289] The second issue deals with the problem of implicit waivers. Mostly, there are several contracts between the state and the investor (not the BIT), which might include special forum selection clauses.[290] In both cases, the investor might not be able to use ICSID arbitration anymore.

Yet, some ICSID tribunals have concluded that a forum selection clause does not waive the right of ICSID jurisdiction.[291] Other tribunals have mentioned that a forum selection clause only waives the rights concerning contract claims, but not treaty claims.[292] As other ICSID tribunals have mentioned, there is an option to waive ICSID jurisdiction even in the case of treaty claims.[293] The waiver must be formulated in an explicit way and an ICSID tribunal must deal with the waiver and its consequences within the respective case.[294] Furthermore, one ICSID tribunal mentioned that it is important to: '(...) give effect to any valid choice of forum clause in the contract.'[295] The literature does not support that a forum selection clause waives arbitration.[296] Some authors favor the idea to exclude a forum selection clause to avoid a complicated aftermath.[297] To avoid this dispute, the utilization of national courts or other arbitration tribunals should only be an option in the state contract, and not an obligation.[298]

There can be a problem if a BIT encompasses different dispute settlement regulations than the state contract. In the case of the G/M-BIT, Art. 11(2) stipulates that all disputes related to an investment are subject to ICSID regulation. Therefore, both parties should not agree on other dispute settlement forums (national or international). An investor in particular would risk losing ICSID jurisdiction, since even an implicit waiver (e.g. forum selection clause) might be interpreted as an exclusion of ICSID jurisdiction. The choice of the ICC or CAM should be clearly connected to questions of commercial disputes, since these cannot be subject to ICSID jurisdiction.

---

[289] Rosenberg (2010), 8 (8).

[290] Liebscher (2009), 105 (115).

[291] Specifically for concessions and BITs, in: ICSID [2011] ARB/07/17—Award, 42 para 180; ICSID [2002] ARB/97/3—Annulment, 89 (131) para 101; ICSID [1998] ARB/97/6—Preliminary Decision, 457 (466) paras 25–26.

[292] ICSID [2003] ARB/01/13, 307 (360–361) paras 161 and 162; ICSID [2002] ARB/97/3—Annulment, 89 (130) para 98 and (131) para 101.

[293] ICSID [2005] ARB/02/3, 29 para 118.

[294] ICSID [2005] ARB/02/3, 29 paras 118–119.

[295] ICSID [2002] ARB/97/3—Annulment, 89 (130) para 98.

[296] Vinuesa (2005), 331 (354); cf Rosenberg (2010), 8 (11).

[297] Dolzer and Schreuer (2008), 220.

[298] Griebel (2008), 103 and 104.

2.3.2.4 Arbitrator

Art. 14 of the ICSID Convention mentions that arbitrators must be independent and impartial. According to Art. 37(2)(a) of the ICSID Convention, the tribunal has a sole arbitrator, provided that the parties do not agree otherwise. If the parties do not agree, the number of arbitrators will be three (Art. 37(2)(b) of the ICSID Convention). If the tribunal is not constituted within 90 days, the chairman may appoint the missing arbitrators (Art. 38 of the ICSID Convention). Further questions of appointment are within Rules 3 to 5 of the Arbitration Rules. An arbitration tribunal according to the Arbitration Proceeding is not a real institution of the ICSID, because each case involves the foundation of a new tribunal.[299]

2.3.2.5 Arbitration Process

As with international commercial arbitration, consent is necessary. According to Art. 26 of the ICSID Convention, consent to ICSID arbitration leads to the exclusion of other remedies. Thus, the ICSID Convention includes an avoidance of contradictory verdicts on the same issue.[300] According to Art. 62 of the ICSID Convention, the seat of arbitration is at the seat of the Centre. Art. 63(a) and (b) of the ICSID Convention offer alternatives to the choice of seat. Chapter IV of the ICSID Convention deals with the arbitration procedure. Art. 36(1) of the ICSID Convention stipulates that the claimant must file a request to initiate the process. Paragraphs 2 and 3 of Art. 36 of the ICSID Convention deal with the necessary requirements of the request and the possible consequences of non-compliance.

Art. 43 to 45 of the ICSID Convention regulate specific question of the arbitration process. Art. 44 mentions the conduct of an arbitration process, provided that the parties do not agree otherwise. Furthermore, Chapters II and III of the Arbitration Rules stipulate the procedural and working issues of the tribunals. Rule 1(1) of the Rules of Procedure for the Institution of Conciliation and Arbitration Proceedings (Institution Rules) highlights that the request for arbitration must be in one of the official languages of the Centre. Other than that, the parties can agree on a language, based on Rule 22(1) of the Arbitration Rules. If the language is not an official language of the Centre, the tribunal must approve. Finally, Art. 56 to 58 of the ICSID Convention deal with the possibility to challenge an arbitrator.

---

[299] Hobe and Müller (2009), 65 (67); Constituting every time newly, in: Schöbener and Markert (2006), 65 (73).

[300] Griebel (2008), 120.

2.3.2.6 Applicable Law, Amendments and Timetable

According to Art. 42(1) of the ICSID Convention, the parties can agree on the applicable law and if they do not, the tribunal decides. Art. 46 of the ICSID Convention deals with the tribunal's ability to consider amendments or supplements to the claim. In addition, Rule 40 of the Arbitration Rules deals with ancillary claims and the possibility to present additional claims (Rule 40(1) of the Arbitration Rules).

Rule 26 of the Arbitration Rules offers the possibility to set up a fixed timetable for the arbitration process.

2.3.2.7 Multi-Party Arbitration

There are no explicit provisions concerning consolidation and multi-party arbitration within the ICSID Convention. In practice, there is the possibility to consolidate and create multi-party arbitration. Both can be seen in one case, where the tribunal consolidated two cases at the request of the respondent and the first and second claimants.[301] Although the ICSID Convention does not mention the option of multi-party arbitration, this possibility has already been discussed in the *travaux préparatoires* of the ICSID Convention.[302] Down to the present day, there have been several cases where ICSID tribunals have handled multi-party arbitrations.[303]

2.3.2.8 Evidence, Hearings, Witnesses, Provisional Measures and Confidentiality

Rule 34 of the Arbitration Rules deals with evidence, and Rules 35 and 36 of the Arbitration Rules contain provisions on the hearing of witnesses and experts. The possibility of provisional measures is addressed in Art. 47 of the ICSID Convention and Rule 39 of the Arbitration Rules. Finally, Rule 6(2) of the Arbitration Rules requires the arbitrators to sign a declaration that they will keep all information confidential.

2.3.2.9 Award and Costs

Finally, there are regulations concerning awards and costs. The provisions on costs are similar to the articles of the other arbitration institutions. Art. 59 to 61 of the

[301] ICSID [2006] ARB/03/13 and [2006] ARB/04/8, 3–5 paras 1–4 and 7.

[302] Lamm et al. (2009), 54 (60).

[303] Lamm et al. (2009), 54 (60).

ICSID Convention and Rule 28 of the Arbitration Rules deal with the costs of ICSID arbitration.

2.3.3 Awards Under the ICSID Convention

Section 6 of the ICSID Convention deals with the recognition and enforcement of an ICSID award (Art. 53 to 55 of the ICSID Convention). Art. 53(1) of the ICSID Convention stipulates that '(. . .) (t)he award shall be binding on the parties and shall not be subject to any appeal or to any other remedy except those provided for in this Convention (. . .).' Therefore, national courts cannot review ICSID tribunal decisions very easily.[304] Art. 54(1) of the ICSID Convention points out that there is no need for a declaration of execution by the local courts. As the paragraph emphasizes:

> (. . .) [A]ll Contracting states recognize an award rendered pursuant to this Convention as binding and enforce the pecuniary obligations imposed by that award within its territories as if it were a final judgment of a court in that State (. . .).

According to Art. 54(3) of the ICSID Convention, the execution of the award is according to the laws of the state where such an execution is sought. These obligations of international law take precedence regulations of the respective national arbitration law.[305] Besides the fact that national courts cannot review ICSID awards, other authors also point out that this applies to a review by international courts.[306] Interestingly, Art. 54(1) and (2) of the ICSID Convention refer to enforcement, whereas Art. 54(3) of the ICSID Convention mentions execution. Neither the French nor Spanish versions of the Convention have the same difference.[307] It is therefore proposed that enforcement and execution be treated in the same way.[308]

There is one little "stain" on the recognition and enforcement of ICSID awards. Art. 55 of the ICSID Convention specifically mentions that immunity from execution is untouched by the ICSID Convention. Also, section VI No. 43 of the Report of the Executive Directors states that doctrine of sovereign immunity may prevent execution of the award. The issue of state immunity will be assessed later on in this thesis.

---

[304] Mortenson (2010), 257 (266); Hauschka (2005), 1550 (1559); Krajewski (2009), 213; Hunter (2007), 165 (170); cf Herdegen (2003), 13 (33); Domestic courts are only capable of reviewing the authenticity of the award, in: Tietje (2003), 5 (11); The actual review of the ICSID award is done by an ICSID tribunal and not a national court, in: Hauschka (2005), 1550 (1559).

[305] Hauschka (2005), 1550 (1559).

[306] Tietje (2003), 5 (11); Hobe and Müller (2009), 65 (74).

[307] Bjorklund (2009), 302 (306).

[308] Bjorklund (2009), 302 (306).

Furthermore, the question of the relationship between Art. 53 and 54 of the ICSID Convention may cause difficulties. The respondent (Argentina) claimed that an award is only binding according to Art. 53(1) of the ICSID Convention if the investor has previously resorted to local enforcement proceedings according to Art. 54 of the ICSID Convention.[309] One tribunal however declined this approach and highlighted that Art. 53 and 54 address different subjects and obligations.[310] It especially underlined that:

> (. . .) [I]f the interpretation were accepted that there is no obligation to comply with an award unless and until the judgment creditor avails itself of enforcement mechanisms established pursuant to Article 54, the result could be that there would never be an obligation to comply with non-pecuniary obligations in an award.[311]

This decision was later widely favored and supported.[312]

2.3.4 *Difference to Enforcement of International Commercial Arbitration Awards*

Based on the articles of the ICSID Convention, there are three main differences concerning the enforcement of international commercial awards. First of all, the execution of the ICSID award is solely due to the ICSID Convention and not according to the NY Convention.[313] Due to the non-application of the NY Convention, the ICSID Convention prevents that the state uses public policy as a reason to prevent ICSID award enforcement.[314] Lastly, awards of other arbitration institutions (e.g. ICC) need recognition by the local courts (e.g. § 1061 of the German code of civil procedure), because of Art. 5 of the NY Convention, unlike ICSID awards.[315] However, this is only valid for ICSID awards according to the Arbitration Rules and not to the ICSID Additional Facility.[316]

The ICSID Convention benefits from Art. 54(1) of the ICSID Convention. It is a unique regulation which separates the ICSID Convention from other arbitration rules. If the award grants financial benefits, the ICSID judgment is automatically enforceable, without further recognition by domestic courts.[317] Thus, ICSID

---

[309] ICSID [2008] ARB/01/3—Annulment, 28 para 55.

[310] ICSID [2008] ARB/01/3—Annulment, 30–34 paras 61–70.

[311] ICSID [2008] ARB/01/3—Annulment, 33 para 66; Alexandrov (2009), 322 (336–337).

[312] ICSID [2007] ARB/97/3—Annulment, 1 (13–14) para 34.

[313] Saunders and Salomon (2007), 467 (469); cf Schöbener and Markert (2006), 65 (106–107).

[314] Reinisch (2008), 107 (112); Delaume (1983), 784 (801); Tietje (2003), 5 (11).

[315] In Germany ICSID awards are awards sui generis and they need no recognition according to § 1061 ZPO, in: Semler (2003), 97 (99); cf Griebel (2008), 120.

[316] Semler (2003), 97 (101).

[317] Schöbener and Markert (2006), 65 (107); Hobe and Müller (2009), 65 (74); cf Hunter (2007), 165 (170); McIlwrath and Savage (2010), 393 para 7-058; No scrutiny of domestic courts, in: Delaume (1983), 784 (785).

tribunals are able to award '(. . .) public law remedy (. . .)', because the award relates to compensation due to damage committed by the state.[318]

This does not alter the fact that the ICSID Convention cannot supersede the immunity from execution because this depends on the local immunity rules.[319] The domestic courts are obligated not to interfere with the ICSID proceedings.[320] The "only" purpose of national courts under the ICSID Convention is to recognize and enforce an ICSID award.[321] If the host state claimed immunity from execution, this would restore the option of diplomatic protection again.[322] This is also laid down in Art. 27(1) of the ICSID Convention. Furthermore, the contracting state whose national is involved could rely on Art. 64 of the ICSID Convention and submit the question of application of the ICSID Convention to the ICJ.

The risk that the state claims immunity from execution of the awards is relatively low because this would damage its credibility immensely, leading to fewer investments.[323] All arbitration courts face this problem, whereby the ICSID Convention deals with less issues of sovereignty than others.[324] It is also mentioned within the literature that ICSID awards are enforced more often than awards from other arbitration institutions.[325] The ICSID Convention and therewith investment protection benefit from the special provisions concerning the enforcement. Although the risk of immunity of execution remains, it is beneficial that public policy is excluded. As mentioned above, the provisions on public policy within the NY Convention are loopholes. Hence, there is a risk of abuse, which does not exist within the ICSID Convention. The issue of state immunity and its effect on arbitration awards will be discussed later in this chapter.

2.3.5 *Interpretation, Revision and Annulment*

Unlike other arbitration institutions, the ICSID Convention offers special provisions dealing with the interpretation, revision and annulment of awards. Within the ICSID Convention, Section 5 deals with interpretation, revision and annulment, encompassing Art. 50 to 52 of the ICSID Convention. Art. 50(1) of the ICSID

---

[318] Harten and Loughlin (2006), 121 (131); Individualized damage claims are rare in international law, in: Harten and Loughlin (2006), 121 (131).

[319] Nmehielle (2001), 21 (30–31); Moses (2008), 227; cf McIlwrath and Savage (2010), 393 para 7-058.

[320] Delaume (1983), 784 (785).

[321] Delaume (1983), 784 (785).

[322] Schöbener and Markert (2006), 65 (107–108); Delaume (1983), 784 (801).

[323] Hauschka (2005), 1550 (1559); Possibility to claim sovereignty was not used often, in: Tietje (2003), 5 (11).

[324] Mortenson (2010), 257 (266).

[325] McIlwrath and Savage (2010), 393–394 para 7-059; Reinisch (2008), 107 (112).

Convention stipulates that any request for interpretation of the award by the parties shall be submitted to Secretary-General. If possible, the original tribunal will deal with the award again. Art. 51 of the ICSID Convention includes the rules and requirements for a revision of the award. The deadline for a revision is 3 years after the award was rendered (Art. 51(2) of the ICSID Convention).

Art. 52 of the ICSID Convention deals with the question of annulment. Art. 52 (1) lists all the possible reasons for an annulment request in letters (a) to (e). In a normal case, an application for annulment must be made no later than 120 days after the award was rendered, Art. 52(2) ICSID Convention. In the case of corruption, the deadline is 120 days after the discovery of corruption, according to Art. 52(2) of the second alternative ICSID Convention. Pursuant to Art. 52(3) of the ICSID Convention, a new tribunal (annulment tribunal) is going to be set up. The annulment tribunal can decide that the award stays in force (Art. 52(5) ICSID Convention) or annuls it and submits it to a new tribunal, if requested by the parties (Art. 52(6) of the ICSID Convention).

Parties frequently resubmit the same case that one ICSID tribunal has already finished to "ICSID tribunals" again for new investment arbitration.[326] However, interpretations of annulment reasons in Art. 52 ICSID Convention happen in a narrow and reluctant way.[327] There are increasing numbers of annulment cases, which illustrates that the parties were not happy with the outcome.[328] Others authors claim that only 7.4 % of all annulment cases are successful.[329] This makes it quite difficult to talk about an increasing amount of annulments. There is support for this view.[330]

Despite certain drawbacks, there is a wide recognition and enforcement of ICSID awards among states.[331] The only comparable feature of other arbitration tribunals is the award review within the ICC. Nevertheless, it does not quite compare to the possibility of annulment. If parties use the annulment option in a reasonable way and ICSID tribunals decide in an understandable and correct way, it is an enrichment of international arbitration law. Yet, it can be a double-edged sword if parties abuse the annulment option. This causes the arbitration to last longer, thus creating a lot of additional costs. It is also difficult to prove that a party is abusing its rights. Down to the present day, there have only been a few cases repeatedly subject to annulment processes. Abusive behavior from any party would automatically cast doubt on their credibility. Both parties could not risk jeopardizing their credibility within the international community.

---

[326] Jagusch et al. (2010), 75 (98).

[327] Tietje (2005), 47 (57).

[328] UNCTAD (2010b), 15.

[329] Saunders and Salomon (2007), 467 (474); Comparable numbers for 2006, in: Schöbener and Markert (2006), 65 (105).

[330] Jagusch et al. (2010), 75 (98).

[331] Alexandrov (2009), 322 (323) and (329).

2.3.6 Evaluation of ICSID Jurisdiction

Overall, ICSID arbitration offers a better investment protection than other arbitration institutions, mostly due better enforceability of awards and the option of annulments. ICSID tribunals have gained a vast experience of negotiating investment disputes over the past few decades. The still rising number of cases also underlines the acceptance of ICSID as an international arbitration institution. The practical benefits of international investment arbitration are obvious (impartiality and effectiveness as just two examples).[332] It is repeatedly emphasized that there is no arbitration court as powerful as the ICSID, which makes it very successful.[333] Another benefit of ICSID jurisdiction is the above-mentioned neutrality of the ICSID tribunals. The ICSID process is completely self-contained (respectively self-sufficient) and therefore "delocalized".[334] Other arbitration convention, e.g. the NY Convention, leaves enforcement to domestic law or other applicable treaties, unlike the ICSID Convention.[335]

Furthermore, ICSID cost schedules are transparent and structured.[336] It is sometimes pointed out that the voluntary execution of ICSID verdicts is due to it being a part of the World Bank's institution.[337] The World Bank might interpret disobedience as a loss of political credibility.[338] Publication of all proceedings and core elements (e.g. parties to the disputes) are done by the Secretary-General online or in an ICSID Annual Report.[339] It is beneficial to publicize the arbitration process so that the state cannot "cloak" proceedings.[340]

---

[332] Walter (2006), 815 (816).

[333] Mortenson (2010), 257 (265) and (267).

[334] Reed et al. (2004), 8; Delaume (1983), 784 (784); Griebel (2008), 119; Moses (2008), 225–226; cf Brower et al. (2009), 843 (848); Horn (2008), 587 (592); Jagusch et al. (2010), 75 (98); Therewith the investor is not subject to political uncertainties, in: Vannieuwenhuyse (2009), 115 (119); Dolzer and Schreuer (2008), 223; cf Metje (2008), 167; Depoliticized, in: UNCTAD (2010b), XXII; Depoliticized, because no politicians are members of ICSID tribunals, in: Nmehielle (2001), 21 (24–25).

[335] Griebel (2008), 119; Reed et al. (2004), 8; cf Harten and Loughlin (2006), 121 (135).

[336] Reed et al. (2004), 9; Griebel (2008), 120–121; A more transparent ICSID proceeding today, in: Egonu (2007), 479 (483).

[337] Reed et al. (2004), 9; cf Hunter (2007), 165 (170).

[338] Adverse political and financial consequences, in: Brower et al. (2009), 843 (847); McIlwrath and Savage (2010), 393–394 para 7-059; Hunter (2007), 165 (170); cf Jagusch et al. (2010), 75 (97–98); Reed et al. (2004), 9; Pressure of international financial institutions, in: Harten and Loughlin (2006), 121 (134); Griebel (2008), 121.

[339] Egonu (2007), 479 (482); Lörcher (2005), 11 (16); cf Partasides and Fullelove (2010), 1 (10); Jagusch et al. (2010), 75 (97).

[340] There is an public interest in the publication of awards, because it often involves public services, in: Egonu (2007), 479 (487–488); Public interest, in: Tietje (2005), 47 (60); Great transparency for international arbitration, in: McIlwrath and Savage (2010), 391 para 7-051; Griebel (2008), 120; Reed et al. (2004), 9; Disclosure because of public policy grounds, in: Trakman (2002), 1 (5).

There are also some drawbacks to ICSID arbitration. Some claim that the obligation of publicity foils with the above-mentioned privacy.[341] ICSID tribunals are in favor of publishing cases and awards, even if one party is against this.[342] This is against to Rule 15 of the Arbitration Rules, which highlights in paragraph one that all '(...) deliberations of the (t)ribunal shall take place in private and remain secret.' Other authors mention that ICSID proceedings are as private as other international arbitration proceedings.[343] In addition, the public interest, which is higher than in cases of international commercial arbitration, creates a certain pressure to make proceedings public. The higher interest frequently relates to the possible involvement of tax money or important domestic projects. Another drawback of ICSID arbitration rests in the fact that treaty interpretation, the mixture of national and international as applicable law, and the dealings with state parties need specialists, which complicates the process.[344] Furthermore, the long duration of ICSID trials is a negative factor.[345] The average ICSID tribunal arbitration period is between 2 and 3 years.[346] However, it is important that arbitration tribunals reach a verdict within a reasonable amount of time.[347]

Overall, the benefits outweigh the drawbacks. There is no doubt that treaty interpretation and the mixture of laws require specialists, which applies to all forms of international arbitration. The duration of ICSID proceedings greatly depends on the behavior of the parties and their will to cooperate. In addition, the size of the project also determines the length of the arbitration. This is comparable to international commercial arbitration. It would be wrong solely limit this argument to ICSID arbitration. Most of the negative perceptions connected to ICSID arbitration also apply to other forms of international arbitration. One last benefit of ICSID jurisdiction remains. Even in the worst case scenario that Morocco (or any other MENA state) denounces the ICSID Convention, the state would still be bound by its previous consent to ICSID arbitration related to all for pending or predicted cases.[348] Hence, an investor should avoid waiving its right to ICSID jurisdiction, as this is a functional tool to protect an investor's rights. The same is valid for Morocco/ONE. Only in the case of ICSID jurisdiction do they have the possibility to annul an award. Most importantly, both parties must try to settle all disputes amicably and not to abuse any rights. Since it is a long-term project, it is necessary that there is a constantly positive climate.

---

[341] Vannieuwenhuyse (2009), 115 (121); Egonu (2007), 479 (487); Tietje (2010), 5 (17–18); cf Trakman (2002), 1 (5); For more publicity in investment proceedings, in: Lörcher (2005), 11 (16).

[342] Egonu (2007), 479 (485–486).

[343] Privacy to protect sensitive business or government information, in: Egonu (2007), 479 (482); Tietje (2010), 5 (17); Reed et al. (2004), 8.

[344] Jagusch et al. (2010), 75 (98).

[345] Hunter (2007), 165 (170); cf UNCTAD (2010b), 18.

[346] Hauschka (2005), 1550 (1558).

[347] Fiebiger (2010), 269 (270).

[348] Fouret (2008), 71 (76); UNCTAD (2010a), 2.

2.3.7 New Development Within Investment Arbitration

As mentioned above the ICSID forum is the best way to handle any investment disputes due to its specialization on international investment. However, there have been new developments within international investment arbitration, which are quite remarkable. On the 01.01.2014 the UNCITRAL Rules on Transparency in Treaty-based Investor-State Arbitration (UNCITRAL Rules on Transparency) entered into force. The main idea was to make the whole process public (Art. 1(4) UNCITRAL Rules on Transparency), as there is tax money involved.[349] The UNCITRAL Rules on Transparency can be applied to all bilateral and multilateral investment treaties, although it is restricted to all forms of ad hoc investment arbitration (Art. 1 (9) UNCITRAL Rules on Transparency).

It applies to all investment disputes after the day it entered into force, but parties can exclude it according to Art. 1(1) UNCITRAL Rules on Transparency. It can also be used concerning all pre-existing investment disputes, if the parties agree (Art. 1(2) UNCITRAL Rules on Transparency). Art. 3–5 and 8 UNCITRAL Rules on Transparency stipulate, how relevant data is made public (e.g. names of the parties, the legal claim, the defense etc.). Art. 6 UNCITRAL Rules on Transparency deals with public hearing, with exceptions in Art. 7 UNCITRAL Rules on Transparency.

In addition, starting from the 01.04.2014 the UNICITRAL Rules were added and Art. 1(4) UNCITRAL Rules on Transparency was introduced. It mentions that the UNCITRAL Rules on Transparency are now part of the UNICITRAL Rules, if the scope of Art. 1 UNCITRAL Rules on Transparency is met.

This illustrates that ad hoc investment arbitration is on a good way to become an equivalent tool of investment arbitration. It is very important to note that the UNCITRAL Rules on Transparency take the need of transparency into account, as most people are unsatisfied with procedure behind closed doors. The UNCITRAL Rules on Transparency are a good step towards a greater acceptance of international investment arbitration.

2.4 Important Prerequisites for International Arbitration

This section deals with important prerequisites which affect international commercial and investment arbitration. Hence, if there is not an explicit reference to either international commercial arbitration or investment arbitration, the following applies to both.

[349] Germany Trade and Invest (2013).

2.4.1 Double Litigation

There is no hierarchy of courts on an international level.[350] Thus, two different arbitration tribunals of a domestic court could deal with the same issue, i.e. double litigation. This can have several reasons, starting from an unclear arbitration agreement or a different interpretation up to a deliberate attempt to avoid undesirable arbitration institutions. Besides the fact that multiple litigations are costly, inconsistent verdicts are undesirable.[351]

There are two important legal doctrines in the case of multiple litigation, namely *lis pendens* and *res judicata*.[352] *Lis pendens* is concerned with several courts dealing with the same issue at the same time and codifies that all courts must wait for the verdict of the court, which dealt with the legal matter at first.[353] *Res judicata* deals with subsequent court proceedings, emphasizing that courts are '(. . .) bound to (. . .) recognize and give effect to foreign judgments which determine the matter in issue before it(. . .).'[354] Nevertheless, both doctrines are not able to solve all related problems, as issues like the legality of an arbitration provision would lead to *res judicata* becoming inapplicable.[355] Another tool to avoid multiple litigation is an anti-suit injunction, which prevents parties from escaping their arbitration obligations by initiating a different process.[356] Overall, an investor and Morocco/ONE should avoid getting different courts involved in the same issue. Various outcomes are not desirable for a successful long-term cooperation.

2.4.2 Applicable Law

Traditionally, the place of arbitration has determined the applicable law. However, this has started to change in recent years.[357] Some authors mention that it would be better not to choose domestic law if takes place in the MENA region.[358] If the applicable law for a concession is the host state law, there is a risk that the state will use its sovereignty and interrupt the contractual equilibrium.[359] In question of recognition and enforcement in particular, it is important to find alternative ways to the Shari'a.[360]

---

[350] Shookmann (2010), 361 (362).

[351] Shookmann (2010), 361 (362).

[352] Shookmann (2010), 361 (363).

[353] Shookmann (2010), 361 (363); Further requirements, in: Vinuesa (2005), 331 (354).

[354] Shookmann (2010), 361 (363–364).

[355] Weiler (2005), 516, Shookmann (2010), 361 (364).

[356] Sattar (2010), 51 (53) and (54); Shookmann (2010), 361 (364–365).

[357] McIlwrath and Savage (2010), 24–25 para 1-032.

[358] Vogl (2010), 32 (34).

[359] Herdegen (2003), 13 (17).

[360] Vogl (2010), 32 (34).

The choice of the applicable law is a highly sensitive issue for both contracting parties.[361] There are several laws important for international arbitration. There are the law of the seat of arbitration (*lex arbitri*), procedural rules, the law governing the arbitration agreement, the performance of the agreement, the law governing the contract, the law governing the recognition and enforcement of the award, and other applicable rules or guidelines.[362] In the following section, the focus will be on the law governing the contract. This section mainly deals with the categorization of contracts between Morocco and an investor. In addition, an analysis of the relationship between international and national law as applicable law will be carried out. Finally, stabilization clauses will be discussed.

First of all, there is a difference between *lex arbitri* (the law of the seat of arbitration) and the law governing the contract.[363] In most cases, the parties chose a "neutral" place of arbitration, without intending to apply *lex arbitri* to their contracts.[364] Besides the choice of national law, there are also other possible legal frameworks. The parties can also choose the *lex mercatoria* as the applicable law.[365] It is rare that the *lex mercatoria* is chosen as the only applicable law, as it is rather useful as assistance of contract negotiation or to supplement the law of the contract.[366] In practice, the choice of a third state law as the proper law did not pay off.[367] Lastly, the choice of international law as the applicable law remains. A choice of international law might be irrelevant if the contract between the investor and the state is subject to international law. Thus, both parties must be subject to international law. Traditionally, only states and international organizations are subject to international law.[368] However, international law does not have a numerus clausus of legal subjects.[369] Legal personality depends whether there is a possibility to assign the "party" their own legal position (rights and obligations).[370] As mentioned above, a possible investor, like Dii was, has the potential of being a TNC. After World War II, TNCs became increasingly important, which has to do with their growing economic power.[371] Hence, the question arises whether TNCs are subjects to international law.[372]

---

361 Dolzer and Schreuer (2008), 73–74; Important choice, in: Wood (2007), 575 (588).

362 Moses (2008), 63–75; Partasides and Fullelove (2010), 1 (4–5).

363 Redfern et al. (2004), 78 para 2-05.

364 Redfern et al. (2004), 78–79 para 2-05.

365 Moss (2009), 782 (785); Ouerfelli (2008), 241 (242); Moses (2008), 70; Buchanan (1988), 511 (529).

366 Moses (2008), 70.

367 Herdegen (2003), 13 (32); Especially in the case of the law of a third state, an application of labor law, tax law or administrative law would be difficult and impractical, in: Schreuer (2009), 559 para 27.

368 Schöbener et al. (2010), 59, Kap. 1 §5 para 194.

369 Dahm et al. (2002), 195 et seq.; Tietje (2005), 47 (61); Tietje (2003), 5 (16).

370 Schöbener et al. (2010), 60, Kap. 1 §5 para 197; Shaw (2008), 195; Verdross and Simma (2010), 221–222 para 375; Weh (2008), 143; Tietje (2005), 47 (61); Tietje (2003), 5 (16).

371 Dahm et al. (2002), 246; Schweisfurth (2006), 48–49 para 165.

372 Tietje (2005), 47 (61); Tietje (2003), 5 (16); Dahm et al. (2002), 245.

2.4.2.1 TNCs as Subjects of International Law

There are different views on the international law subjectivity of TNCs. First of all, there are some authors who completely oppose this idea. Contracts between a TNC and a state are not subject to international law, according to the dualistic approach.[373]

Nevertheless, there are also other authors arguing in favor of TNCs being subjects of international law. The first one is the international agreement doctrine. Their point of view starts with the idea of equating TNC/state contracts with international agreements under certain circumstances (e.g. completion of the contract by the highest state organs).[374] In these cases, the TNC gains a 'functional statehood.'[375] Especially if the state contract includes an internationalization or stabilization clause, some arbitration courts accepted that the contract is partially an international contract.[376] It must be noted that this doctrine is not customary international law.[377] In the case of investment protection (ICSID Convention), authors claim that TNCs have a partial legal subjectivity[378] According to the ECT companies have the right to file claims under ICSID jurisdiction because all member states have given their unconditional consent.[379] Art. 26(3) of the ECT includes consent to ICSID arbitration, which makes it comparable to the ECHR. Therefore, the investor can file a claim without the consent of the state. All this is connected to the derived legal subjectivity of international law established for international organizations.[380]

Yet, there are some arguments against the point of view that TNC are legal subjects of international law. First of all, legal personality to international law does not depend on state recognition, as the founding document of the international organization establishes legal subjectivity of an international organization.[381] Recognition can also be revoked by a state, without violating the pacta sunt servanda doctrine, but leading to a negation of the partial legal personality.[382] There are no international customs concerning the international agreement doctrine.[383]

[373] Epping in Ipsen (2011), §9, para 8 et seq.; Doehring (2004), para 397; Weh (2008), 158.

[374] Schöbener et al. (2010), 61, Kap. 1 §5 paras 199 and 246; cf Krajewski (2009), 182; cf Schweisfurth (2006), 51 para 166; Muchlinski (2009), 341 (342); Including further references, in: Weh (2008), 158–159.

[375] Weh (2008), 159; cf Verdross (1958), 635 (639 et seq.).

[376] If there is a clear expression that both parties want to "denationalize" the contract, in: Herdegen (2003), 13 (19); Krajewski (2009), 183.

[377] Schöbener et al. (2010), 62, Kap. 1 §5 para 199; Dahm et al. (2002), 257; Gruss (1979), 782 (785); Tribunal accepted that a private can be partially subject of international law but is genuine subject, in: Seidl-Hohenveldern (1977), 502 (502).

[378] Schöbener et al. (2010), 60, Kap. 1 §5 para 197; Schöbener and Markert (2006), 65 (69).

[379] Tietje (2005), 47 (62); Tietje (2003), 5 (16).

[380] Epping in Ipsen (2011), § 8 para 18.

[381] Weh (2008), 159.

[382] Epping in Ipsen (2011), § 8 para 18; Weh (2008), 159–160.

[383] Liemen (1980), 150 et seq.; Weh (2008), 160.

Furthermore, it does not sufficiently answer the complex question of international law subjectivity.[384] The ICSID Convention argument also does not work. The investor is only capable of using ICSID as an arbitration platform, because the state previously agreed that the ICSID Convention is applicable (in other words: only the state can trigger ICSID jurisdiction).[385] Finally, there is the fact that being a subject of international law includes rights and obligations. It is difficult to argue that TNCs have an international obligation of any kind.[386]

There is also an attempt to grant TNCs international legal personality because of the already existing power of TNCs.[387] One author argues that the main purpose of international law is peaceful cooperation.[388] Hence, it should actually accept and implement the accumulated power of TNCs.[389] However, this approach does not work because it does not define what "influential power" actually means and requires.[390]

As the PCIJ mentioned '(a)ny contract which is not a contract between States in their capacity as subjects of international law is based on the municipal law of some country.'[391] The ICJ mentioned that a contract cannot relate to both international and municipal law at the same time.[392] In the case, the ICJ decided that a contract between a state and a private company is not a source of international law because it is a mere contract.[393]

Based on the literature and the rulings of the PCIJ and the ICJ, TNCs do not have legal personality within international law at the moment and the contracts of TNCs are not subject to international law. The only exception could be internationalization and stabilization clauses. However, these do not transform TNCs into subjects of international law. Instead, they "merely" lead to an application of international law. There is no intention that the partial international contract leads to functional subjectivity of a TNC.

---

384 Dahm et al. (2002), 257; Not accepted in international law, in: Schweisfurth (2006), 51 para 167.

385 Hobe (2002), 249 (251); Metje (2008), 106; Tietje (2005), 47 (61–62); Tietje (2003), 5 (16); However the feature that a sovereign state is sued by a private investor leads to special classification of the ICSID system within international law, in: UNCTAD (2010a), 10.

386 Metje (2008), 106–107; No subject of international law, in: Muchlinski (2009), 341 (342).

387 Including further references, in: Weh (2008), 160–162.

388 Dahm et al. (1989), 40 et seq; Weh (2008), 160–162.

389 Weh (2008), 161.

390 Weh (2008), 161–162.

391 PCIJ [1929] Judgement (Case Concerning the Payment of Various Serbian Loans Issued in France), 31.

392 ICJ [1952] Anglo-Iranian Oil Co. Case, Preliminary Objection—Judgment, 93 (111–112).

393 ICJ [1952] Anglo-Iranian Oil Co. Case, Preliminary Objection—Judgment, 93 (111–112).

2.4.2.2 Applicable Law According to Art. 42 of the ICSID Convention

According to Art. 42(1) sentence 1 of the ICSID Convention, the ICSID tribunal decides the respective case based on the law agreed upon by the parties. Art. 42 (1) sentence 2 of the ICSID Convention mentions that: '(. . .) (i)n the absence of (. . .) agreement, the Tribunal shall apply the law of the Contracting State party to the dispute (. . .) and such rules of international law as may be applicable.' As illustrated in No. 10 of the ICSID Model Clauses and its commentary, the parties are free to choose any law and can refer to 'national law, international law, a combination of national and international law, or a law frozen in time or subject to certain modifications.' Thus, Art. 42(1) of the ICSID Convention is an expression of the freedom of party doctrine, which is characteristic of arbitration proceedings.[394]

A mere choice of domestic law is rare as applicable law of the contract between the state and the investor.[395] On the other side, the choice of the investor's home law would create a lot of difficulties.[396] The investor is closely linked to domestic regulations, e.g. administrative law, labor law, and the tax law of the host state.[397] It is also not advisable to detach the contract from domestic and international law and to treat it as self-contained.[398]

Art. 42(1) of the ICSID Convention talks about 'rules of law', which means that it is generally accepted that parties are free to combine rules or sets of rules from different legal origins as applicable law.[399] In addition, there is no doubt that the choice of law must be reasonable and connected to the investment.[400] In practice, in most investment related cases domestic law is chosen as applicable law.[401] However, the application of international law is chosen as well.[402]

2.4.2.3 Internationalization of Contracts

An internationalization clause is a contract clause stipulating the application of international law for the respective contract.[403] Such a clause can also be

[394] Hobe and Müller (2009), 65 (73); Böckstiegel (1999), 141 (144–145); Griebel (2008), 138; Schöbener and Markert (2006), 65 (101).

[395] Dolzer and Schreuer (2008), 265; Schreuer (2009), 558 para 24.

[396] Schreuer (2009), 559 para 27.

[397] Dolzer and Schreuer (2008), 265; Schreuer (2009), 559 para 27.

[398] Schreuer (2009), 562 para 37.

[399] Schreuer (2009), 563 para 39.

[400] Schreuer (2009), 564 para 43.

[401] Tietje (2003), 5 (11).

[402] Tietje (2003), 5 (11).

[403] Metje (2008), 66; Schreuer (2009), 560–561 para 33; Krajewski (2009), 183; Schöbener et al. (2010), 244–245, Kap. 4 § 16 para 90.

incorporated into the concession agreement[404] as they are generally accepted.[405] Although it is not advisable because international law does not have the regulation density of domestic laws, e.g. no regulation for technical questions.[406] If it is chosen, both parties must rely on the common legal principles of international law and the arbitrators have a lot of leeway regarding their decisions.[407] In practice the application of international law, governing contracts between the state and the investor (e.g. concession contract) decreased.[408] This is also due to the fact that international law, including minimum standards, always applies and therefore tribunals consider the latter.[409] One reason for the decrease is that the relationship between developing and developed countries has changed and moved from an emotional to a professional platform.[410] Today, developing countries have better legal frameworks to host a project, which is leading to a "renationalization" of contracts (e.g. concessions).[411]

Despite the improving national legal frameworks within developing countries, de facto international law still plays an important role. Instead of using international law as proper law, it is frequently applied as a 'regulative' of the host state law.[412] Therefore, the parties should include a section in their agreement which states that international law functions as a back-up.[413] This was best illustrated in the *Aminoil case*, as the tribunal applied Kuwait law ("as the most directly involved") and international law, 'which formed a constituent part of Kuwait law.'[414] The tribunal did not decide the relationship between international and Kuwait law, but mentioned that they do not conflict with each other.[415] International law is often also incorporated into national law, so an investor could rely on domestic regulations.[416] Nevertheless, the status of international law within domestic law varies from state to state, which requires a specific inclusion of international law.[417]

With regard to the Desertec Concept, it is very important that both parties can rely on a detailed legal framework. The size and relevance of the project almost

404 Schöbener et al. (2010), 244–245, Kap. 4 § 16 paras 90–91; Concerning concession agreements, in: Herdegen (2003), 13 (18).

405 Schöbener et al. (2010), 245, Kap. 4 § 16 para 93.

406 Schreuer (2009), 562 para 36; Herdegen (2003), 13 (19).

407 Herdegen (2003), 13 (19).

408 Metje (2008), 103 et seq.; Nevertheless there are still some contracts applying international law as proper law, in: Herdegen (2003), 13 (20).

409 Metje (2008), 104–105; Herdegen (2003), 13 (22).

410 Herdegen (2003), 13 (20).

411 Metje (2008), 105 and 107–108; Herdegen (2003), 13 (20).

412 Herdegen (2003), 13 (20); cf Metje (2008), 72–73.

413 Maniruzzaman (2001), 309 (325); Metje (2008), 108.

414 Ripinsky and Williams (2008), 4.

415 Ripinsky and Williams (2008), 4.

416 Schreuer (2009), 582 para 100.

417 Schreuer (2009), 582–583 para 103.

rules out the application of an "incomplete" system of law. Hence, the direct application of international law as the applicable law of the contract, e.g. concession and state contract, is not advisable. Both parties should agree on a domestic law as the applicable law, but should include a section that international law functions as a back-up.

The "corrective function" of international law is also accepted nowadays.[418] International commercial[419] and investment[420] arbitration include the principle that parties can freely choose the applicable law. If they decide to choose domestic law, it is a clear expression of party autonomy, which is a prevailing feature of international arbitration. However the influence of international law on domestic law is nothing new. One example is jus cogens rules, which always apply. As expressed in Art. 53 of the VCLT, jus cogens is a general principle of law which influences all contracts. Without a doubt, it is part of international customary law.

2.4.2.4 Stabilization Clauses

To prevent a change of national legislation to the disadvantage of the investor, there is a possibility to conclude a stabilization clause.[421] If a concession is subject to host state law, the investor faces the risk that the state will use its sovereignty to interfere with the contractual balance. In countries with a French legal background, it is not uncommon that the executive authority (unilaterally) changes, terminates or modifies a concession contract if it is within the interest of the public.[422] It is possible to apply stabilization clauses only to certain parts of the contract.[423] As long as changes to the law are within the constitutional framework of the host state, they are not forbidden.[424]

There is a wide acceptance of stabilization clauses if they are not permanent and irrevocable, and thus do not limit state sovereignty too much.[425] Long stabilization clauses in contracts can lead to problems as BITs are not insurance policies against

418 Schreuer (2009), 620–627 paras 214–235.

419 ICC supports it, in: Maniruzzaman (2001), 309 (318).

420 ICSID [2002] ARB/98/4—Annulment (2002), 933 (941) para 40, (941–942) para 42 and (942) paras 44–45; ICSID [2007] ARB/01/3—Award, 65 para 207; ICSID [2007] ARB/02/16—Award, 69 para 237; ICSID [2006] ARB/01/12—Award, 20–21 paras 66–68; ICSID [2000] ARB/96/1—Award, 169 (191) paras 64–67; ICSID [1999] ARB/94/2—Award, 197 (216–217) para 69; ICSID tribunals against this view, in: Schreuer (2009), 627–630 paras 236–244.

421 Joffé (2000), 33 (42); Metje (2008), 66–67 and 111; Hobe and Müller (2009), 65 (73); Lörcher (2005), 11 (17); Krajewski (2009), 182; Herdegen (2003), 13 (23); Dolzer and Schreuer (2008), 75; Schreuer (2009), 588, para 117; cf Berger (2003), 65 (76).

422 Herdegen (2003), 13 (17).

423 Schreuer (2009), 588–589 para 119.

424 Schreuer (2009), 588 para 116.

425 Aaken (2006), 544 (551); cf Schöbener et al. (2010), 245, Kap. 4 § 16 para 93.

bad business decisions.[426] It is possible to limit the length of time that the stabilization clause applies.[427] Nevertheless, according to the Iran–US Claims Tribunal, a 35-year-long stabilization clause is a permanent limitation of sovereignty.[428] The principle of fair and equitable treatment also serves as a stabilization clause as it protects legitimate expectations of the investor.[429] ICSID tribunals treat stabilization clauses in a positive way if both parties freely agreed to its inclusion.[430]

2.4.2.5 Effect and Problems of Stabilization Clauses

The stabilization clause only limits sovereignty concerning the respective contract (e.g. concession).[431] In addition, stabilization clauses always interfere with state sovereignty[432] as they freeze the status quo.[433] This means that the state party cannot change or terminate the contract at an early stage, except when it uses administrative measures or enacts a law.[434] When a state concludes a stabilization clause, it is an expression of its sovereignty.[435] Sovereignty also encompasses the possibility of limitation due to a sovereign act.

The mere breach of contract does not automatically constitute a breach of international law.[436] Thus, international law does not guarantee the contractual position, but compensation on a secondary level.[437] Some authors point out that only a combination of stabilization and internationalization clauses lead to an international law obligation to not change the domestic law.[438] The application of both clauses seems to have decreased in recent years.[439]

It is important to locate the stabilization clause in the international law, if the host state law is the contract's proper law, so as to prevent the host from changing the stabilization clause.[440] A possible solution would be that international law's

---

[426] Aaken (2006), 544 (550–551).

[427] Metje (2008), 110.

[428] Aaken (2006), 544 (551).

[429] Tietje (2010), 5 (12).

[430] Schreuer (2009), 589 para 120; A stabilization clause is a compromise between the state and the investor, in: Dolzer and Schreuer (2008), 75.

[431] Berger (2003), 65 (77); Schreuer (2009), 588 para 117; Herdegen (2003), 13 (23).

[432] Dolzer and Schreuer (2008), 75; Berger (2003), 65 (76); Metje (2008), 67–68.

[433] Aaken (2006), 544 (551); Schreuer (2009), 588 para 117.

[434] Aaken (2006), 544 (551).

[435] Metje (2008), 68; A legal expropriation is not prohibited in international law, in: Shaw (2008), 828; cf Karl (2003), 37 (49–50); Herdegen (2003), 13 (17).

[436] Herdegen (2003), 13 (17).

[437] Herdegen (2003), 13 (17).

[438] Metje (2008), 67.

[439] Schöbener et al. (2010), 245, Kap. 4 § 16 para 94; Dolzer and Schreuer (2008), 75.

[440] Herdegen (2003), 13 (23).

public policy (minimum standards for investment protection) prohibits the unilateral disregard of a stabilization clause.[441] The best way to conclude the stabilization clause would be an international agreement between the host and home state, comparable to the UK–France channel project.[442] Nevertheless, tribunals have used the stabilization clause restrictively because of its ability to limit state sovereignty.[443] If the host state does not want to include a stabilization clause in the PPP to protect its sovereignty, an adjustment clause could solve this problem.[444] Adjustment clauses can also be the opposite of stabilization clauses in that they bring flexibility into the contract.[445]

Some questions remain concerning stabilization clauses. It is still questionable if a stabilization clause can really hinder a determined attack by the host state.[446] Even today, tribunals are still discussing the relationship between a stabilization clause and expropriation, and whether a stabilization clause can prevent an expropriation.[447] Consequently, the investor should ensure that there is an explicit inclusion of expropriation and its effect within the stabilization clause.[448]

2.4.2.6 Evaluation

There is no doubt that international law should not be the governing law of the contracts. Due to the corrective nature of international law, this does not seem to be a big problem. Hence, an investor and Morocco should choose a domestic system of law. Since the project will be implemented in the MENA region, it would be best to apply a MENA law. For the benefit of both parties, the applied law must have connections to European legal traditions, e.g. Morocco.

Although a stabilization clause limits the sovereignty of the host state, it does not lead to unacceptable results. First of all, international tribunals and the literature support the utilization of stabilization clauses. Secondly, each state is free to limit its own sovereignty. It is only necessary that this limitation does not extend to core elements of the state's function. Both parties should previously assess, if the stabilization clause is within the constitutional framework of the host state and in accordance with international law. At the moment, there is no indication that Germany and Morocco plan to conclude an international agreement related to a stabilization clause. As mentioned above, the investor should not include an

---

[441] Herdegen (2003), 13 (23–24).

[442] Herdegen (2003), 13 (24).

[443] Ripinsky and Williams (2008), 4.

[444] Griebel (2008), 33.

[445] Griebel (2008), 33; Berger (2003), 65 (76); Importance of flexibility, in: Herdegen (2003), 13 (28–29).

[446] Joffé (2000), 33 (42).

[447] Dolzer and Schreuer (2008), 75–76.

[448] Metje (2008), 112.

internationalization clause. The need for a stabilization clause exists, which is highlighted by recent revolutions in the MENA region.

2.4.3 *Compensation*

Furthermore, there is the issue of compensation in a case of expropriation. The issue of compensation has dominated international law for a long time. The PCIJ has already dealt with the issue of expropriation and compensation in the *Chorzów Factory case*, which has its roots in the Treaty of Versailles, signed in 1919.[449] In the *Chorzów Factory case*, it pointed out that there must be a distinction between legal and illegal acts and that the remedy is appropriate.[450] Concerning the question of appropriate compensation, the PCIJ stated that

> reparation must, as far as possible, wipe out all the consequences of the illegal act and reestablish the situation which would, in all probability, have existed if that act had not been committed.[451]

A lawful expropriation required only a restitution of the undertaking's value at the day of expropriation plus 'interest to the day of payment.'[452] As mentioned, historically, two approaches govern the dispute concerning compensation—the Hull Formula and the Calvo Doctrine.

2.4.3.1 Calvo and Hull

The Calvo Doctrine says that an investor's compensation is equal to a the national treatment, not better or worse.[453] This includes the exhaustion of local remedies and the surrender of diplomatic protection.[454] The Calvo Doctrine is an expression of the principle of national treatment and the developing countries' fear of neocolonialism and the protection of local private companies through public policy.[455] The Calvo Doctrine experienced a new boost especially after the Argentinean crisis in

---

[449] The Jean Monnet Center for International and Regional Economic Law and Justice (2013).

[450] PCIJ [1928] Judgment (*The Factory At Chorzów* (*Claim for Indemnity*) (*The Merits*), 39–40; There are also some author, who claim that the Chorzów case is broadly misunderstood and the judgement includes also ex post compensation, in: Abdala and Spiller (2008), 103 (104 et seq.).

[451] PCIJ [1928] Judgment (*The Factory At Chorzów* (*Claim for Indemnity*) (*The Merits*), 125.

[452] PCIJ [1928] Judgment (*The Factory At Chorzów* (*Claim for Indemnity*) (*The Merits*), 39–40.

[453] Schöbener et al. (2010), 230, Kap. 4 §15 para 30; Garcia-Bolivar (2010), 3–4; Krajewski (2009), 195; Metje (2008), 74.

[454] Garcia-Bolivar (2010), 5.

[455] Garcia-Bolivar (2010), 3; Fear of interference in internal affairs, in: Shaw (2008), 824.

2001 and the financial crisis in 2008.[456] Nevertheless, authors point out that there is no actual comeback of the application of the Calvo Doctrine.[457]

According to the Hull Formula, in a case of expropriation a 'prompt, adequate (and) effective compensation' must take place.[458] Hence, the actual value (market price) of the expropriated company is the amount of compensation.[459] These different approaches are mostly used by either capital-importing states (Calvo) or capital-exporting states (Hull).[460] The Hull Formula was direct result of Mexican expropriations of various properties (owned by US citizen) during 1915–1940.[461]

Depending on the habits of the respective state, there can be a major difference between the Hull Formula and the Calvo Doctrine. If the state does not pay any expropriations to its own citizens, the investor is treated equally under the Calvo Doctrine. Since the Desertec Concept involves a total investment sum of 400 billion euros, it is vital to know prior to the start of the project, which kind of compensation applies.

In 1974, the Charter of Economic Rights and Duties of States Resolution (no. 3281) was passed by the UN General Assembly as part of the "New International Economic Order." This Charter was a compromise settlement which stated that there should be an appropriate compensation.[462] Despite the fact that it was adopted by the majority of the UN Assembly, almost all major economies did not approve, which led to the result that the Charter is not customary international law.[463] Similarly, the 1962 proposed Declaration on Permanent Sovereignty over Natural Resources (Resolution 1803) failed to achieve a consensus between the developing and developed states.[464]

Furthermore, the OECD offers guidelines and codes for foreign investments. One of these is the Guidelines on the Treatment of Foreign Direct Investment (GTFDI). Section IV No. 2 and 3 GTFDI talk about an appropriate compensation

[456] Markert and Wilske (2010), 62 (66); Garcia-Bolivar (2010), 9–10.

[457] This is supported by facts that Cuba keeps on ratifying new BITs with investor rights and even Venezuela ratified a new BIT with Russia including more than the Calvo doctrine, Ecuador and Venezuela remain active in front of ICSID despite denouncing several BITs, Nicaragua issued a case against an investor after threatening to leave ICSID Argentina official remaining party to ICSID Haiti just ratifying the ICSID Convention and other South American countries entering into new BITs, in: Garcia-Bolivar (2010), 17–21; Furthermore the increased amount of South American companies operating abroad and there is a desire to remain in an investment protection framework, in: Garcia-Bolivar (2010), 22–23 and 25; Metje (2008), 74.

[458] Desbordes and Vicard (2009), 372 (373); Metje (2008), 116; Hobe (2002), 249 (249); Krajewski (2009), 194; cf Joffé (2000), 33 (42); Schöbener et al. (2010), 230 Kap. 4 §15 para 31; Schäfer (1998), 199 (201); Norton (1991), 474 (475–476).

[459] Hobe (2002), 249 (249); Norton (1991), 474 (476).

[460] Harten (2007), 91.

[461] The Jean Monnet Center for International and Regional Economic Law and Justice (2013).

[462] Schöbener et al. (2010), 231, Kap. 4 §15 para 35; Krajewski (2009), 195.

[463] Lowenfeld (2008), 492–493; Krajewski (2009), 195, Karl (2003), 37 (39); Schöbener et al. (2010), 231–232, Kap. 4 §15 para 37.

[464] Reasons for failure, in: Lowenfeld (2008), 489.

based on the adequate fair market value determined on the day of the decision of expropriation. Compensation must be effective and prompt according to Section IV No. 7 and 8 of the GTFDI.

2.4.3.2 Prevailing Approach and the Relevance of the BIT

Authors frequently view the Hull Formula as being part of customary international law.[465] However, the Hull Formula is not undisputed.[466] Nevertheless, it seems to prevail in international law.[467] In practice this discussion is irrelevant because many BITs include a more precise definition of compensation and codifies a full compensation obligation of the host state, in the case of nationalization.[468] The calculation of compensation is mostly according to the fair market value.[469] To avoid problems concerning the calculation of the compensation, some authors propose that compensation be included 'in a lump sum' in the contract.[470]

Art. 4(2) of the G/M-BIT regulates questions of compensation. Compensation must be prompt, according to the actual value (market price) of the investment. This compensation needs to be immediately before an actual or imminent expropriation, nationalization or a comparable action is made public. Furthermore, there is an explicit citation of direct and indirect expropriations. Therefore, the G/M-BIT reflects the idea of the Hull Formula. This means that an investor has a guarantee that it would get adequate compensation in the event of an expropriation.

2.5 State Immunity

Furthermore, the question of state immunity is discussed, since it is a general issue concerning all forms of international arbitration. The effects of public policy will be explored later on in this thesis because it is closely connected to arbitration in MENA countries and only touches on international commercial arbitration.

---

465 Desbordes and Vicard (2009), 372 (373); Hobe (2002), 249 (249); Metje (2008), 115; Krajewski (2009), 194; cf Herdegen (2003), 13 (17); Case law is not as clear, but an increasing amount of BITs (aside of American BITs) include the Hull Formula, in: Shaw (2008), 834.

466 Krajewski (2009), 195; Brazil, Ecuador and Bolivia supporting the Calvo Doctrine, Venezuela and Nicaragua threatening to abandon the Hull formula, in: Garcia-Bolivar (2010), 8; Schäfer (1998), 199 (204); Norton (1991), 474 (476).

467 Although Resolution 1803 talks about an "appropriate compensation", most arbitration institution award a full compensation, in: Schöbener et al. (2010), 239, Kap. 4 §15 para 66; Krajewski (2009), 195.

468 Schöbener et al. (2010), 238, Kap. 4 §15 para 67.

469 Schöbener et al. (2010), 238–239, Kap. 4 § 15 para 68.

470 Berger (2003), 65 (84); Concerning the problem to estimate the correct sum of compensation, in: Horn (1981), 255 (287).

2.5.1 Setup of State Immunity

State immunity can be described as '(. . .) a hybrid of customary international law, municipal law, and to a very limited extent treaty law.'[471] In the past, the theory of absolute immunity prevailed, which led to no possible legal actions against the state.[472] Nevertheless, the state had the right to waive immunity.[473] The doctrine of absolute immunity has changed over time and been replaced by the doctrine of restricted immunity.[474] According to the new doctrine, the state waived its immunity in some cases because of the new and more intense cooperation in commercial activities (e.g. GATT and WTO).[475] It is also important how the state acts. The state (or its entity) can act as the sovereign (*jure imperii*) or according to private law (*jure gestionis*).[476] In the case of *jure gestionis*, the state cannot invoke state immunity.[477] Thus, the investor could try to conclude with the state (or its entity) that all acts related to the projects are private acts.[478] Courts tend to define the term "private act" narrowly.[479] Frequent cases have involved the distinction between property severing sovereign or non-sovereign purposes.[480] It is problematic to draw a distinctive line between commercial property and property serving public or governmental purposes.[481] The German constitutional court (BVerfG) mentioned in one case that there can be no execution against the property of a foreign state if the property is devoted to sovereign purposes.[482]

There are different forms of immunity, all of which are very distinct from each other.[483] Whereas immunity from jurisdiction (limitation of the power of national courts) is treated in a more open way, immunity of execution (restricting the

---

[471] Bjorklund (2009), 302 (309).

[472] Pengelley (2009), 859 (859).

[473] Pengelley (2009), 859 (859).

[474] Recent development, in: Pengelley (2009), 859 (860); Differently, in: Nmehielle (2001), 21 (35).

[475] Pengelley (2009), 859 (860).

[476] Pfeiffer (2003), 141 (167); Delaume (1985), 319 (321) and (322); McIlwrath and Savage (2010), 359 para 6-091.

[477] Saunders and Salomon (2007), 467 (470); Pfeiffer (2003), 141 (167); cf Delaume (1985), 319 (339–340).

[478] Delaume (1985), 319 (339).

[479] Saunders and Salomon (2007), 467 (471); Dolzer and Schreuer (2008), 215.

[480] Bjorklund (2009), 302 (303); Reinisch (2006), 803 (808); Dolzer and Schreuer (2008), 289; Reinisch (2008), 107 (113); European courts differ between sovereign purpose property or not, in: Turck (2001), 327 (342).

[481] Reinisch (2008), 107 (113).

[482] BVerfG [1983] BVerfGE 64, 1 (1), (16) and (45).

[483] Reinisch (2006), 803 (803).

enforcement power of national courts) continues to be very strict.[484] There have been several attempts to deal with the issue of immunity. In 1972 the European Convention on State Immunity was signed in Basle. As only eight states ratified it (e.g. Germany, Switzerland and Austria), it never lived up to its expectations. Even the ILC dealt with the question of state immunity in the 1982 Montreal Draft Articles for a Convention on State Immunity.[485] Up until today the convention did not enter into force. With regard to monetary claims, Art. 18(2) mentioned that there is a possibility of waiver, but this does not include the execution immunity.[486]

In 2004, the UN General Assembly issued Resolution A/59/38, called the United Nations Convention on Jurisdictional Immunities of States and Their Property. It is very similar to the ILC Draft. Resolution A/59/38 includes in Art. 18 and 19 provisions concerning the immunity related to pre- and post-judgment measures. However, the issue of execution immunity is quite problematic, as it is treated in very strict way. Nevertheless, Art. 18 and 19 are exempt from the immunity in cases of state waivers as paragraphs a in both articles illustrate. Art. 5 of Resolution A/59/38 mentions that each country has immunity, with regard to itself and its property, from the jurisdiction of another state. As Art. 7 and 19 of Resolution A/59/38 illustrate, the Convention also differentiates between jurisdictional and execution immunity. Resolution A/59/38 admits execution against property may be used '(. . .) governmental non-commercial purposes (. . .)', as long as the property has a connection with the entity against which the proceeding is directed (Art. 19(c) and 21 of Resolution A/59/38). However, until today, Resolution A/59/38 has not entered into force due to it lacking the required amount of signatures and it is not legally binding. Resolution A/59/38 can only serve as an interpretation instrument.

Immunity also depends on the signed international conventions and autonomous rules of each country.[487] Success of arbitration and its awards depends on the enforceability against the losing party.[488] An award without the backing of the national court orders does not have the power to impose sanctions on the losing party.[489] In most cases, the losing party will automatically perform the award.[490]

---

[484] Mixed results of claims, in: Bjorklund (2009), 302 (309); Reinisch (2006), 803 (803–804); cf Nmehielle (2001), 21 (31); Bjorklund (2009), 302 (303–304); cf with different background, in: Pengelley (2009), 859 (861) and (866); cf Reinisch (2008), 107 (112).

[485] The 1994 Buenos Aires Revised Draft Articles for a Convention on State Immunity did not change the articles relating enforcement immunity.

[486] Heß (1993), 269 (277).

[487] Lörcher (2001), 275 (277); The implementation of international convention is also highly important, in: Sattar (2010), 51 (56).

[488] Nmehielle (2001), 21 (29).

[489] Sattar (2010), 51 (55).

[490] Badmus-Busari (2013), 47–48; Saunders and Salomon (2007), 467 (467); Partasides and Fullelove (2010), 1 (11); Concerning ICSID awards, in: Moses (2008), 227; Euro-Mediterranean Association for Cooperation and Development e.V. (2010), 1; Enforcement is at 98 % of all awards, in: Moses (2008), 208; Harten and Loughlin (2006), 121 (134); Sattar (2010), 51 (54–55); Different opinion, in: McIlwrath and Savage (2010), 343 para 6-042.

Although most awards are enforced today,[491] there is still some reluctance if it comes down to state property and entities.[492]

2.5.2 *Immunity from Execution and Waiver*

Due to the fact that states increasingly participated within commercial transactions, there idea came up that immunity must be restricted related to these activities.[493] Immunity from execution makes an arbitration process highly unpredictable for an investor.[494] There is a difference between setting an award aside and resisting its execution. In the case of setting aside, '(. . .) the award ceases to exist (. . .) and cannot be enforced (. . .).'[495] On the other hand, resisting execution means that the award is not recognized and enforced, whereby the award itself is untouched.[496] Some European courts have declared and states recognized that is not an absolute form of immunity anymore.[497] This alone does not help because the immunity from execution still exists, which means that there is a need for state waivers. The state could waive its immunity from execution. The investor should demand a waiving of immunity with regard to execution.[498] Frequently host states waived there right to plea on immunity.[499] However, this is difficult to negotiate with the host state,[500] yet it is a common practice.[501] The chance to conclude a waiver depends on the economic power of the investor.[502] Some mention that concluding an arbitration agreement normally encompasses that the state waives all typical "support" from its national courts.[503] If the state

[491] Reinisch (2008), 107 (111); cf Ziegler (2012), 4.

[492] Swedish and Hong Kong courts, in: Loya (2012); Bono et al. (2012), 2–3.

[493] Ziegler (2012), 2.

[494] Bjorklund (2009), 302 (304).

[495] McIlwrath and Savage (2010), 31 para 1-047.

[496] McIlwrath and Savage (2010), 31 para 1-047.

[497] Reinisch (2006), 803 (804) and (807).

[498] Badmus-Busari (2013), 19–20; Moses (2008), 227; Nmehielle (2001), 21 (39); Waiver is the '(. . .) ultimate technique of (. . .) immunity avoidance (. . .)', in: Delaume (1985), 319 (344); Hunter (2007), 165 (174); Dolzer and Schreuer (2008), 289; Also citing other authors supporting this claim, in: Pengelley (2009), 859 (871); Concerning different possibilities of waivers, in: McIlwrath and Savage (2010), 331 para 6-012.

[499] Ziegler (2012), 2.

[500] Dolzer and Schreuer (2008), 289.

[501] Ziegler (2012), 4.

[502] Nmehielle (2001), 21 (39).

[503] Ouerfelli (2008), 241 (255); Western position, in: McIlwrath and Savage (2010), 359 para 6-090; For investment arbitration, in: Harten and Loughlin (2006), 121 (128); Herdegen (2003), 13 (33); cf Hunter (2007), 165 (166); Sometimes it is pointed out that the stipulation of an arbitration agreement includes a waiving of exequatur, in: Herdegen (2003), 13 (34).

has agreed to arbitration, it must live with the award.[504] This is a known practice in international agreements.[505]

The host state cannot resort to immunity if the host state's law is the proper law of the contract.[506] In one case, the French *Cour de Cassation* extended the state's waiver (due to the arbitration agreement) to execution immunity, because of ICC rules.[507] The court applied the 1988 Art. 24 of the ICC Rules of Arbitration (today Art. 34(6) of the ICC Rules of Arbitration), which mentioned that every arbitration award should be final and binding.[508] Furthermore, it pointed out that the signature of the arbitration clause included a waiver of execution immunity.[509] This verdict was later on restricted by the *Cour d'appel* (Paris), which mentioned that ICC Rules of Arbitration do not apply to accounts of diplomatic missions.[510] In 2000, the decision of the *Cour d'appel* was revised by the *Cour de Cassation* again, after the claimant appealed the appellate decision and applied the old verdict again.[511] Overall, French courts tend to be reluctant if it comes down to a waiver of execution immunity.[512] The successful party also sued in the US against the same defendant (seeking to execute the judgment), but failed, as the US Court of Appeals of the District of Columbia refused to apply the French courts' reasoning.[513] This raises the problem that there is no clear judicial consensus concerning the application of Art. 34(6) ICC Rules.[514]

Most agree that an arbitration agreement does not include a waiving of immunity concerning execution.[515] Courts tend to limit the scope of waivers to avoid conflicts with other immunities stemming from consular or diplomatic law.[516] Some courts

---

504 Nmehielle (2001), 21 (35).

505 Russia waiving its immunity of execution, in: Turck (2001), 327 (331).

506 However if host state law is not the proper law, there is not an automatically waive of immunity by the host state, in: Herdegen (2003), 13 (34).

507 Pengelley (2009), 859 (864); McIlwrath and Savage (2010), 359 para 6-092; Herdegen (2003), 13 (34).

508 Reinisch (2006), 803 (819–820); Herdegen (2003), 13 (34); Pengelley (2009), 859 (864).

509 Pengelley (2009), 859 (864); Differently in other French court decisions, in: Turck (2001), 327 (327–328).

510 Turck (2001), 327 (329–330); Herdegen (2003), 13 (34–35).

511 Turck (2001), 327 (330); *Cour de Cassation* maintaining the waiver of jurisdiction immunity in the past, in: Nmehielle (2001), 21 (32).

512 French courts see enforcement immunity as absolute in most cases, in: Reinisch (2006), 803 (807); Other cases, in: Turck (2001), 327 (328); Opposing opinion, in: Turck (2001), 327 (341).

513 Badmus-Busari (2013), 33–34.

514 Badmus-Busari (2013), 34.

515 McIlwrath and Savage (2010), 359 para 6-090; Ouerfelli (2008), 241 (255); Hunter (2007), 165 (166); Herdegen (2003), 13 (34); cf Reinisch (2006), 803 (819); Cited from another case, in: Pengelley (2009), 859 (866); Opposing this claim, in: Pengelley (2009), 859 (867); Bjorklund (2009), 302 (302).

516 Reinisch (2006), 803 (818).

have proposed (e.g. in Switzerland) that: '(...) immunity from enforcement measures should be treated in the same way as restrictive immunity from jurisdiction.'[517] In total it is possible to observe that the plea of immunity is slowly pushed back by national courts.[518] State immunity remains an important issue as globalization keeps on and states try to protect itself against unjustified claims of third parties.[519] At the same time is noticeable that there is a movement granting immunity to global PPPs. This might be interesting as an investor, like Dii could have been, has the potential to become a TNC.[520] If this would be the case immunity issue might arise on both side of the cooperation.

2.5.2.1 ICSID and Immunity

As mentioned above, there is the same problem within investment arbitration. The ICSID Convention also differentiates between immunity from jurisdiction and immunity from execution.[521] Some authors point out that the solution rests with the domestic courts and their application of Art. 54(3) and 55 of the ICSID Convention.[522] The proposed solution mentions that domestic courts should do a combined reading of both articles to give the claimant a valid title for execution '(...) provided, (...) that where such measures are directed at the assets of a state, execution is possible under the law of the (host state) in which execution is sought.'[523] However, this is not applied by ICSID tribunals and therefore can only serve as an idea.

The ICSID Model Clause offers a possible solution to the problem of immunity according to Art. 55 of the ICSID Convention. The ICSID Model Clause 15[524] proposes a waiver of immunity from execution of the award. In reality it might be difficult for an investor to conclude a comparable waiver with Morocco or ONE because the host state might understand it as an infringement of its immunity. The investor would be faced again with the difficult task of negotiating a waiver. Nevertheless, the obligation under the ICSID Convention remains, independent of the problems concerning from execution.[525]

517 Reinisch (2006), 803 (809); Pengelley (2009), 859 (865); Other courts also pointed out that a waiver of immunity encompasses all forms of immunity, in: Pengelley (2009), 859 (861); Opposite behaviour of courts, in: Reinisch (2008), 107 (112).

518 Krieger (2012), 1.

519 Krieger (2012), 1–2.

520 Krieger (2012), 2.

521 Saunders and Salomon (2007), 467 (469).

522 Nmehielle (2001), 21 (36).

523 Nmehielle (2001), 21 (36); cf Delaume (1983), 784 (800).

524 Clause 15 ICSID Model Clause: The Host State hereby waives any right of sovereign immunity as to it and its property in respect of the enforcement and execution of any award rendered by an Arbitral Tribunal constituted pursuant to this agreement.

525 Nmehielle (2001), 21 (40); cf Delaume (1983), 784 (799) and (801).

2.5.2.2 Consequences for the Investor

As Art. 20 of Resolution A/59/38 and the strict wording of the other provisions imply, there is a need for an explicit waiver. Thus, a mere arbitration clause is not sufficient.[526] If the state has not explicitly declared a waiver of execution immunity, enforcement by writ can only be against non-sovereign assets.[527] Some authors point out that the 'cloak of immunity' extends over almost every asset of the state.[528] It is therefore very important that the waiver is clearly worded, so as to avoid any interpretation and hence rejection of it.[529] Waivers of immunity are rare outside the financial field.[530] It must be kept in mind that a concluded waiver does not solve all issues concerning award enforcement, as some risks remain.[531]

Although there are some courts in favor of a wide interpretation of waivers, this is not the general rule. The prevailing opinion demands a separate waiver for execution immunity. As mentioned above, only a powerful investor has a chance to successfully agree to a waiver of execution. Since an investor has the potential to become a TNC, they might have the competence to agree to waiver with Morocco. State sovereignty allows Morocco to conclude this kind of waiver. To prevent any misunderstandings, both parties should conclude separate waivers for the contracts (e.g. concession contract and state contract). In this way, they will clarify that the waivers affect international commercial and investment arbitration. If there are no separate waivers, an investor runs the risk of facing immunity from execution claims in all cases of international arbitration. This is not an uncommon risk, and does not only relate to the Desertec Concept. The fact remains that the state complies with the arbitration award in most cases. Otherwise it would jeopardize its international credibility. Hence, the risk of immunity objection is relatively low. A missing waiver should not lead to the abandonment of the complete project.

---

[526] Reinisch (2006), 803 (819).

[527] French and US courts also differ between explicit and implicit waivers of immunity, in: Turck (2001), 327 (341); Semler (2003), 97 (102); McIlwrath and Savage (2010), 359 para 6-091; cf Saunders and Salomon (2007), 467 (469); cf Reinisch (2006), 803 (804); Herdegen (2003), 13 (34); cf Hunter (2007), 165 (166).

[528] Saunders and Salomon (2007), 467 (469); In most cases execution of awards within the territory of the host state imposed by home state courts are going to fail, in: Semler (2003), 97 (102).

[529] Delaume (1985), 319 (344); Nmehielle (2001), 21 (39).

[530] Delaume (1985), 319 (344); Nmehielle (2001), 21 (40).

[531] Ziegler (2012), 4.

2.6 Arbitration and Islamic Countries

After finding the suitable arbitration courts for the Desertec Concept, the acceptance of international arbitration within Morocco and the MENA region is important. If MENA countries do not accept the verdict of international arbitration institutions, it would be useless to conclude any such agreement. Furthermore, the problem of public policy remains. Public policy can be a serious problem for a functional arbitration, especially in countries with an Islamic background.

2.6.1 Morocco and Arbitration

The UNCITRAL CAR led to a harmonization of arbitration rules within the Maghreb countries.[532] In 2007, Morocco adopted the UNCITRAL CAR, with only few changes.[533] Morocco was also one of the first Maghreb countries to adopt the NY Convention.[534] Furthermore, in 1978 Morocco was the first country to be sued according to the ICSID Convention in the case *Holiday Inns v Morocco* (ICSID, ARB/72/1). In general, Maghreb countries try to solve their disputes with foreign investors in an amicable manner.[535] As mentioned above, Moroccan law is closely related to French law. An increase in strict interpretations of the Koran and Shari'a makes it still important to understand the concepts of the Shari'a.[536] This can be done by examining the Moroccan arbitration laws and the Shari'a to fully understand Morocco's attitude towards international arbitration.

2.6.2 Arbitration and Shari'a

First of all, a closer look at the Shari'a is necessary to find the general attitude to arbitration. Although the Shari'a is not directly applicable to the law in Morocco, it still plays a major role because of its sacred status. Furthermore, some MENA states which may be possible Desertec partners in the future rely heavily on the Shari'a.

---

[532] Ouerfelli (2008), 241 (243).

[533] Ouerfelli (2008), 241 (243); Taarji et al. (2011), 11.

[534] Taarji et al. (2011), 11; Mernissi (2002), 179 (179); Ouerfelli (2008), 241 (244).

[535] Ouerfelli (2008), 241 (247).

[536] Kutty (2006), 565 (568).

2.6.2.1 The Meaning of the Shari'a

The word Islam derives from the word salaam (meaning peace) and means submission to God, whereby '(t)he Shari'a is the path to achieve this submission.'[537] According to the Islam, God is the sole source of authority and thus the sovereign lawgiver.[538] Hence, the Shari'a is divine and eternal,[539] and often the source of legislation.[540] For most Muslims, the Qur'an is '(. . .) the actual verbatim revelation sent by God.'[541] The Qur'an deals with legal issues only in 350 out of 6616 verses and thus "only" lays down guidelines and general principles for the realization of a civilized society.[542] Most Islamic legal scholars view the Qur'an as the primary source of the Shari'a.[543] Furthermore, there is the *sunnah*, which encompasses '(. . .) behaviors, decisions, actions, tacit approvals and disapprovals of the Prophet (. . .) (, which were) heard, witnessed, memorized, recorded, and transmitted from generation to generation (. . .).'[544] Many Islamic scholars view the *sunnah* as the second legal source of the Shari'a.[545] The Shari'a is explained by different schools of thought.[546]

Furthermore, three important legal tools are *qiyas* (the reasoning by analogy in cases of new legal problems), *ijtihad* (establishing rules in relation to the principles of Islam), and *ijma* (consensus by a community concerning the acceptance of *qiyas* and *ijtihad*.[547] Based on *ijtihad*, Muslim jurists cannot create new law as God created all rules.[548] If arbitrational tribunals were allowed to include principles as the *ijtihad*, some claim that a lot of enforcement obstacles would be avoided.[549] It is important to realize that customs cannot change rules of the Shari'a,[550] but the Shari'a views customs as binding.[551] The Shari'a is often used as a back-up to fill judicial loopholes.[552] It is also problematic that the Shari'a deals with legal issues case-by-case, instead of establishing general rules.[553] It would be wrong to think that there are no similarities between Western laws and the Shari'a.

---

537 Kutty (2006), 565 (578).

538 Kutty (2006), 565 (578) and (580).

539 Kutty (2006), 565 (578).

540 Ayad (2009), 93 (94).

541 Kutty (2006), 565 (583).

542 Kutty (2006), 565 (584).

543 Ayad (2009), 93 (99); Kutty (2006), 565 (584).

544 Kutty (2006), 565 (585).

545 Kutty (2006), 565 (585).

546 Darwazeh and El-Kosheri (2008), 203 (204), Kutty (2006), 565 (581–582).

547 Kutty (2006), 565 (586–588).

548 Ayad (2009), 93 (105).

549 Ayad (2009), 93 (93) and (103–104).

550 Kutty (2006), 565 (589).

551 Ayad (2009), 93 (109).

552 Ayad (2009), 93 (98).

553 Ayad (2009), 93 (111).

> [L]ike most Western legal systems, the Shari'a is a positive system of law and not merely religious law. Additionally, both systems apply judge-made law using the case law method in their own particular ways.[554]

2.6.2.2 Acceptance of Arbitration

Traditional Islamic law accepts arbitration and the Qur'an approves it.[555] Even the Prophet Muhammad called for arbitration in a conflict with the tribe of *Banu Qurayza*.[556] Yet, down to the present day, in many Islamic countries, women still cannot be arbitrators.[557]

One partial driving factor is globalization because self-isolation is not possible anymore.[558] Nevertheless, in Middle Eastern countries, people have the feeling that international arbitration does not consider Islamic culture, values and legal traditions.[559] It is also important to remember that Islamic countries share the same language, but also have different cultural backgrounds.[560] They share a close cultural connection through Islam.[561] The distrust of international arbitration is also deeply rooted in the MENA region.[562] There is also evidence that Islamic legal tradition accepts *tahkim* (arbitration).[563] While most Western scholars view arbitration as binding, Islamic scholars are still discussing the extent to which *tahkim* is binding.[564] The main problem is that the four important Islamic law schools interpret *tahkim* differently.[565] Furthermore, the scope of *tahkim* is questionable.[566] For example, Turkey excluded state concessions for public services and investments from arbitration for a long time.[567] The parliament changed this in 2001,

---

[554] Kutty (2006), 565 (579); Similarities of both systems, in: Ayad (2009), 93 (109).

[555] Vogl (2010), 32 (33); Ahdab and Stackpool-Moore (2008), 275 (283); Brower and Sharpe (2003), 643 (643); Kutty (2006), 565 (589–590).

[556] Brower and Sharpe (2003), 643 (643); Darwazeh and El-Kosheri (2008), 203 (204).

[557] Darwazeh and El-Kosheri (2008), 203 (204–205); Kutty (2006), 565 (606).

[558] Brower and Sharpe (2003), 643 (646–647).

[559] Sayed (2008), 289 (292); Brower and Sharpe (2003), 643 (645–646); Kutty (2006), 565 (592); Historically reinforced by Western attitudes in oil concession contracts, in: Brower and Sharpe (2003), 643 (644).

[560] Hammoud and Houerbi (2008), 231 (231).

[561] Darwazeh and El-Kosheri (2008), 203 (203).

[562] Nariman (2004), 123 (123).

[563] It will also expand in the upcoming years, in: Darwazeh and El-Kosheri (2008), 203 (209); Ahdab and Stackpool-Moore (2008), 275 (283–284); Brower and Sharpe (2003), 643 (643); Kutty (2006), 565 (596).

[564] Darwazeh and El-Kosheri (2008), 203 (204); Kutty (2006), 565 (596–597).

[565] Kutty (2006), 565 (597).

[566] Prohibition of arbitration of state entities in some Islamic states, in: Brower and Sharpe (2003), 643 (651–652); Kutty (2006), 565 (598–601).

[567] Brower and Sharpe (2003), 643 (652).

which led to a rapid increase in investments.[568] Islamic countries had bad experiences with international arbitration during the first oil crises in the 1970s.[569] Western arbitrators who were not able to speak any Arabic denied the application of Islamic host state law as applicable law.[570] A lot of Moslems were annoyed by the: '(. . .) ignorance, carelessness or unjustified psychological superiority complexes (. . .) of the arbitrators and the (. . .) humiliating nature' of Islamic law exclusion.'[571]

Most Islamic countries have prejudices concerning international arbitration. Yet, there is an overall acceptance of arbitration as a legal tool. Within the last few decades, the acceptance of international arbitration has grown. Since Mediterranean states are not as Islamic as Middle Eastern states, the influence of the Shari'a is significantly lower.

2.6.2.3 Acceptance of International Arbitration

The arbitration rules within Moroccan law are Art. 306 to 327 loi n° 1-74-447 (Code de procédure civile). In general, the state and its entities cannot be subject to arbitration, as stated in Art. 310 of the Code de procédure civile. Although Art. 310 of the Code de procédure civile was introduced in 2007, there is no reference towards an exception of international arbitration. Interestingly, the *Projet du Code de l'Arbitrage* included an exception for international commercial relationships in Art. 2(3). The new law does not include this proposal. Furthermore, Art. 323 of the Code de procédure civile sets forth a comprehensive list of grounds for the rejection of arbitration. Yet, the newly introduced Art. 327-39 to 327-69 of the Code de procédure civile deal with international arbitration. Art. 327-29 of the Code de procédure civile explicitly mention that this section deals with the issue of international arbitration, acknowledging all international conventions ratified by Morocco. Under the old Code de procédure civile, there was only a section dealing with national arbitration. Besides existing legal obligations, e.g. the ICSID, the former law and practice already included an exception from the general rule of Art. 310 of the Code de procédure civile. There is Moroccan case law, mentioning that the restriction does not apply to '(. . .) industrial, public or commercial entities when the contract has an international scope.'[572] The principle that arbitration is not allowed for matters involving acts or property governed by administrative law should not apply to international arbitration.[573] Furthermore, there is an increased

---

[568] Brower and Sharpe (2003), 643 (652).

[569] Ahdab and Stackpool-Moore (2008), 275 (278).

[570] Nariman (2004), 123 (124); Ahdab and Stackpool-Moore (2008), 275 (278–279).

[571] Ahdab and Stackpool-Moore (2008), 275 (279); Problematic treatment by Western scholars and arbitrators, in: Darwazeh and El-Kosheri (2008), 203 (205).

[572] Mernissi (2002), 179 (180).

[573] Mernissi (2002), 179 (183).

usage of arbitration in Morocco and the public authorities seem to promote this trend.[574]

Nevertheless, the wording of Art. 310 of the Code de procédure civile is clear and does not allow the state and its entities to be part of arbitration. There is the assumption that there was no intention to make the state subject to international arbitration. As Moroccan case law reveals, there was an awareness of the issue that the state could not by law be subject to international arbitration. The Projet du Code de l'Arbitrage included a provision which would have solved this problem once and for all. However, the legislator did not include this section in the new provisions of the Code de procédure civile. This could be a clear expression of rejection. However, the legislator included several new provisions on international arbitration and mentioned its obligation under international conventions. Morocco also continued to ratify international treaties, e.g. the 2004 BIT with Germany, which includes an obligation of international arbitration. International trade and commerce require the acceptance of international arbitration. Morocco would be isolated by not participating in international investment and commercial arbitration. Therefore, the intention of the legislator cannot be to exclude the possibility of international investment and commercial arbitration for the state. Consequently, the historical interpretation and the intention of the legislator result in an acceptance of the international arbitration subjectivity of the state and its entities.

Since the Moroccan system has close ties with the French legal system, it is also beneficial to analyze it. Taking a look at the French system, Art. 2060 of the Code Civil arbitration is not possible if a public collective or public establishment is involved. However, today there is no doubt that this does not apply in cases of international arbitration. French courts have highlighted that public entities cannot escape their international obligations by invoking national law,[575] making the situation comparable. It is clear that the state of Morocco can be subject to international arbitration. Globalization and international cooperation (e.g. WTO) do not allow any different interpretation. The new section concerning international arbitration in the Code de procédure civile would be pointless. The missing proposal can only be a mistake of the legislator. Furthermore, the old case law remains in place, which means the state could be subject to international arbitration, allowing Morocco/ONE to be subject to international commercial and investment protection.

2.6.3 *ICC Arbitration and Morocco*

As mentioned above, Morocco has been party to ICSID arbitration several times. In addition, the state of Morocco has already been party to ICC proceedings.[576]

---

[574] Taarji et al. (2011), 12–13.

[575] Hanotiau and Caprasse (2008), 721 (724–725).

[576] In the MENA region, Moroccan parties submitted, among others, the most cases to the ICC, in: Hammoud and Houerbi (2008), 231 (232).

Under ICC proceedings, Morocco has been chosen as a seat of arbitration only once.[577] The statistics of the ICC reveal that Western parties do not tend to choose an Arab country as the seat of arbitration.[578] When an Arab country is involved, the most popular seat of arbitration is Paris.[579] The choice of Arab law as applicable law in ICC cases has varied over the years.[580] In the majority of ICC cases, Western laws were the applicable law, whereby French and Swiss law were frequent choices.[581] Moroccan law was rarely chosen.[582]

2.7 Public Policy

Finally, there is the issue of public policy. Public policy and *ordre public* are comparable, whereas public policy relates more to states with a common law tradition and *ordre public* to states with a civil law tradition.[583] Therefore *ordre public* and public policy will be treated alike. In Islamic and Western countries, there is a big difference in the understanding of public policy.[584] One of the major problems of public policy is that it varies from state to state.[585] Public policy has a kind of supervisory role within arbitration proceeding, as it should correct awards, which are against fundamental principles of the host states legal framework.[586] As the ILC proposed:

> [T]he international public policy of any State includes: (i) fundamental principles, pertaining to justice or morality, that the State wishes to protect even when it is not directly concerned (ii) rules designed to serve the essential political, social or economic interests of the State, these being known as "*lois de police*" or "public policy rules" and (iii) the duty of the State to respect its obligation towards other States or international organisations.[587]

---

577 Hammoud and Houerbi (2008), 231 (236); Not as attractive as other Arabic nations, e.g. Egypt, Tunisia or Lebanon, in: Darwazeh and El-Kosheri (2008), 203 (207).

578 Darwazeh and El-Kosheri (2008), 203 (207); Hammoud and Houerbi (2008), 231 (237).

579 Also Switzerland and the UK, in: Hammoud and Houerbi (2008), 231 (238); Paris is a reliable forum, in: Darwazeh and El-Kosheri (2008), 203 (207).

580 Hammoud and Houerbi (2008), 231 (238–239).

581 Hammoud and Houerbi (2008), 231 (239).

582 Hammoud and Houerbi (2008), 231 (239).

583 UNCITRAL (1985), 3 (36) para 296; Both terms used interchangeably, in: Ma (2005), 72; Both terms are the same, in: Saunders and Salomon (2007), 467 (471).

584 Kutty (2006), 565 (603–604).

585 Fei (2010), 301 (301); Saunders and Salomon (2007), 467 (471); Ozumba (2009), 4; Partasides and Fullelove (2010), 1 (11–12).

586 Buchanan (1988), 511 (513).

587 Recommendation 1: No. 1(d) of the Resolution of the ILA on Public Policy as a Bar to Enforcement of International Arbitral Awards 2002 (2002 ILC Public Policy Recommendations).

The UNCITRAL also dealt with the issue of public policy as well. It mentioned explicitly (in relation to the UNCITRAL Model Law and the NY Convention) that public policy only encompasses fundamental principles of law and justice.[588] Arbitrators are obligated to apply issues of public policy already during the arbitration.[589] Nevertheless, the amount of number enforced under the NY Convention is very high.[590] Furthermore, the ILC Public Policy Recommendations offer some guidance as well. For example, Recommendation 1 No. 1(e) 2002 of the ILC Public Policy Recommendations gives examples of the term "fundamental principle", e.g. the prohibition of abuse of rights. Recommendation 2 No. 2(b) 2002 of the ILC Public Policy Recommendations offers a guideline of possible reasons of rejection before award enforcement is denied.

Mandatory rules are different to public policy rules. As the ILA recommends in Recommendation 3(a) of ILA Resolution, mere mandatory rules cannot deny enforcement of an arbitral award, unless the mandatory rule is part of international public policy. However, the ILA does not define mandatory rules, which makes the application quite difficult. One author proposes that:

> [M]andatory rules are commonly understood as rules which cannot be excluded by the parties' agreement, and which may apply alternatively (or at least additionally) to the parties' chosen law These rules are so important that they are applicable in spite of, and even independent of, any choice of law process.[591]

The main differences between both concepts are that public policy is not regarded as law affecting enforcement and choice of law issues, and has a wide variety of sources. This is different concerning mandatory rules, which are status of law, mainly affecting the choice of law and have predominantly statutory sources.[592]

2.7.1 *Types of Public Policy*

Sometimes there is a distinction between domestic and international public policy.[593] Thus, not everything which constitutes a breach of domestic public policy violates international public policy.[594] Some authors mention that both forms of public policy often overlap, but until today there are still tensions because of the difficult delimitation of both concepts.[595] Domestic public policy encompasses

---

[588] UNCITRAL (1985), 3 (36) para 297.

[589] Otherwise there would be a high risk that a national court overturns the decision afterwards, in: Hanotiau and Caprasse (2008), 721 (726).

[590] Saunders and Salomon (2007), 467 (467).

[591] Ma (2005), 122.

[592] Ma (2005), 124–127.

[593] Further theories concerning public policy, in: Ma (2005), 34 and 80 et seq.

[594] Ozumba (2009), 6; Hanotiau and Caprasse (2008), 721 (730).

[595] Ma (2005), 82–87.

local standards, which limit the principle of freedom to contract.[596] On the other hand, international public policy includes more flexible standards and rules of the state, which are applied in an international context.[597] Art. 1502(5) of the Code Civil in France also stipulates that an appeal against a decision which grants recognition and enforcement is only possible if either one is contrary to international public policy. Even English courts also accepted that foreign arbitral awards might violate British public policy, but not international public policy.[598]

Other authors claim that there is a transnational public policy superseding domestic public policy to serve the higher interests of the world community.[599] The main difference to international public policy lies in the fact that transnational public policy is that international public policy is subject to the choice of law process.[600] International and transnational public policy have the trans-boundary application in common, which illustrates that they also overlap to a certain extent.[601] Consequently, it is difficult to distinct between the different forms and within the next few years a distinction might be irrelevant, due to an even closer connection between national and international law.[602] In practice this does not matter because courts will only apply public policy issues if they are part of the international public policy of the state, even though they may also be transnational public policy.[603]

2.7.2 *Problems Related to Public Policy*

The interpretation of public policy is particularly difficult in Islamic countries because there is no unified approach.[604] The unpredictable issue of how the Shari'a might influence public policy underlines this.[605] Nevertheless, the Shari'a is independent of public policy, but can support the later.[606] Interpretation of the Shari'a is

[596] Buchanan (1988), 511 (513).

[597] Buchanan (1988), 511 (514); Example of international public policy in Article 2(4) UN Convention (e.g. prohibition of slavery, piracy, drug trade, terrorism and genocide and the protection of basic human rights), in: Schreuer (2009), 566 para 49.

[598] Saunders and Salomon (2007), 467 (472).

[599] Hanotiau and Caprasse (2008), 721 (731–732); Buchanan (1988), 511 (514) and (529–531).

[600] Ma (2005), 87; cf Buchanan (1988), 511 (514) and (529).

[601] Ma (2005), 88; Transnational public policy is independent of the chosen law, whereas international public policy adjust according to the chosen law, in: Ma (2005), 89; Buchanan (1988), 511 (514).

[602] Ma (2005), 91.

[603] Hanotiau and Caprasse (2008), 721 (732); cf ILA recommendations on public policy as a bar to enforcement of arbitral awards, in: Hanotiau and Caprasse (2008), 721 (732); Cases applying transnational public policy are rare, in: Buchanan (1988), 511 (530).

[604] Ayad (2009), 93 (96) and (112).

[605] Ayad (2009), 93 (100).

[606] Ayad (2009), 93 (100).

not done by the average lawyer, '(. . .) but rather to discovery of the intent behind the divine law and its appropriate application.'[607] This means that it is difficult for arbitrators to interpret the law because the wording is less important than the intent.[608] Furthermore, the risk that utilization of public policy undermines arbitration awards has been around for a long time.[609]

Cases in different countries provide additional examples of the problems of public policy. In Pakistan, was a case where the Supreme Court stopped an arbitration process in front of the ICC because it decided retroactively that the issue could not have been subject to arbitration due to public policy.[610] The verdict of the court is regarded to be arbitrary and a great step backwards within the international arbitration community.[611] In India, the Supreme Court purposely lowered the standard of public policy to 'patently illegal' to be able to review an arbitration award.[612] This means that an arbitral award can now be challenged on the mere fact that it is '(. . .) in (. . .) contravention with the law and thus in conflict with public policy in India.'[613] The afore-mentioned examples show how public policy can be raised to prevent the enforcement of the complete contract or specifically on the arbitration clause.[614] Although public policy is not an issue under the ICSID Convention, the GTFDI shows that it also can cause difficulties concerning investments. The GTFDI mentions in section 2 No. 5 that all restrictions because of *ordre public* also apply to investments.

The refusal of arbitration awards for public policy reasons is rare and thus often unsuccessful.[615] A state cannot agree to apply a certain law to the contract and afterwards claim that this law breaches its public policy.[616] If the part which violates international public policy can be separated from the award, the rest should be enforced according to Recommendation 1 No. 1(h) 2002 of the ILC Public Policy Recommendations. Recommendation 4 2002 of the ILC Public Policy Recommendations highlights that the refusal of enforcement should be ultima ratio. In France, only the fundamental notions of the legal system are regarded as public policy.[617] French courts tend to have a strict approach towards the application of public policy, [618] resulting in public policy only being utilized in a narrow

---

[607] Ayad (2009), 93 (102).

[608] Ayad (2009), 93 (103).

[609] Concerning Chinese courts, in: Fei (2010), 301 (302).

[610] Sattar (2010), 51 (60).

[611] Sattar (2010), 51 (60).

[612] Also not excluding international awards, in: Sattar (2010), 51 (62) and (63).

[613] Sattar (2010), 51 (62).

[614] Buchanan (1988), 511 (515); In theory also the arbitral tribunal could raise public policy, in: Buchanan (1988), 511 (515).

[615] Saunders and Salomon (2007), 467 (471).

[616] Schreuer (2009), 566 para 48.

[617] Hanotiau and Caprasse (2008), 721 (730) and (736).

[618] Hanotiau and Caprasse (2008), 721 (733–734).

way. Nevertheless, it must be reviewed how Morocco deals with the issue of public policy.

It is very difficult to define public policy in Maghreb countries.[619] Some mention that arbitration is generally forbidden in cases of public policy.[620] Another problem is to determine whether the Shari'a is part of the public order or not. In Maghreb countries there is a clear distinction between the Shari'a and law, whereby all countries emphasis their Islamic background. Nevertheless, some authors say that it does not make a difference if international arbitration or national legislation is chosen, because both solutions are problematic.[621]

2.7.3 *Moroccan and Arbitration Awards*

To understand the relationship between Moroccan law and the Shari'a, a closer look at the (1996) Dahir n° 1-96-141 du 8 rabii II 1417 (Moroccan Constitution) is necessary. The Preamble of the Moroccan Constitution highlights that Morocco is a sovereign and Islamic country. Secondly, Art. 1 of the Moroccan Constitution points out that it is a democratic and social constitutional monarchy. Art. 82 stipulates that the judicial sector is independent from the executive and legislative branch. Furthermore, all judgments are executed in the name of the king, according to Art. 83. Overall Morocco also makes a clear distinction between the Shari'a and law. All articles of the Constitution underline the fact that there is a clear distinction between law and religion. This is also illustrated by Art. 6 of the Constitution, which separately addresses the question of religion.

There are also problems of arbitral award enforcement in Africa and the Middle East.[622] The compliance with awards is rather the exception than the rule.[623] E.g. the mere fact that Saudi Arabia accepted the NY Convention does not mean that all awards are automatically enforceable.[624] There are some courts in Islamic nations which have refused the enforcement of an award on the grounds of domestic public policy, although the conventions referred to international public policy.[625] Comparable to Algeria, '(. . .) the presiding judge of the court of first instance grants enforcement (. . .) (and this) decision is subject to appellate recourse, and thereafter, to cassation before the High Court', this also applies to Morocco.[626] Most Maghreb

[619] Ouerfelli (2008), 241 (254).

[620] Aljazy (2000), 1 (1).

[621] McIlwrath and Savage (2010), 364 para 6–105.

[622] McIlwrath and Savage (2010), 364 para 6–105; cf Ouerfelli (2008), 241 (241).

[623] McIlwrath and Savage (2010), 364 para 6–105.

[624] Kutty (2006), 565 (602).

[625] Kutty (2006), 565 (603).

[626] Ouerfelli (2008), 241 (247–248).

laws do not include sections on how to act in cases of conflict between national laws and international conventions.[627]

Finally, Moroccan national law also deals further with international arbitration. Art. 327-46 of the Code de procédure civile highlights that international arbitration awards are generally accepted, if they are in accordance with national and international public policy. It is important to note that this Article mention domestic (national) and international public policy. However, as mentioned above, in cases of international arbitration, only international public policy should be the deciding factor. Other jurisdictions, e.g. Algeria, Lebanon, Tunisia, want to implement an exclusive application of international public policy.[628] As the Moroccan law was changed a couple of years ago, it is doubtful that it will be changed soon again. As domestic public policy applies in addition to international law, the investor should be aware that this might cause unforeseeable difficulties. As far as it is possible, the investor should try to negotiate that only the international standard of public policy applies in the case of the Desertec Concept.

Taking everything into account, there is a high risk that an investor will face issues of public policy. In Morocco, there were frequent problems with execution of contracts.[629] This especially relates to international commercial arbitration. The application of domestic public policy makes it highly unpredictable for an investor. Moroccan law is very clear on the application of domestic public policy. Combined with the "Islamic revitalization" the understanding of public policy is hard to predict.

2.7.4 Conclusion

In the case of Morocco, the risk of it using public policy to escape its obligations is relatively low. This does not necessarily apply for other states in the MENA region. As a basic rule, it comes down to more: the more the Shari'a applies, the higher is the risk of public policy application. As mentioned above, the Desertec Concept should be set up as a concession. Within a lot of Islamic countries, oil concessions have been a day-to-day practice. Furthermore, there have already been a lot of arbitration proceedings related to these concessions. Hence, they might offer some guidance on the topic. Oil concessions were one of the first cases of international arbitration that Islamic states were faced with in modern times.[630] Interestingly, oil concessions never violate public policy within Islamic countries because the profit is for the benefit of the community.[631] These benefits '(...) must be economic and financial, contributing to the material enrichment and profit the Islamic community.'[632]

[627] Ouerfelli (2008), 241 (249).

[628] Brower and Sharpe (2003), 643 (649).

[629] Brenton et al. (2006), 9.

[630] Brower and Sharpe (2003), 643 (644).

[631] Ayad (2009), 93 (115).

[632] Ayad (2009), 93 (115).

Taking this as a worst case scenario, the Desertec Concept would not violate public policy if these benefits prevailed. The risk of public policy claims remains rather small in the case of an investor and Morocco partnership.

2.8 Remarks on Desertec and Arbitration

International commercial and investment arbitration offer a good possibility to secure the Desertec Concept and to resolve disputes between the parties. It is however not advisable that national courts be used. Due to widespread corruption, they do not offer an effective and impartial alternative to international arbitration. The problems which the project might face are not particular problems connected to this project, but exist generally. There are tools like the stabilization clause and the waiver of immunity from execution, which offer additional security. Both are subject to negotiation and should not be demanded by the investor. It is necessary that there is balance between all interests in order to successfully implement this project and to avoid unnecessary disputes. Besides the choice of the ICC as arbitration institution, the CAM, which closely cooperates with Arab arbitration institutions, might function as a "mediator" between European and MENA stereotypes. In addition, the parties should also carefully choose each arbitrator and pay attention to his or her cultural background. Then no party would feel underrepresented. All this must be balanced. Morocco should not "play the cultural card" too often as this might cause discontent and harm the cooperation.

In addition, arbitration is common legal tool in the MENA region. Therefore, it is likely that these verdicts will be accepted. Despite the revolutions, even liberalization groups adhered to international contracts previously negotiated by the government (e.g. oil exports from Libya). Hence, there seems to be a general interest to fulfill contractual obligations which are beneficial for the country in question. This means that an investor should focus on Morocco and not Europe with regard to energy production and the benefits resulting from the Desertec Concept. In doing so, it will gain a lot of credit and to some extent immunity if difficult political times occur.

References

Aaken A (2006) Zwischen Scylla und Charybdis: Völkerrechtlicher Staatsnotstand und Internationaler Investitionsschutz. ZVglRWiss 544–569

Abdala MA, Spiller PT (2008) Chorzów's standard rejuvenated. J Int Arbitr 25:103–120

Ahdab JE, Stackpool-Moore R (2008) Arab arbitration v. international arbitration? The case for a reconciliation. J Int Arbitr 25:275–288

Alexandrov SA (2009) Enforcement of ICSID awards: Articles 53 and 54 of the ICSID convention. In: Binder C, Kriebaum U, Reinisch A, Wittich S (eds) International investment law for the 21st century – essays in honour of Christoph Schreuer. Oxford, pp 322–337

Aljazy OMH (2000) Arbitration in Islamic law. The Lebanese review of arbitration 16 (received via Email 12 July 2010)
Aust A (2007) Modern treaty law and practice, 2nd edn. Cambridge et al
Ayad MB (2009) Harmonisation of international commercial arbitration law and Sharia. Macquarie J Bus Law 6:93–118
Badmus-Busari F (2013) Sovereign immunity and enforcement of awards in international commercial arbitration. Available via international commercial arbitration/Tulane Law School. http://papers.ssrn.com/sol3/papers.cfm?abstract_id=2336664. Accessed 08 April 2014
Bagner H, Dandenell H (2010) SCC. In: Rowley JW (ed) Arbitration world, 3rd edn. London, pp 123–138
Bellhouse J, Capper P, Connellan C, Sabharwal D (2010) LCIA. In: Rowley J W (ed) Arbitration world, 3rd edn. London, pp 101–112
Berger K-P (2003) Äquivalenzstörungen, Neuverhandlungsklauseln und Vertragsanpassung bei internationalen Konzessionsverträgen sowie Probleme der Streitschlichtung. In: Baur J F, Hobe S (eds) Rechtsprobleme von Auslandsinvestitionen, Baden-Baden, pp 65–97.
Birnie P, Boyle A, Redgwell C (2009) International law & the environment, 3rd edn. Oxford
Bjorklund AK (2009) State immunity and the enforcement of investor-state arbitral awards. In: Binder C, Kriebaum U, Reinisch A, Wittich S (eds) International investment law for the 21st century – essays in honour of Christoph Schreuer. Oxford, pp 302–321
Böckstiegel K-H (1999) Die Anerkennung der Parteiautonomie in der internationalen Schiedsgerichtsbarkeit. In: Geimer R (ed) Wege zur Globalisierung des Rechts – Festschrift für Rolf A. Schütze, München, pp 141–151
Bono S, Charles P, Gharibian A, Hasofer MM, Hilgard M, Kraayvanger J, Mafra R, Reynolds S, Reynolds S, Sarles J, Telles E (2012) Sovereign immunity and enforcement of arbitral awards: navigating international boundaries. Available via Mayer Brown LLP. http://www.mayerbrown.com/files/Publication/2e0f7077-9b25-430e-8b70-6a8f8c1a9d76/Presentation/PublicationAttachment/1be4c54c-bfc2-4d78-9403-85e37c560f43/12270.PDF. Accessed 08 Apr 2014
Brenton P, Baroncelli E, Malouche M (2006) Trade and investment integration of the Maghreb, Middle East and North Africa. The World Bank working paper series no. 44
Brower CN, Sharpe JK (2003) International arbitration and the Islamic World: third phase. Am J Int Law 97:643–656
Brower CN, Ottolenghi M, Prows P (2009) The saga of CMS: Res Judicata, precedent, and the legitimacy of ICSID arbitration. In: Binder C, Kriebaum U, Reinisch A, Wittich S (eds) International investment law for the 21st century – essays in honour of Christoph Schreuer. Oxford, pp 843–864
Buchanan MA (1988) Public policy and international commercial arbitration. Am Bus Law J 26:511–531
CAM (2014a) Arbitration. http://www.camera-arbitrale.it/en/Arbitration/index.php?id=5. Accessed 05 May 2014
CAM (2014b) Mediterranean project. http://www.camera-arbitrale.it/en/International+Projects+and+Activities/Mediterranean+project.php?id=19. Accessed 05 May 2014
Cicogna M (2010) Italy. In: Rowley JW (ed) Arbitration world, 3rd edn. London, pp 347–364
Clapham J (2009) Finality of investor-state arbitral awards: has the tide turned and is there a need for reform? J Int Arbitr 26:437–466
CRCICA (2014a) History – Board of Trustees. http://www.crcica.org.eg/boardoftrustees.html. Accessed 05 May 2014
CRCICA (2014b) History. http://www.crcica.org.eg/history.html. Accessed 05 May 2014
Dahm G, Delbrück J, Wolfrum R (1989) Völkerrecht – vol I/1. Berlin
Dahm G, Delbrück J, Wolfrum R (2002) Völkerrecht – vol I/2, 2nd edn. Berlin
Darwazeh N, El-Kosheri AS (2008) Arbitration in the Arab world: an interview with Professor Ahmed Sadek Wl-Kosheri. J Int Arbitr 25:203–210
Delaume GR (1983) ICSID arbitration and the courts. Am J Int Law 77:784–803
Delaume GR (1985) Economic development and sovereign immunity. Am J Int Law 79:319–346

Desbordes R, Vicard V (2009) Foreign direct investment and bilateral investment treaties: an international political perspective. J Comp Econ 37:372–386

Doehring K (2004) Völkerrecht. Heidelberg

Dolzer R, Schreuer C (2008) Principles of international investment law. Oxford

Ebner T (2005) Streitbeilegung im Welthandelsrecht. Tübingen

Egonu MI (2007) Investor-state arbitration under ICSID: a case for presumption against confidentiality? J Int Arbitr 24:479–489

Ergan A O (2012) The new [2012] arbitration rules of International Chamber of Commerce ["ICC"]. Ankara Bar Rev 83–92

Euro-Mediterranean Association for Cooperation and Development e.V. (2010) Schiedsgerichtsbarkeit und Streitschlichtung – nützliche Instrumente für Investoren (nicht nur) im Mittelmeerraum. http://ema-hamburg.org/media/download_gallery/Mediterranes/Heft%201/Schiedsgerichtsbarkeit_MR.pdf. Accessed 05 May 2014

Fei L (2010) Public policy as a bar to enforcement of international arbitral awards: a review of the Chinese approach. Arbitr Int 26:301–311

Fiebiger G (2010) Vienna arbitration days 2010. SchiedsVZ 269–271

Fouret J (2008) Denunciation of the Washington convention and non-contractual investment arbitration: "manufacturing consent" to ICSID arbitration. J Int Arbitr 25:71–87

Garcia-Bolivar OE (2010) Sovereignty v. investment protection: Back to Calvo?. http://works.bepress.com/cgi/viewcontent.cgi?article=1011&context=omar_garcia_bolivar. Accessed 05 May 2014

Griebel J (2008) Internationales Investitionsrecht. München

Gruss HJ (1979) Enteignung und Aufhebung von Erdölkonzessionen: Der Schiedsspruch im libyschen Erdölstreit. ZaöRV 39:782–806

Hammoud L, Houerbi S (2008) ICC arbitration in the Arab world. J Int Arbitr 25:231–240

Hanotiau B, Caprasse O (2008) Arbitrability, due process, and public policy under Article V of the New York convention. J Int Arbitr 25:721–741

Harten G (2007) Investment treaty arbitration and public law. Oxford

Harten G, Loughlin M (2006) Investment treaty arbitration as a species of global administrative law. Eur J Int Law 17:121–150

Hauschka CE (2005) Bauprojekte im Ausland – Durchsetzung der Ansprüche des Bauunternehmers gegen den Staat vor ICSID-Schiedsgerichten. BauR 1550–1560

Herdegen M (2003) Rechtsprobleme des internationalen Konzessionswesens. In: Baur JF, Hobe S (eds) Rechtsprobleme von Auslandsinvestitionen. Baden-Baden, pp 13–35

Herdegen M (2010) Völkerrecht, 9th edn. München

Herrmann C, Weiß W, Ohler C (2007) Welthandelsrecht, 2nd edn. München

Heß B (1993) The International Law Commission's draft convention on the jurisdictional immunities of state and their property. Eur J Int Law 4:269–282

Hilf M, Oeter S (eds) (2010) WTO-Recht, 2nd edn. Baden-Baden

Hobe S (2002) Enteignung zu Zwecken des Umweltschutzes: Zur ICSID Compania del Desarollo de Santa Elena, S.A. und Republik von Costa Rica (ILM 39 [2000], 1317). IPRax 249–251

Hobe S, Müller A (2009) Die Schiedsgerichtsbarkeit des International Centre for Settlement of Investment Disputes. WiVerw 1:65–76

Hobér K (2008) State responsibility and investment arbitration. J Int Arbitr 25:545–568

Hobér K, Mckechnie W (2007) New rules of the Arbitration Institute of the Stockholm Chamber of Commerce. Arbitr Int 23:261–292

Hodges P, Kirby J, Parker C, Fulton S (2010) International Chamber of Commerce (ICC). In: Rowley JW (ed) Arbitration world, 3rd edn. London

Hofmann R (2012) Völkerrecht II. http://www.jura.uni-frankfurt.de/43681020/Volkerrecht-II-Teil-3_SoSe-2012.pdf. Accessed 03 Apr 2014

Horn N (1981) Neuverhandlungspflicht. AcP 181:255–288

Horn N (2008) Current use of the UNCITRAL arbitration rules in the context of investment arbitration. Arbitr Int 24:587–601

Hughes A (2010) The UNCITRAL arbitration rules. In: Rowley JW (ed) Arbitration world, 3rd edn. London, pp 149–164
Hunter R (2007) Investitionsschutz und Streitbeilegung bei BOT-Projekten im Ausland. In Nicklisch F (ed) Betreibermodelle – BOT/PPP-Vorhaben im In- und Ausland. München, pp 165–174
ICC (2014) The merchants of peace. http://www.iccwbo.org/about-icc/history/. Accessed 05 May 2014
ICSID (2014) About ICSID. http://icsid.worldbank.org/ICSID/FrontServlet?requestType=CasesRH&actionVal=ShowHome&pageName=AboutICSID_Home. Accessed 05 May 2014
Germany Trade and Invest (2013) Rechtsnews. http://www.gtai.de/GTAI/Content/DE/Trade/_SharedDocs/Pdf/Recht/2013-10-rechtsnews.pdf. Accessed 04 Apr 2014
Ipsen K (ed) (2011) Völkerrecht, 6th edn. München
Jagusch S, Gearing M, Sinclair A (2010) ICSID. In: Rowley JW (ed) Arbitration world, 3rd edn. London, pp 75–100
Joffé G (2000) Foreign investment and the rule of law. Mediterr Polit 5:33–49
Karl J (1994) Neue Entwicklungen beim internationalen Investitionsschutz. RIW 809–817
Karl J (2003) Aktuelle Probleme des Investitions- und Enteignungsschutzes am Beispiel des Internationalen Konzessionsvertrages. In: Baur JF, Hobe S (eds) Rechtsprobleme von Auslandsinvestitionen. Baden-Baden, pp 37–61
Kausch K (2009) The European Union and political reform in Morocco. Mediterr Polit 14:165–179
Kouris S (2005) Confidentiality: is international arbitration losing one of its major benefits? J Int Arbitr 22:127–140
Kovacs RB (2008) Challenges to international arbitral awards. J Int Arbitr 25:421–431
Krajewski M (2009) Wirtschaftsvölkerrecht, 2nd edn. Heidelberg
Krieger H (2012) Immunität: Entwicklung und Aktualität als Rechtsinstitut. Available via Deutsche Gesellschaft für internationales Recht. http://www.dgfir.de/workspace/media/documents/krieger-thesen.pdf. Accessed 09 Apr 2014
Kutty F (2006) The Shari'a factor in international commercial arbitration. Loyola Los Angel Int Comp Law Rev 28:565–624
Lamm CB, Pham HT, Bay AKM (2009) Consent and due process in multiparty investor-state arbitrations. In: Binder C, Kriebaum U, Reinisch A, Wittich S (eds) International investment law for the 21st century – essays in honour of Christoph Schreuer. Oxford, pp 54–74
LCIA (2014a) Organisation. http://www.lcia.org/LCIA/Organisation.aspx. Accessed 05 May 2014
LCIA (2014b) History of the LCIA. http://www.lcia.org/LCIA/Our_History.aspx. Accessed 05 May 2014
Liebscher C (2009) Monitoring of domestic courts in BIT arbitrations: a brief inventory of some issues. In: Binder C, Kriebaum U, Reinisch A, Wittich S (eds) International investment law for the 21st century – essays in honour of Christoph Schreuer. Oxford, pp 105–128
Liemen E (1980) Erdöl-Produktionsverträge des Iran. Bochum
Loncle J-M (2005) The arbitration option in protection of investment treaties: treaty claim vs. contract claims. Int Bus Law J 1:3–12
Lörcher G (2001) Enforceability of agreed awards in foreign jurisdictions. Arbitr Int 17:275–285
Lörcher T (2005) ICSID-Schiedsgerichtsbarkeit. SchiedsVZ 11–21
Lowenfeld AF (2008) International Economic Law, 2nd edn. Oxford
Loya K (2012) State immunity as bar to execution of awards in international investment arbitrations. Available via KSLR commercial law. http://www.kslr.org.uk/blogs/commercial/2012/01/20/state-immunity-2/#_ftn1. Accessed 09 Apr 2014
Ma W (2005) Public policy in the judicial enforcement of arbitral awards: lessons for and from Australia, Bond University
Maier B (2010) How has international law dealt with the tension between sovereignty over natural resources and investor interests in the energy sector? Is there a balance? Int Energy Law Rev 95–109

Maniruzzaman AFM (2001) State contracts in contemporary international law: monist versus dualist controversies. Eur J Int Law 12:309–328

Markert L, Wilske S (2010) Entwicklungen in der internationalen Schiedsgerichtsbarkeit im Jahr 2009 und Ausblick auf 2010. SchiedsVZ 62–67

McIlwrath M, Savage J (2010) International arbitration and meditation. Alphen aan den Rijn

Mernissi ME (2002) Arbitration in Morocco. J Int Arbitr 19:179–183

Metje TM (2008) Der Investitionsschutz im internationalen Anlagenbau. Tübingen

Mortenson JD (2010) The meaning of investment: ICSID's travaux and the domain of international investment law. Harv Int Law J 51:257–318

Moses ML (2008) The principles and practice of international commercial arbitration. Cambridge

Moss GC (2009) Commercial arbitration and investment arbitration: fertile soil for false friends? In: Binder C, Kriebaum U, Reinisch A, Wittich S (eds) International investment law for the 21st century – essays in honour of Christoph Schreuer. Oxford, pp 782–797

Muchlinski P (2009) The diplomatic protection of foreign investors: a tale of judicial caution. In: Binder C, Kriebaum U, Reinisch A, Wittich S (eds) International investment law for the 21st century – essays in honour of Christoph Schreuer. Oxford, pp 341–362

Nariman F (2004) East meets west: tradition, globalization and the future of arbitration. Arbitr Int 20:123–138

Nathan KVSK (1995) Submissions to the International Centre for Settlement of Investment Disputes in breach of the convention. J Int Arbitr 1:27–52

Nicklisch F (1994) BOT-Projekte: Netzwerk komplexer Langzeitverträge und Fragen der Streitbeilegung. In: Nicklisch F (ed) Rechtsfragen privatfinanzierter Projekte. Heidelberg, pp 53–64

Nmehielle VOO (2001) Enforcing arbitration awards under the International Convention for the Settlement of Investment Disputes. Annu Surv Int Comp Law 7:21–48

Norton PM (1991) A law of the future or a law of the past? Modern tribunals and the international law of expropriation. Am J Int Law 85:474–505

Ouerfelli A (2008) Enforcement of foreign arbitral awards in Maghreb countries. J Int Arbitr 25:241–256

Ozumba O (2009) Enforcement of arbitral awards: does the public policy exception create inconsistency? http://www.dundee.ac.uk/cepmlp/gateway/files.php?file=cepmlp_car13_8_127246631.pdf. Accessed 05 May 2014

Partasides C, Fullelove G (2010) Global overview. In: Rowley JW (ed) Arbitration world, 3rd edn. London, pp 1–16

Pengelley N (2009) Waiver of sovereign immunity from execution: arbitration is not enough. J Int Arbitr 26:859–872

Pfeiffer T (2003) Zahlungskrisen ausländischer Staaten im deutschen und internationalen Rechtsverkehr. ZVglRWiss 141–194

Plaßmeier H (2010) Ende des "Doppelexequatur" bei ausländischen Schiedssprüchen. SchiedsVZ 82–85

Redfern A, Hunter M, Blackaby N, Fartasides C (2004) Law and practice of international commercial arbitration, 4th edn. London

Reed L, Paulsson J, Blackaby N (2004) Guide to ICSID arbitration. The Hague

Reinisch A (2006) European court practice concerning state immunity from enforcement measures. Eur J Int Law 17:803–836

Reinisch A (2008) The proliferation of international dispute settlement mechanisms: the threat of fragmentation vs. the promise of a more effective system? Some reflections from the perspective of investment arbitration. In: Buffard I, Hafner G (eds) International law between universalism and fragmentation – Festschrift in honour of Gerhard Hafner. Boston, pp 107–125

Ripinsky S, Williams K (2008) Case summary – The Government of the State of Kuwait v The American Independent Oil Company. Available via BIICL. http://www.biicl.org/files/3938_1982_kuwait_v_aminoil.pdf. Accessed 05 May 2014

Rosenberg CB (2010) Avoiding waiving a right to ICSID arbitration in the negotiation of a concession agreement. Int Arbitr Law Rev 13:8–13
Ruchay D (1996) Partnerschaftliche Entsorgungsprojekte. In: Nicklisch F (ed) Partnerschaftliche Infrastrukturprojekte. Heidelberg, pp 69–87
Sattar S (2010) National courts and international arbitration: a double-edged sword? J Int Arbitr 27:51–73
Saunders M, Salomon C (2007) Enforcement of arbitral awards against states and state entities. Arbitr Int 23:467–476
Sayed A (2008) Towards a reflexive sociology of the arbitration field in the Arab world. J Int Arbitr 25:289–294
SCC (2014a) About the SCC. http://www.sccinstitute.com/hem-3/om-oss-3.aspx. Accessed 05 May 2014
SCC (2014b) SCC Arbitration. http://www.sccinstitute.com/skiljeforfarande-2.aspx. Accessed 05 May 2014
Schäfer M (1998) Der Entschädigungsstandard im allgemeinen Völkerrecht. RIW 199–204
Schöbener B, Markert L (2006) Das International Centre for Settlement of Investment Disputes (ICSID). ZVglRWiss 65–116
Schöbener B, Herbst J, Perkams M (2010) Internationales Wirtschaftsrecht. Heidelberg
Schreuer C (2009) The ICSID convention – a commentary, 2nd edn. Cambridge
Schütze RA (ed) (2011) Institutionelle Schiedsgerichtsbarkeit, 2nd edn. Köln
Schweisfurth T (2006) Völkerrecht. Tübingen
Seidl-Hohenveldern I (1977) Schiedsspruch über die Nationalisierung von Erdölkonzernen der Texaco under Calasiatic durch Libyen. RIW/AWD 502–504
Semler, F-J (2003) Schiedsverfahren im Rahmen von Investitionsschutzabkommen der Bundesrepublik Deutschland. SchiedsVZ 97–102
Shaw MN (2008) International law, 6th edn. Cambridge
Shookmann J (2010) Too many forums for investment disputes? ICSID illustrations of parallel proceedings and analysis. J Int Arbitr 27:361–378
Sornarajah M (2010) The international law on foreign investment, 3rd edn. Cambridge
Taarji R, Hamzi ZI, Essakalli Y (2011) Morocco country chapter. Rome
The Jean Monnet Center for International and Regional Economic Law and Justice (2013) The Hull rule. Available via The NYU Institutes of the Park. http://www.jeanmonnetprogram.org/archive/papers/97/97-12-III.html#fn17. Accessed 09 Apr 2014
Tietje C (2003) Grundstrukturen und aktuelle Entwicklungen des Rechts der Beilegung internationaler Investitionsstreitigkeiten. In: Tietje C, Kraft G (eds) Arbeitspapiere aus dem Institut für Wirtschaftsrecht, vol 10. Halle (Saale), pp 5–20
Tietje C (2005) Die Beilegung internationaler Investitionsstreitigkeiten. In: Marauhn T (ed) Streitbeilegung in den internationalen Wirtschaftsbeziehungen. Tübingen, pp 47–62
Tietje C (2010) Internationales Investitionsschutzrecht im Spannungsverhältnis von staatlicher Regelungsfreiheit und Schutz wirtschaftlicher Individualinteressen. In: Tietje C, Kraft G, Lehmann M (eds) Beiträge zum Transnationalen Wirtschaftsrecht, vol 93. Halle (Saale), pp 5–22
Trakman LE (2002) Confidentiality in international commercial arbitration. Arbitr Int 18:1–18
Tschanz P-Y, Viñuales JE (2009) Compensation for non-expropriatory breaches of international investment law. J Int Arbitr 26:729–743
Turck NB (2001) French and US courts define limits of sovereign immunity in execution and enforcement of arbitral awards. Arbitr Int 17:327–342
Ulfstein G, Marauhn T, Zimmermann A (2007) Introduction. In: Ulfstein G (ed) Making treaties work. Cambridge, pp 3–12
Ulusoy A, Oguz F (2007) The privatization of electricity distribution in Turkey: a legal and economic analysis. Energy Pol 35:5021–5034

UNCITRAL (1981) Report of the secretary-general: possible features of a model law on international commercial arbitration – A/CPARA9/207. http://daccess-dds-ny.uparaorg/doc/UNDOC/GEN/NL8/102/47/PDF/NL810247.pdf?OpenElement. Accessed 07 Jan 2011
UNCITRAL (1985) Report of the United Nations Commission on international trade law on the work of its eighteenth session – A/40/17. http://www.uncitral.org/pdf/english/yearbooks/yb-1985-e/vol16-p3-46-e.pdf. Accessed 05 May 2014
UNCITRAL (2014a) About UNCITRAL. http://www.uncitral.org/uncitral/en/about_us.html. Accessed 05 May 2014
UNCITRAL (2014b) FAQ – UNCITRAL and private disputes/Litigation. http://www.uncitral.org/uncitral/en/uncitral_texts/arbitration_faq.html#difference. Accessed 05 May 2014
UNCTAD (2009) Selected recent developments in IIA arbitration and human rights – IIA MONITOR No. 2. http://www.unctad.org/en/docs/webdiaeia20097_en.pdf. Accessed 05 May 2014
UNCTAD (2010a) Denunciation of the ICSID convention and BITS: impact on investor-state claims. http://www.unctad.org/en/docs/webdiaeia20106_eparapdf. Accessed 30 Dec 2010
UNCTAD (2010b) Investor–state disputes: prevention and alternatives to arbitration. http://www.unctad.org/en/docs/diaeia200911_en.pdf. Accessed 05 May 2014
Vannieuwenhuyse G (2009) Bringing a dispute concerning ICSID cases and the ICSID convention before the International Court of Justice. Law Pract Int Courts Tribunals 8:115–141
Verdross A (1958) Die Sicherung von ausländischen Privatrechten aus Abkommen zur wirtschaftlichen Entwicklung mit Schiedsklauseln. ZaöRV 18:635–651
Verdross A, Simma B (2010) Universelles Völkerrecht, 3rd edn. Berlin
Vinuesa RE (2005) Jurisdictional objections to ICSID arbitration under bilateral investment treaties. In: Bröhmer J, Bieber R, Calliess C, Langenfeld C, Weber S, Wolf J (eds) Internationale Gemeinschaft und Menschenrechte – Festschrift für Georg Ress. Köln, pp 331–355
Vogl T (2010) Schiedsgerichtsbarkeit im Mittelmeerraum. SchiedsVZ 32–41
Walter A (2006) Die Reichweite von Schirmklauseln in Investitionsschutzabkommen nach der jüngsten ICSID-Spruchpraxis. RIW 815–824
Weh M (2008) Transnationale Unternehmen im System kollektiver Sicherheit. Regensburg
Weiler T (2005) International investment law and arbitration. London
Wood DP (2007) Private dispute resolution in international economic law. In: Guzman AT, Sykes AO (eds) Research handbook in international economic law. Cheltenham, pp 575–597
Zampetti AB, Sauvé P (2007) International investment. In: Guzman A, Sykes AO (eds) Research handbook in international economic law. Cheltenham, pp 211–270
Ziegler AR (2012) Immunität – Immunität der Staaten. Available via Deutsche Gesellschaft für internationales Recht. http://www.dgfir.de/workspace/media/documents/ziegler-thesen.pdf. Accessed 09 Apr 2014

Cases

German Courts

BVerfG, *National Iranian Oil Company* [1983], [1983] BVerfGE 64, 1

PCIJ

France v Kingdom of the Serbs, Croats and Slovenes [1929] Judgement (Case Concerning the Payment of Various Serbian Loans Issued in France), Series A – No. 20/21
Germany v Poland [1928] Judgment (*The Factory At Chorzów (Claim for Indemnity) (The Merits)*, Series A - No. 17

International Court of Justice

Islamic Republic of Iran v United States of America [1996] Oil Platforms, Preliminary Objection – Judgment [1996] ICJ Reports, 803

Islamic Republic of Iran v United States of America [1996] Oil Platforms, Separate Opinion of Judge Higgins [1996] ICJ Reports, 847
Lichtenstein v Guatemala [1955] Nottebohm Case (second phase) – Dissenting Opinion Judge Read, [1955] ICJ Reports, 34
Republic of Guinea v Democratic Republic of the Congo [2007] Case concerning Ahmadou Sadio Diallo, Preliminary Objections – General List No. 103, 17
United Kingdom v Iran [1952] Anglo-Iranian Oil Co. Case, Preliminary Objection – Judgment [1952] ICJ Reports, 93
Yugoslavia v Italy [1999] Legality of Use of Force, Provisional Measures (Order of 2 June 1999) [1999] ICJ Reports, 481

ICSID Tribunals
Aguas del Tunari, S.A. v Republic of Bolivia [2005] ARB/02/3
Aguaytia Energy LLC v Republic of Peru [2008] ARB/06/13 – Award
Azurix Corp. v The Argentine Republic [2006] ARB/01/12 – Award
Bayindir Insaat Turizm Ticaret Ve Sanayi A.S. v Islamic Republic of Pakistan [2005] ARB/03/29
Burimi SRL and Eagle Games SH.A v Republic of Albania [2013] ARB/11/18 – Award
Compañiá de Aguas del Aconquija S.A. and Vivendi Universal S.A. v Argentine Republic [2002] ARB/97/3 – Annulment [2002] ICSID Review, 89
Compañia del Desarrollo de Santa Elena S.A. v Republic of Costa Rica [2000] ARB/96/1 – Award [2000] ICSID Review, 169
El Paso Energy International Company v The Argentine Republic [2006] ARB/03/15
Enron Corporation and Ponderosa Assets, L.P. v Argentine Republic [2007] ARB/01/3 – Award
Enron Corporation and Ponderosa Assets, L.P. v Argentine Republic [2008] ARB/01/3 – Annulment
Glamis Gold, Ltd. v United States of America [2009] In accordance with the UNCITRAL Arbitration Rules
Impregilo S.p.A. v Argentine Republic [2011] ARB/07/17 – Award
Jan de Nul N.V. Dredging International N.V. v Arab Republic of Egypt [2006] ARB/04/13
Lanco International Inc. v The Argentine Republic [1998] ARB/97/6 – Preliminary Decision (2001) 40 ILM, 457
Pan American Energy LLC, and BP Argentina Exploration Company v The Argentine Republic [2006] ARB/03/13 and *BP America Production Company, Pan American Sur SRL, Pan American Fueguina, SRL and Pan American Continental SRL v The Argentine Republic* [2006] ARB/04/8
PantechnikiS.A.Contractors&Engineers(Greece)vTheRepublicofAlbania[2009]ARB/07/21–Award
Phoenix Action Ltd. v The Czech Republic [2009] ARB/06/5 – Award
Saipem S.p.A. v The People's Republic of Bangladesh [2007] ARB/05/07
Sempra Energy International v The Argentine Republic [2007] ARB/02/16 – Award
SGS Société Générale de Surveillance S.A. v Islamic Republic of Pakistan [2003] ARB/01/13 [2003] ICSID Review, 307
Siemens v The Argentine Republic [2004] ARB/02/8 – Decision
Toto Costruzioni Generali S.P.A. v The Republic of Lebanon [2009] ARB/07/12
Tradex Hellas S.A. v Republic of Albania [1999] ARB/94/2 – Award, [1999] ICSID Review, 197
Wena Hotels Ltd. v Arab Republic of Egypt [1999] ARB/98/4 (2002) 41 ILM, 881
Wena Hotels Ltd. v Arab Republic of Egypt [2002] ARB/98/4 – Annulment (2002) 41 ILM, 933

Chapter 3
The Desertec Concept and the ICSID Convention

This chapter deals with special questions of investment protection under the ICSID Convention. The focus will be on the scope of *ratione personae* and *ratione materiae* of Art. 25(1) of the ICSID Convention. Especially the requirement of "investment" will be analyzed in depth as this may cause difficulties in the case of the Desertec Concept. In addition, the applicability of the umbrella clause will be assessed.

3.1 Investment and Africa

First of all, some general aspects concerning investments in Africa require discussion. This includes the investment environment within Africa, renewable energy projects as investments, and possible investment legal frameworks. In general, projects encompassing more than one billion USD are called 'mega-deals.'[1] There are two types of investments, namely FDIs and portfolio investments. FDIs aim to invest for a longer period in the host state, and portfolio investments usually include cooperation in terms of shares and stocks.[2] However, there is no single definition of portfolio investment.[3] In cases where the investor builds a new project in a foreign country, the FDI is also known as a 'Greenfield investment.'[4] Since there is no company in the MENA region involved with solar thermal energy, portfolio investments are not an option. Consequently, the Desertec Concept is a FDI. As it requires construction of a new plant, it is also a Greenfield investment.

---

[1] Krajewski (2009), 167.

[2] Herrmann et al. (2007), § 17 para 785; Krajewski (2009), 168.

[3] Michaelis, Salomon in Hilf and Oeter (2010), § 15 para 2.

[4] Krajewski (2009), 168.

R. Gömmel, *Investing into North African Solar Power*,
DOI 10.1007/978-3-319-15756-6_3

3.1.1 The Desertec Concept and Investment Security

Over the last few decades, several different projects have been brought up before ICSID tribunals. These included everything from small to large scale projects. Despite the fact that almost none of these projects had something to do with renewable energy, the ICSID dealt with numerous projects related to energy.

As mentioned above, the ICSID is part of the World Bank group as one of the five institutions. The World Bank was founded in 1944 to rebuild Europe after World War II.[5] During the 1950s, this changed and the World Bank focused more on development aid for "third world countries".[6] In 2009, four institutions of the World Bank group expressed their support of renewable energies in a special report.[7] These four institutions are the International Bank for Reconstruction and Development (IBRD), the International Development Association (IDA), the International Finance Corporation (IFC), and the Multilateral Investment Guarantee Agency (MIGA).[8] Nevertheless, the ICSID was not part of this group of supporting World Bank institutions.[9] Within a different Annex to the 2009 report, it was pointed out that the ICSID does not directly support any renewable energy or energy efficiency activities.[10] It would be wrong to assume that the Desertec Concept will receive special treatment just because of its benefits for the environment and the people.

3.1.2 Investment in Africa

Africa is still considered a high risk area of the world for foreign investment.[11] In most cases, only the strong African countries are supported, whereby there are only scattered activities to promote and develop science and technology in Africa.[12] Overall, only 3 % of all FDIs take place in Africa.[13] Investments are very important for developing states as they get more money from investments than from development aid.[14] African states support the improvement of tertiary education, the improvement of the local environment, cooperation concerning environmental

[5] Schöbener and Markert (2006), 65 (66).
[6] Schöbener and Markert (2006), 65 (66–67).
[7] IBRD (2009), 53.
[8] IBRD (2009), 53.
[9] IBRD (2009), 53.
[10] IFC (2010), 47 (47).
[11] Abraham (2009), 206 (212).
[12] Saß (2009), 27 (28) and (29).
[13] Krajewski (2009), 167.
[14] Tietje (2003), 5 (7).

protection, and the upgrade of health and security.[15] It is important for the investor to recognize human rights and social and environmental standards.[16]

The stability of the political system and the political relationship between the respective countries are essential for a successful investment.[17] New empirical studies revealed that a good interstate political relationship increases FDIs, and historical ties between the countries influence the political relationship.[18]

Over the past years FDIs decreased within the MENA region, illustrated by a drop of over 50 % between 2007 and 2011.[19] Especially the Arabian Spring led to a wait-and-see attitude towards investors.[20] However it is expected that the MENA region is going to offer a good investment field in the future due to the ongoing democratization.[21] Especially investments within the field of energy have great potential.[22] Despite these positive views it is expected that the investments risks are going to be significantly high until 2020–2030.[23] It is necessary to avoid harmful investments within the MENA region as these could lead to a dependence of the local economy. This underlines that projects like solar thermal power plants are well suited for the MENA region as they help to decrease the dependence on fossil fuels like gas and oil.[24]

Morocco is an exception concerning the abovementioned decrease as FDIs started to rise again by 2011.[25] Within the Doing-Business-Ranking of the World Bank Morocco also proceeded to No. 94 (advancement of 21 places)[26] Morocco has an Investment Charta which stipulates benefits for FDIs that e.g. include an unlimited transfer of profit.[27] There are also certain requirements to get these benefits, such as a minimum FDI of 17.5 million Euros, creating 250 new jobs, settlement in a priority region, transfer of technology and a contribution to environmental protection.[28]

---

[15] Saß (2009), 27 (32).

[16] Herdegen (2003), 13 (30).

[17] Desbordes and Vicard (2009), 372 (374).

[18] Desbordes and Vicard (2009), 372 (379) and (383).

[19] Rang (2011), 29 (30); cf BDI (2011).

[20] Rang (2011), 29 (31).

[21] Rang (2011), 29 (35).

[22] Rang (2011), 29 (36–37).

[23] Schulz (2012).

[24] Schulz (2012).

[25] Rang (2011), 29 (30).

[26] Rang (2011), 29 (33).

[27] Deutsche Industrie- und Handelskammer in Marokko (2008), 7.

[28] Deutsche Industrie- und Handelskammer in Marokko (2008), 7.

3.2 Multilateral Investment Protection

As mentioned above, the ICSID Convention offers the best investment arbitration platform within international law, this does not completely answer the question of the best suited investment protection framework. Besides the already discussed BITs, there are also multilateral investment frameworks. The most important multilateral and regional treaties are the Energy Charter Treaty (ECT) and treaties within the World Trade Organization (WTO).

3.2.1 Energy Charter Treaty

Comparable to the North American Free Trade Agreement (NAFTA), the ECT is a regional investment protection agreement and thus different from BITs.[29] Different from NAFTA, the ECT is limited in its *ratione materiae*, which sometimes leads to the description of a sectoral agreement.[30] It must be kept in mind that the ECT is only a declaration of intent, which sets out 'broad policy objectives' and only the agreements of the signatory states to go beyond this status to negotiate a legal framework led to an investment protection in 1994 by creating the binding ECT.[31]

The idea of the ECT dates back to the collapse of the Soviet Union as there was a desire to unify Europe (especially the energy sector).[32] The ECT's intention was to assist Eastern European states and Russia to develop real economic activity instead of relying on economic aid.[33] The treaty itself is a multilateral international treaty which aims to facilitate cooperation in the field of energy.[34] The ECT is a Charta of basic rights in the field of energy,[35] and it encompasses regulations concerning the trade of energy, environmental protection, transmission and rules of investment protection.[36] The ECT mainly protects investments from political risks.[37] The provisions of the ECT protect all direct or indirect investments of an investor, if he is a national of an ECT contracting party.[38] Furthermore Western European

[29] Krajewski (2009), 175–176; Schöbener et al. (2010), 250, Kap. 4 §17 para 112.

[30] Griebel (2008), 59.

[31] Marauhn (1996), 301 (328).

[32] Kemper (2003), 504 (504); Schöbener et al. (2010), 251, Kap. 4 §17 para 115.

[33] Marauhn (1996), 301 (328).

[34] Kemper (2003), 504 (505); Krajewski (2009), 176; Schöbener et al. (2010), 251, Kap. 4 §17 para 115.

[35] Griebel (2008), 59; Happ (2007), 129 (129).

[36] Schöbener et al. (2010), 252, Kap. 4 §17 paras 116–117; Happ (2007), 129 (129); Kemper (2003), 504 (505); cf Krajewski (2009), 176; Griebel (2008), 59–60; cf Hunter (2007), 165 (168); cf Harten (2007), 81–82.

[37] Happ (2007), 129 (129); Kemper (2003), 504 (505).

[38] Happ (2007), 129 (130).

states shared common interests in '(. . .) capital yields security of supply an environmental improvements (. . .)' and Eastern European states desired '(. . .) economic recovery, security of supply through diversified forms of energy production, energy efficiency, and again environmental improvements'.[39] The ECT is the only treaty which guarantees a free transfer of capita,[40] and does not infringe on state sovereignty in respect of national energy supply.[41] The ECT is not limited to investment and trade regulations as environmental standards play also a major role, which is not surprising as '(. . .) efficiency and environmental protection are considered to be essentially economics-based'.[42] However, Morocco is not a party to the ECT, as most MENA countries only have an observer status and the ECT cannot serve as an investment protection framework.

3.2.2 Multilateral Agreements

There are also multilateral agreements, which differ from regional agreements and BITs. Multilateral agreements mostly complement regional agreements and BITs, but they do not offer a comprehensive investment protection. Besides the regulations of the WTO, the OECD offers interesting approaches to investment protection.

3.2.2.1 WTO Regulation

The settlement of disputes under the auspices of the WTO is done according to the Dispute Settlement Understanding (DSU).[43] Pursuant to Art. 23(1) of the DSU, the DSU is in charge to solve disputes related to WTO law.[44] Art. 1(1) of the DSU regulates that the DSU shall apply to all disputes settlement provisions of what are known as "covered agreements" (Appendix 1 to the DSU). Appendix 1 of the DSU mentions several agreements within its paragraphs, whereas in paragraph B, the Agreement on Trade-Related Investment Measures (TRIMs), the Agreement on Trade-Related Aspects of Intellectual Property Rights (TRIPs) and the General Agreement on Trade in Services (GATS) are of high relevance for investments. The World Trade Organization (WTO) law includes some regulations concerning

[39] Marauhn (1996), 301 (327).

[40] Karl (1994), 809 (813).

[41] Kemper (2003), 504 (506).

[42] Marauhn (1996), 301 (328–329).

[43] Ebner (2005), 107–108.

[44] Zampetti and Sauvé (2007), 211 (261–262); No parallel processes are allowed, in: Ebner (2005), 108; Article 23(1) DSU is not intended to be the exclusive norm, forbidding any other dispute resolution, in: Neumann (2001), 529 (554).

investment protection within TRIMs.[45] However, TRIMs only include single parts of investment protection.[46] A clarification is also delivered in Art. 1 of the TRIMs, as it mentions that investment measures only relate to trade activities. Other frameworks, like the GATS, are limited to services and FDIs,[47] which is irrelevant for the implementation of the Desertec Concept. In addition, TRIPs only partially include investment regulations and are limited to intellectual property.[48] Further attempts of the WTO to include investment issues into the negotiation at the Doha round failed.[49] Consequently, neither covered agreement offers a platform for investment protection for the Desertec Concept at the moment.

3.2.2.2 OECD

The OECD offers guidelines and codes for foreign investments. First of all, there is the GTFDI. There are several relevant regulations within the GTFDI. Right from the beginning of a GTFDI investment, it is important there is a prompt issue of all licenses (e.g. concessions) according to Section III No. 5(a). Section III No. 6(b) of the GTFDI mentions that there also should be free transfer of revenues. In Section III No. 8 of the GTFDI, the prevention of corruption is highlighted as an important tool to facilitate foreign investments. However, the GTFDI is soft law and thus not as relevant in practice.[50] With regard to the Desertec Concept, it can only severe as a non-committal guideline, even though it mostly repeats important well-known requirements.

Furthermore, there are also OECD codes, which deal with the issue of investment (e.g. Code of Liberalisation of Capital Movement and Code of Liberalisation of Current Invisible Operations). These codes are not "binding" international law, but they are binding on all OECD members.[51] It must be noted that they do not encompass regulations concerning protection against expropriation,[52] which makes them less important for the implementation of the Desertec Concept.

[45] Zampetti and Sauvé (2007), 211 (249) and 252–254); Krajewski (2009), 177; Herrmann et al. (2007), § 17 paras 791–792; Expansion of investment security due to the WTO, in: Schöbener et al. (2010), 49, Kap. 1 §4 paras 162 and 321, Kap. 4 §20 para 411.

[46] Michaelis, Salomon in Hilf and Oeter (2010), § 15 para 19; Krajewski (2009), 177; Griebel (2008), 60; cf Schöbener et al. (2010), 312, Kap. 4 §20 paras 412–413.

[47] Harten (2007), 72; Karl (2003), 37 (41); Schöbener et al. (2010), 322–323, Kap. 4 §20 paras 419–420; cf Griebel (2008), 60; Investment regulations within GATS, in: Zampetti and Sauvé (2007), 211 (249) and (254–257).

[48] Griebel (2008), 60; Investment regulations within TRIPs, in: Zampetti and Sauvé (2007), 211 (249) and (261).

[49] Schöbener et al. (2010), 253, Kap. 4 §17 para 124.

[50] Schöbener et al. (2010), 320, Kap. 4 §20 paras 406–408.

[51] Krajewski (2009), 180; Not legally binding, in: Karl (2003), 37 (40).

[52] Krajewski (2009), 180.

Members of the OECD also tried to negotiate the Multilateral Agreement on Investments (MAI) from 1995 to 1998.[53] The MAI was the first attempt to lift the investment protection from a bilateral to a multilateral level.[54] The MAI's main idea was to unify different the protection standards of several BITs and to raise the level of investment protection in some cases.[55] Nevertheless, the MAI never entered into force because developing countries did not feel correctly represented during the negotiation process,[56] which leads to the fact that MAI does not help concerning the Desertec Concept investment. Overall, there is no multilateral or regional investment protection regime which could apply in the case of the Desertec Concept, leaving protection to BITs

3.3 The Applicability of the G/M-BIT

As mentioned above, Germany and Morocco concluded a new BIT in 2001, which would apply in the case of the Desertec Concept. Nevertheless, there are some doubts concerning the applicability of the G/M-BIT, due to a shift of competence on the European level. This question not only deals with the G/M-BIT, but rather all BITs of EU member states. Until the Treaty of Lisbon, there was no doubt that the competence to conclude BITs was located in the EU member states,[57] although there were cases before the Treaty of Lisbon which dealt with the issue of BITs concluded before EU membership.[58] According to Art. 351(2) of the TFEU (formerly Art. 307(2) of the TEC), all '(...) (m)ember (s)tate(s) (...) concerned shall take all appropriate steps to eliminate the incompatibilities (...)' if prior treaties are not compatible with the European Treaties. In 2009, the ECJ dealt with a case concerning Austrian and Swedish BITs and concluded that even hypothetical incompatibilities with the European Treaties are against community law.[59]

[53] Zampetti and Sauvé (2007), 211 (249) and (250); Karl (2003), 37 (37); Krajewski (2009), 180; Schöbener et al. (2010), 252–253, Kap. 4 §17 para 121; Kim (2007), 55–59.

[54] Karl (2003), 37 (37).

[55] Kim (2007), 53–55; Schöbener et al. (2010), 252–253, Kap. 4 §17 para 121; Krajewski (2009), 180; cf Herdegen (2003), 13 (27); Zampetti and Sauvé (2007), 211 (250).

[56] Herrmann et al. (2007), § 17 para 789; Karl (2003), 37 (41–42); Kim (2007), 57–59; Schöbener et al. (2010), 252, Kap. 4 §17 para 120 and 253, Kap. 4 §17 para 122; Krajewski (2009), 181; Herdegen (2003), 13 (27); Detailed reasons for MAI failure, in: Zampetti and Sauvé (2007), 211 (250–251).

[57] Griebel (2009), 469 (469); Schöbener et al. (2010), 227, Kap. 4 § 14 para 20; Due to the new European competence there is need for coordination and adjustments, in: Tietje (2010), 5 (19).

[58] Potestà (2009), 225 (238).

[59] Potestà (2009), 225 (241); The cases dealt with transfer clauses, which do occur in almost all BITs, in: Potestà (2009), 225 (239) and (243).

3.3.1 Change Due to the Treaty of Lisbon

Since the Treaty of Lisbon came into force, there have been doubts about the BIT competence of the EU member states. Art. 207(1) of the TFEU deals with the common commercial policy of the EU. Art. 207(1) of the TFEU stipulates that: 'die Handelsaspekte des geistigen Eigentums, die ausländischen Direktinvestitionen'[60] are part of the common commercial policy. This is a major shift of competence within investment law as, for the first time, a supranational organization can set up investment agreements with third states for its members. It was surprising that this change of competence involved very little prior preparation by the EU and its member states.[61] The EU member states even seemed to be surprised about the range of the transfer of competence.[62] Based on the Treaty of Lisbon, the EU already states that it has exclusive competence concerning investments.[63] The EU also actively uses its new competence as it has been working towards an EU investment policy for the past few years and it includes investment sections within border agreements.[64] The new Transatlantic Trade and Investment Partnership with the USA (negotiations started in 2013) illustrates that the EU is actively participating in the field of investment.[65]

Besides the problematic consequences of this change, the scope of the new European competence is also unclear. Based on the wording of Art. 207 (1) of the TFEU, the EU only has competence regarding FDIs, which would exclude all portfolio investments. This opinion is not shared by the EU institutions, which consider all kinds of investments to be within their new competence. This is not undisputed as some legal authors claim that decisions concerning portfolio investments need to be ratified by EU member states,[66] but at the same time it is highly likely that the ECJ will adopt the view of the EU institutions.[67] Other legal authors even claim that Art. 207(1) of the TFEU does not cover the protection of property.[68] This also corresponds to Art. 345 of the TFEU, which stipulates that: '(t)he treaties shall in no way prejudice the rules in Member States governing the system of

---

[60] Not as clear in the English version of the contract, where it is mentioned that "commercial aspects of intellectual property, foreign direct investment" are part of the common commercial policy.

[61] Seattle to Brussels Network (2010), 12.

[62] Reinisch (2014), 111 (115).

[63] European Commission (2014) Commission to consult European public on provisions in EU–US trade deal on investment and investor-state dispute settlement; There are different perceptions within literature, in: Reinisch (2014), 111 (117–118 and 125).

[64] Tams (2014).

[65] European Commission (2014) Commission to consult European public on provisions in EU–US trade deal on investment and investor-state dispute settlement.

[66] Seattle to Brussels Network (2010), 13.

[67] Reinisch (2014), 111 (136–141).

[68] Krajewski (2005), 91 (114); Already Article 133 ECT did not include FDIs, in: Osteneck in Schwarze (2009), Art. 133 para 16; Johannsen (2009), 5 (16–17); Griebel (2009), 469 (471).

property ownership.'[69] The EU therefore could only stipulate the case of expropriation by an EU organ.[70]

3.3.1.1 EU Investment Policy

The EU is unwilling to negotiate with all non EU member states separately as negotiations should be limited to the bigger states (e.g. USA, India, China, Russia, etc.) and free trade agreements.[71] There is no plan to create a 'one-size-fits-all model.'[72] This was already highlighted in 2010 when the European Commission mentioned that after notifying an EU member state BIT, the BIT should stay in force.[73] There was no plan to replace all existing BITs. The European Commission's aim is to introduce a common and comprehensive investment policy which addresses the needs of the investor (European and from other states).[74] The main idea of the European Commission is to clarify and improve investment protection.[75] Standards of an EU-BIT could encompass non-discrimination, fair and equitable treatment, umbrella clauses, cultural aspects of the third party, and provision of adequate compensation.[76] It could also include provisions to ensure an effective dispute settlement.[77] It is stressed that '(t)he Union's future action in this field should be inspired and guided by the best available standards, so as to offer a level playing field of a high quality to all EU investors.'[78] Finally, it is important for the European Commission to include the same standards which apply within the EU and its member states (especially concerning human health and the environment).[79] The main goal is to improve investment dispute settlements, to protect the state and the general public, to create transparency and to avoid any investor claims which are unlikely to succeed.[80]

The European Parliament also stresses that investor protection must be the first priority.[81] New BITs must be based on best practice experiences by the EU member states.[82] The European Parliament proposes that non-discrimination, fair and

[69] Johannsen (2009), 5 (16).

[70] Griebel (2009), 469 (471).

[71] Seattle to Brussels Network (2010), 13.

[72] European Commission (2010), 6.

[73] Europäische Kommission (2010), 2 and 4.

[74] European Commission (2010), 5.

[75] European Commission (2013b), 2.

[76] European Commission (2010), 8–9; cf European Commission (2013b), 2–3.

[77] European Commission (2010), 9.

[78] European Commission (2013b), 3; European Commission (2010), 6.

[79] European Commission (2010), 9.

[80] European Commission (2013b), 8–9 and 10–11.

[81] European Parliament (2011).

[82] European Parliament (2011).

equitable treatment, protection against direct and indirect expropriation and umbrella clauses should be part of any EU-BIT.[83] It says that the

> EU's future policy must also promote investment which is sustainable, respects the environment (...) and encourages good quality working conditions in the enterprises targeted by the investment.[84]

In contrast to the European Commission, the European Parliament believes that it will be difficult to maintain the protection level for investors involved in the new EU-BITs as guaranteed under existing member state BITs (e.g. the need to protect the state's right to regulate).[85] The European Parliament also wants to exclude speculative investments.[86] However, the European Commission has pointed out that such a distinction would be difficult and that both forms should be part of the notion of investment.[87]

3.3.1.2 Regulation 1219/2012/EU

Due to the shift in competence, the EU decided to pass Regulation 1219/2012/EU to clarify the future procedure concerning bilateral investment agreements (BIA) (Art. 1(1) of Regulation 1219/2012/EU). Although Regulation 1219/2012/EU covers BIAs, this term encompasses BITs as No. 2 of the guiding principles to Regulation 1219/2012/EU provides a broad definition of BIAs. This is also illustrated by Art. 1 (2) of Regulation 1219/2012/EU, which mentions that BIAs encompass 'any agreement with a third country that contains provisions on investment protection'. Within the guiding principles of Regulation 1219/2012/EU, the EU emphasizes again that it has exclusive competence with respect to common commercial policy (No. 1). It acknowledges that there are several BITs of EU member states (No. 4) and underlines that BIAs remain binding under public international law and 'will be progressively replaced by agreements of the Union relating to the same subject matter' (No. 5). According to No. 6, specifications and guarantees concerning conditions of investments should be maintained in the interests of the investors and member states until they are appropriately replaced. No. 7 highlights that Regulation 1219/2012/EU applies to all BIAs signed prior to 01 December 2009 and lays down conditions for BIAs signed between 01 December 2009 and 09 January 2013 (No. 8). EU member states also have the power to take necessary measures to adjust BITs if they are not in accordance with EU law (No. 11).

If the BIT of an EU member state was signed before 01 December 2009 and should be maintained, the state must notify the respective BIT to the European

[83] European Parliament (2011); Reinisch (2014), 111 (123).

[84] European Parliament (2011).

[85] European Parliament (2011).

[86] European Parliament (2011).

[87] European Commission (2011), 2–3.

Commission (Art. 2 of Regulation 1219/2012/EU). Art. 3 of Regulation 1219/2012/EU clarifies that although the BIT can remain in force, it might eventually be replaced by a newly negotiated EU-BIT. According to Art. 4(1) of Regulation 1219/2012/EU, all notifications of the member states are published. The European Commission then assesses whether all the notified BITs comply with EU law. This is repeated by Art. 12(1) of Regulation 1219/2012/EU and according to Art. 12(2) and Art. 9(1) and (2) of Regulation 1219/2012/EU, there is a list of possible incompatibilities between BITs and EU law. If there are serious obstacles, the European Commission will request that the respective member state adjust the BIT (Art. 6(1) and (2) of Regulation 1219/2012/EU). If the requirements of Art. 9 of Regulation 1219/2012/EU are met, the European Commission must authorize the usage of the BIT (Art. 12(3) of Regulation 1219/2012/EU). The European Commission must decide within 180 days of receipt of notification (Art. 12(4) of Regulation 1219/2012/EU). Finally, Art. 13 of Regulation 1219/2012/EU deals with the conduct of the member state if the BIT falls within the scope of Regulation 1219/2012/EU. This requires that the member state must inform the European Commission of all meetings and their agendas under the provisions of the BIT (Art. 13(a) of Regulation 1219/2012/EU). If there is a risk of EU law infringement, the European Commission can request that the member states positions itself on the respective issue (Art. 13(a) of Regulation 1219/2012/EU). Furthermore, the member state must inform the European Commission of any request for dispute settlement under the auspices of the BIT (Art. 13(b) of Regulation 1219/2012/EU). The European Commission and the member state must cooperate to provide an effective defense (Art. 13(b) of Regulation 1219/2012/EU). According to Art. 13(c) of Regulation 1219/2012/EU, the member state must seek the agreement of the European Commission before activating any relevant mechanisms for dispute settlement against a third party under the BIT. Mechanisms are defined broadly (encompassing any consultation) and the European Commission will actively work together with the member state (Art. 13(c) of Regulation 1219/2012/EU). Art. 7–11 of Regulation 1219/2012/EU deal with questions of new negotiations of BITs and participation of the European Commission.

It must be noted that a prior draft of Regulation 1219/2012/EU accorded a greater screening power to the European Commission, but this was downsized by EU member states to avoid a greater loss of influence.[88] This also indicates that the list of incompatibilities in Art. 9 of Regulation 1219/2012/EU is final and cannot be interpreted more broadly. Finally, it is notable that there is the possibility of 'grandfathering' existing BITs by notifying them to the European Commission.[89] This is also known as a grandfather clause.

[88] Reinisch (2014), 111 (120).

[89] Seattle to Brussels Network (2010), 12.

3.3.1.3 Unresolved Conflicts

Due to the shift in competence, there are some unresolved issues, which mainly deal with the question of dispute settlement. The EU stressed that there must be an effective investment dispute resolution.[90] It is problematic that the ICSID Convention is only 'open to signature and ratification by states members of the World Bank or party to the Statute of the International Court of Justice' and that the EU qualifies under neither.[91] The European Parliament mentioned that it 'is aware that the EU cannot use existing (...) ICSID and UNCITRAL dispute settlement mechanisms since the EU as such is a member of neither organization.'[92] Therefore, it 'calls on the EU to include a chapter on dispute settlement in each new EU investment treaty in line with the reforms suggested in this resolution' and 'requests that the European Commission and the Member States take up their responsibility as major international players to work towards the necessary reforms of the ICSID and UNCITRAL rules.'[93] It is doubtful that the ICSID Convention is going to be changed easily to make the EU a member and it is not clear why the European Parliament also excludes UNCITRAL.[94] The result would be that only UNCITRAL or other arbitration rules are available for all EU-BITs.[95]

Furthermore, the question of who the respondent will be needs to be clarified if an EU member state is sued. The European Commission issued a proposal on how to handle the distribution of roles between the EU and its member states in relation to the conduct of disputes under agreements to which the Union is a party. According to the proposal, the EU would be the respondent, 'where the treatment alleged to be inconsistent with the agreement is treatment afforded by one or several Union institutions', the member state would be the respondent, 'where the treatment in question is afforded by the Member State', and lastly the EU would be again act as the respondent 'in respect of treatment afforded by a Member State (...) (and the member state) (...) has opted not to act as respondent (...) (or) (...) where the European Commission decides that issues of Union law are involved.'[96] This proposal builds on the ECT, where the EU and its member states can decide within 30 days who will be the respondent.[97]

Finally, issues remain on how the ECJ will handle the possibility that tribunals might rule over EU law.[98] The ECJ is very reluctant if it comes down to its

[90] Reinisch (2014), 111 (133).

[91] European Commission (2010), 10; Reinisch (2014), 111 (134).

[92] European Parliament (2011).

[93] European Parliament (2011).

[94] Reinisch (2014), 111 (134).

[95] Reinisch (2014), 111 (134).

[96] European Commission (2012a), 6.

[97] Reinisch (2014), 111 (133).

[98] Reinisch (2014), 111 (154).

exclusive competence to decide on facts of EU Law, especially as it views tribunals as "only" a private form of dispute settlement.[99]

3.3.2 *Effects on the Desertec Concept*

The shift in competence will have an influence on the Desertec Concept as its success mainly depends on a functional ISDS. If there are any doubts concerning the investment security, no investor would be willing to spend any money.

3.3.2.1 Application of the G/M-BIT

Based on the grandfather clause, the G/M-BIT was notified by Germany to the European Commission.[100] As the Federal Ministry for Economic Affairs and Energy (BMWi) issued an overview of all effective BITs concluded by Germany up to 13 February 2014, including the G/M-BIT based on the year 2001,[101] it can be assumed that it is not violating any requirement according to Regulation 1219/2012/EU. Otherwise, the European Commission would have been forced to request that Germany change and renegotiate the relevant parts of the G/M-BIT. It must be kept in mind that due to the shift of competence the EU can lawfully enter into BITs and also establish new ones.[102] The European Commission has already adopted negotiating directives for four MENA countries, including Morocco.[103] It is unclear what would happen to the G/M-BIT if the EU negotiated a new BIT with Morocco.

Interestingly, Regulation 1219/2012/EU also affects all BITs which were concluded prior to the Treaty of Lisbon. As the G/M-BIT is concluded between Germany and Morocco, it is necessary to take a closer look at how German courts treat changes to EU law which affect prior national law. In 2009, the German Constitutional Court mentioned that its judgment had no effect on treaties concluded prior to the 1st of January 1958.[104] This contains the legal principle that the current legal situation within a member state is unchangeable because of later European integration.[105] This legal principle also applies to BITs concluded after the 1st of January 1958.[106] Consequently, the EU must approve all existing BITs

---

[99] Reinisch (2014), 111 (154–157).

[100] European Commission (2013a), 2 (27).

[101] BMWi (2014), 2.

[102] Reinisch (2014), 111 (119–120).

[103] European Commission (2012b).

[104] BVerfG [2009] 2 BvE 2/08, 2267 (2291) para 380.

[105] BVerfG [2009] 2 BvE 2/08, 2267 (2291) para 380.

[106] BVerfG [2009] 2 BvE 2/08, 2267 (2291) para 380.

(comparable to the EC Decision 2001/855/EC).[107] The approval reflects the common and tacit practice of the EU concerning continued applicability of an international law contract.[108]

Taking this into account, a closer look at the G/M-BIT might reveal a solution. First of all, there is customary international law principle pacta sunt servanda, which is referred to in Art. 26 of the VCLT.[109] It is one of the fundamental principles of treaty law.[110] Furthermore, Art. 39 and 54 of the VCLT stipulate that the EU does not replace its member states within BITs. Hence, all contracts which were adopted before a possible change of competence was in sight do not violate European law. A change in competence does not retroactively affect previously concluded agreements.[111] Since the G/M-BIT came into force in 2004, it was concluded before the entry into force of the Treaty of Lisbon, which reorganized EU competences. The G/M-BIT was adopted on the 18th of March 2004, only 3 months before the European Constitutional Contract (ECC). Art. III-315 of the ECC was the first article regulating FDIs. However, the draft of the G/M-BIT dates back to the 6th of August 2001, which was almost 3 years before the ECC. In 2005, the ECC failed after France and the Netherlands refused to ratify it.[112] This means that the ECC never came into force and therefore the competence never shifted due to the ECC. After the ECC failed and before the Treaty of Lisbon, it was unclear how the EC/EU would develop. Since the Treaty of Lisbon came into force in 2009, there were almost 4 years between the ECC and the Treaty of Lisbon,[113] so it would be wrong to claim that Germany certainly knew or predicted that a change in competence might take place. For this reason, it did not violate European law or need EU approval.

For the moment, both the grandfather clause as well as Art. 26 of the VCLT still recognize the G/M-BIT. This might change with a new EU-BIT.

3.3.2.2 Consequences of the Shift in Competence

At the moment, there should be no influence on the G/M-BIT as it was successfully notified by Germany. Therefore, the G/M-BIT is still applicable and the investor can sue the host state according to Art. 25 of the ICSID Convention. If the treaty is replaced, this might cause some issues as it is not clear how and where an ISDS would take place.

---

107 BVerfG [2009] 2 BvE 2/08, 2267 (2291) para 380.

108 BVerfG [2009] 2 BvE 2/08, 2267 (2291) para 380.

109 Binder (2009b), 187 (192); ECJ [1998] Case C-162/96, I-3655 (3670) para 49.

110 Herdegen (2010), § 15 para 4; Binder (2009b), 187 (192).

111 ECJ [2002] Schlussanträge des Generalanwaltes Antonio Tizzano C-466/98, I-9431 (I-9475-I-9476) para 113.

112 Hellmann (2009), 6.

113 Hellmann (2009), 12.

As mentioned above, it is doubtful that the ICSID Convention is going to be changed to allow the EU to become a member. This would be difficult as it would need all the ICSID Convention member states to vote on this unanimously. If the G/M-BIT were replaced, the ICSID Convention may no longer be applicable. As illustrated above, there are other ISDS mechanisms which may be used. A possible solution would be to resort to arbitration under the UNCITRAL Rules or to make use of the ICC Rules. However, at the moment, no one knows how the situation is going to develop as the EU and Morocco are just starting to negotiate a new investment agreement, nor how the EU member states, third states, investors and third state investors are going to react if the ICSID mechanism is not an option anymore. Most European BITs illustrate that the ICSID Convention is the primary choice of the ISID, which makes it difficult to predict how an exclusion of the ICSID would be received. It is noteworthy that the EU member states have already refused to cancel the old BITs which exist between the EU member states, despite an European Commission request.[114] Finally, there also remains the issue of whether the G/M-BIT is still applicable, as mentioned above, due to the fact that it was concluded before the Treaty of Lisbon.[115] This could become an issue if an investor wants to fall back on the ICSID Convention after a new EU-BIT with Morocco is in place. Based on the ruling of the German Constitutional court, it is not clear how other German courts would react to such a claim as they might support the validity of the G/M-BIT.

Until this agreement enters into force, the G/M-BIT will remain fully in force and this thesis will discuss the application of the ICSID Convention according to the G/M-BIT. Questions about the notion of investment under the ICSID Convention might also apply to other arbitration institutions and will also be discussed.

3.4 Precedents in International Law

Over the past few centuries, ICSID tribunals have decided numerous different arbitration cases. These often involved similar topics or facts. Consequently, ICSID tribunals frequently referred to previous ICSID decisions. In some cases, there was even a chain of reference to justify the final decision. The term precedent is mostly used in connection with stare decisis, which means 'to stand by what is decided.'[116] This raises the question if international law recognizes or allows stare decisis.

---

[114] Griebel (2009), 469 (469).

[115] This would not apply, if Germany and Morocco both accept a new form of investment treaty instead of the G/M-BIT, as it is state sovereignty to change existing contracts.

[116] Kaufmann-Kohler (2007), 357 (358); cf Commission (2007), 129 (134).

3.4.1 *Stare Decisis and the ICJ*

Art. 59 of the ICJ Statute stipulates that decisions of the ICJ do not have a binding force except between the parties of the dispute. Most authors view this article of the ICJ Statute as an example that there is no formal stare decisis doctrine within international law.[117] In addition, most arbitral tribunals also confirm that there is no doctrine of binding precedents.[118] Due to Art. 38 of the ICJ Statute, regulating utilization of ICJ decisions as 'subsidiary means for the determination of rules of law', practice reveals that the ICJ frequently relies on its precedent decisions.[119] However, Art. 38 and 59 of the ICJ Statue illustrate the difficulty to balance the desire for consistent judicial decisions and the fear of losing state sovereignty.[120] Concerning its own precedents, the ICJ concluded that:

> [T]here can be no question of holding (parties) to decisions reached by the Court in previous cases. The real question is whether, in this case, there is cause not to follow the reasoning and conclusions of earlier cases.[121]

Comparable to the ICJ, the WTO Appellate Body highlighted that:

> [A]dopted panel reports are an important part of the GATT acquis. (. . .). They create legitimate expectations among WTO Members, and, therefore, should be taken into account where they are relevant to any dispute.[122]

Since there is no comparable regulation like Art. 59 of the ICJ Statute within the WTO Dispute Settlement Understanding (DSU), some authors must admit that a stare decisis does exist in practice.[123] At the same time, practical application of stare decisis is not rejected outright, as it may help to create a coherent judicial framework instead of patchwork decisions.[124] Despite some arbitral tribunals highlighting that they judge independently, they sometimes emphasize that precedents were helpful in reaching the correct decision.[125]

---

[117] Alvarez (2003), 405 (406–407); No binding precedents in international law, in: Brower et al. (2009), 843 (851); Kaufmann-Kohler (2007), 357 (361); cf Sureda (2009), 830 (832).

[118] Sureda (2009), 830 (834).

[119] Commission (2007), 129 (134); Kaufmann-Kohler (2007), 357 (361).

[120] Sureda (2009), 830 (832).

[121] ICJ [1998] Land and Maritime Boundary between Cameroon and Nigeria—Preliminary Objections Judgment, 275 (292) para 28.

[122] WTO Appellate Body [1998] WT/DS58/AB/RW, 30 para 108.

[123] Bhala (1999), 845 (941–942).

[124] Bhala (1999), 845 (955–956).

[125] Sureda (2009), 830 (834).

3.4.2 ICSID Tribunals and Stare Decisis

As indicated in Art. 38 of the ICJ Statute, earlier ICSID decisions are not a primary source of rules.[126] According to Art. 53(1) of the ICSID Convention, the award is only binding on the parties. Recently an ICSID tribunal mentioned that there is no obligation for any ICSID tribunal to establish stare decisis.[127] There can be no stare decisis within ICSID jurisdiction because each arbitration case depends on the respective BIT, which differs depending on the parties involved in the dispute.[128] However, one tribunal said that there is a 'reasonable assumption' that ICSID tribunals take decisions of former tribunals into account.[129] A majority of an ICSID tribunal arbitrators added that tribunals have the duty 'to contribute to a harmonious development of investment law' and hence must take earlier and related cases into account.[130] Nevertheless, there is no doctrine of stare decisis in ICSID jurisprudence, which means precedents are not binding.[131] Decisions only apply inter partes. Despite the fact that ICSID tribunals do not have stare decisis, tribunals take earlier cases into account.[132] ICSID tribunals point out that they try to achieve a harmonious development of investment law by considering and sometimes using precedents.[133]

---

[126] Commission (2007), 129 (130); Andreeva (2008), 161 (170); cf Griebel and Kim (2007), 186 (188).

[127] ICSID [2006] ARB/04/13, 22 para 64; ICSID [2006] ARB/03/15, 14 para 39; ICSID [2005] ARB/02/17, 9 para 23(d); ICSID [2004] ARB/01/3—Decision on Jurisdiction (Ancillary Claim) 8 para 25; ICSID [2004] ARB/02/6, 37 para 97.

[128] ICSID [2005] ARB/02/17, 9 para 23(d).

[129] ICSID [2006] ARB/03/15, 14 para 39; Neither the wording of Article 53(1) ICSID Convention, nor the *travaux préparatoires* exclude any application of stare decisis, in: Kaufmann-Kohler (2007), 357 (368).

[130] ICSID [2010] ARB/08/5, 22–23 para 100.

[131] ICSID [2010] ARB/07/20—Award, 31–32 para 96; ICSID [2010] ARB/01/3—Annulment, 37–38 paras 93–94; ICSID [2008] ARB/05/12, 18 paras 49–50; ICSID [2007] ARB/05/07, 20 para 67; ICSID [2006] ARB/03/15, 14 para 39; ICSID [2005] ARB/02/17, 9 para 23d); ICSID [2004] ARB/01/3—Decision on Jurisdiction (Ancillary Claim) 8 para 25; Bhala (1999), 845 (863); Brower et al. (2009), 843 (845) and (851); Kaufmann-Kohler (2007), 357 (357); Abraham (2009), 206 (209).

[132] ICSID [2013] ARB/09/8—Award, 40–41 paras 154–159; ICSID [2007] ARB/05/07, 20 para 67; Considering another ICSID tribunal decision, in: ICSID [2006] ARB/03/17, 36–37 para 63; "Carefully" considering precedents, in: ICSID [2006] ARB/04/13, 22 para 64; On the contrary, tribunals highlight that precedents are not necessarily relevant to decide a case, in: ICSID [2005] ARB/02/17, 10 para 26; Comparison of other ICSID tribunal's decisions is useful, in: ICSID [2005] ARB/03/10, 23 para 36; Considering different ICSID cases, in: ICSID [2003] ARB/01/8, 788 (799) para 72; ICSID [2002] ARB/00/2—Award (2002) ICSID Review, 142 (158) para 58; Commission (2007), 129 (131); Prawoko (2005), 143 (147).

[133] ICSID [2010] ARB/07/20—Award, 31–32 para 96; cf ICSID [2008] ARB/05/12, 18 para 50; ICSID [2007] ARB/05/07, 20 para 67.

3.4.3 *Evaluation of Stare Decisis Practice in ICSID Arbitration*

ICSID decisions are a subsidiary source of international law as they are not mentioned within Art. 38 of the ICJ Statute. In addition, the ICJ has not cited any ICSID tribunal decisions in its own verdicts.[134] This does not change the fact that ICSID tribunals refer to other ICSID tribunal verdicts within their decisions in the same manner as common law courts.[135] At the same time, ICSID tribunals' decisions are comparable to civil law and its handling of precedents, as key decisions also influence later interpretation of the written law.[136] Nevertheless, there can be no doubt that a de facto stare decisis exists within ICSID case law.[137] Some authors mention that there is a difference between a stare decisis and 'taking precedents into account.'[138] Yet, this distinction does not change the fact that ICSID tribunals did not hesitate in the past to decide in an opposite way than their precedents.[139]

The stare decisis practice is partially accepted. Literature points out that the field of investment arbitration is relatively "young" and still is in the experimental phase.[140] Thus, there are still some inconsistent ICSID tribunal judgments which have lead to a reluctance to accept a stare decisis within investment arbitration.[141] Due to the lack of a formal stare decisis, there is no guarantee that opinions and decisions of other tribunals are considered in future arbitrations.[142] This might end in an annulment process because of the application or non-application of prior ICSID tribunal verdicts.[143] In cases where every decision relates to a different treaty regime (e.g. German, Italian, Moroccan BITs), it is very difficult to use precedents.[144] This also relates to the fact that ICSID tribunals are not set up in the same way, as arbitrators frequently change, making it difficult to achieve a coherent jurisdiction.[145]

---

[134] Commission (2007), 129 (153).

[135] Commission (2007), 129 (148).

[136] Commission (2007), 129 (132).

[137] Jagusch et al. (2010), 75 (98); *De facto* precedents, but without a *jurisprudence constante*, in: Sureda (2009), 830 (833); Brower et al. (2009), 843 (851); Tietje (2010), 5 (18); Reinisch (2008), 107 (123).

[138] Brower et al. (2009), 843 (852–853).

[139] Brower et al. (2009), 843 (853).

[140] Griebel (2008), 58; Johannsen (2009), 5 (27).

[141] Griebel (2008), 58; Johannsen (2009), 5 (27).

[142] Franck (2005), 1521 (1611).

[143] Commission (2007), 129 (156).

[144] Andreeva (2008), 161 (171).

[145] Commission (2007), 129 (135); Over 100 different arbitrators were appointed so far, in: Commission (2007), 129 (138).

On the other hand, there are also arguments supporting stare decisis.[146] First of all, it is a difficult path to establish stare decisis because precedents should not be ignored or undisputedly accepted.[147] Hence, it is vital to assess possible precedents fully prior to utilization.[148] It is not uncommon to use different international legal documents as guidance to solve problems related of a pending case.[149] There is also a risk of fragmentation of international investment law because of the great amount of investment cases.[150] This can undermine the predictability and reliability of the whole investment system.[151] Stare decisis could be a possible way to prevent the fragmentation of international investment law,[152] and help to establish consistency in arbitral proceedings and awards.[153]

Investors and states will examine all relevant decisions to assess whether they have any rights or liabilities,[154] as it is common practice to refer to precedents.[155] Therefore, it would not be surprising if investment tribunals adopted or followed decisions of other investment tribunals, which have dealt with the same BIT.[156] It is undisputed that there is no stare decisis doctrine within international law, and consequently in international investment arbitration. However, a de facto stare decisis exists in practice. There is also no doubt that Morocco/ONE and a possible investor are going to consider suitable precedents to reinforce their arguments. Previous ICSID tribunals' decisions must be guidelines. If one tribunal has already decided a comparable question, there is a great chance that the newly established tribunal will take this into account. There is no guarantee that the tribunal will feel "obliged" to follow any precedent. Hence, it is advisable to take all possible previous cases into account and make a check list of "dos and don'ts".

[146] Commission (2007), 129 (132).

[147] Sureda (2009), 830 (842).

[148] Johannsen (2009), 5 (27); Commission (2007), 129 (154).

[149] A NAFTA tribunal referred to decisions of the European Court of Human Rights (ECHR) to solve a problem of an immunity claim, in: UNCTAD (2009), 4; Other tribunals (including ICSID) referred to the ECHR in a case of definition concerning de facto expropriation or to the Inter-American Courts of Human Rights (IACHR) concerning the analysis of an expropriation claim, in: UNCTAD (2009), 5–6.

[150] Metje (2008), 167; Reinisch (2008), 107 (114).

[151] Reinisch (2008), 107 (119).

[152] Reinisch (2008), 107 (122).

[153] Sureda (2009), 830 (839).

[154] Franck (2005), 1521 (1612).

[155] Gantz (2004), 679 (689); Franck (2005), 1521 (1612); cf ICSID [2007] ARB/05/07, 20 para 67.

[156] Furthermore, if provisions of BITs are treated differently by investment tribunals, this can seriously undermine investment protection, in: Gantz (2004), 679 (689); cf ICSID [2008] ARB/05/12, 18 para 50.

3.5 Jurisdiction According to the ICSID Convention

As mentioned above, Art. 25(1) of the ICSID Convention deals with the issue of jurisdiction.[157] The following section takes a closer look at the question of *ratione personae* and *ratione materiae*. The question of nationality was already discussed earlier and thus will not be part of the following sections. The *ratione temporis* will not be covered because this issue is only relevant at the time the dispute arises.

3.5.1 Ratione Personae *Under the ICSID Convention*

The *ratione personae* encompasses several requirements. Within this section, the focus will be on the *locus standi* and specific questions of the umbrella clause. Parts of the assessment of *locus standi* were conducted in the first chapter under the topic of investor nationality. Thus, the newly formed SPV would be the investor in the case of the Desertec Concept. Since the investor will be the majority shareholder, the following section will refer to Dii as "the investor" for better illustration. This is also due to the fact that Dii would have been the main claimant in a case of expropriation.

3.5.1.1 Locus Standi

Taking a closer look at the present case, there are several possible partners. First of all, there would have been Dii as a private investor. As mentioned above, even a shareholder of an investor company is able to sue according to ICSID jurisdiction and the G/M-BIT. Secondly, there is the contracting state as the other party to the arbitration. As illustrated before, there are two possible contractual partners—ONE and the state of Morocco. Morocco is a contracting state to the ICSID Convention and hence can be a party to a tribunal. In the case of ONE, this is a bit more difficult and thus needs further examination.

It is not unusual that an agency or subdivision, which is legally distinct from the state but exercises public functions, is a party in state contracts.[158] These subdivisions and agencies are not only respondents concerning ICSID arbitration as they may also commence ICSID proceedings.[159] Art. 25(3) of the ICSID

[157] Article 25(1) ICSID Convention: '(. . .) [t]he jurisdiction of the Centre shall extend to any legal dispute arising directly out of an investment, between a Contracting State (or any constituent subdivision or agency of a Contracting State designated to the Centre by that State) and a national of another Contracting State, which the parties to the dispute consent in writing to submit to the Centre. When the parties have given their consent, no party may withdraw its consent unilaterally'.

[158] Schreuer (2009), 149 para 230.

[159] Schreuer (2009), 149 para 231.

Convention requires a designation of the contracting state before any subdivision or agencies can give their consent to ICSID arbitration. Vice versa, consent of a subdivision or agency to ICSID jurisdiction cannot extend to the state,[160] making it is advisable that an investor gets consents from both—the state and the agency or subdivision.[161] Although paragraphs 1 and 3 of Art. 25 of the ICSID Convention are closely related to another, paragraph 1 deals with the *ratione personae* and *locus standi*, while paragraph 3 deals with modalities of the consent to ICSID arbitration. Both are paragraphs include vital requirements for a possible jurisdiction of an ICSID tribunals both requirements must be met.[162] No. 5 of the ICSID Model Clauses offers an example of designation and consent agreements for the investment contract.[163] Hence, in a long-term contract, the investor should demand a written commitment of designation and consent of the agency or subdivision.[164] In a case where this designation is missing, the ICSID tribunal will dismiss the case.[165]

3.5.1.1.1 ONE as an Agency

Consequently, it is essential to assess what ONE is. As highlighted on their homepage, ONE is an "etablissement public",[166] established in 1963.[167] Without a doubt, ONE is not a subdivision, but could be an agency. The concept of agency is functional and not structural.[168] It does not matter if the agency is a corporation, to what extent the government owns it or if it has legal personality, as the most important criterion is if '(...) it performs public functions on behalf of the Contracting State (...).'[169] As mentioned earlier, ONE is the state monopoly for energy production and distribution, which is a service of general interest. Thus, ONE performs functions on behalf of Morocco, which makes it an agency. Finally a

---

160 Schreuer (2009), 170 para 312.

161 Reed et al. (2004), 21; Schreuer (2009), 170 para 312.

162 Schreuer (2009), 150 para 232.

163 No. 5 ICSID Model Clause: The name of constituent subdivision or agency is [a constituent subdivision]/[an agency] of the Host State, which has been designated to the Centre by the Government of that State in accordance with Article 25(1) of the Conventiopara In accordance with Article 25(3) of the Convention, the Host State [hereby gives its approval to this consent agreement]/[has given its approval to this consent agreement in citation of instrument in which approval is expressed]/[has notified the Centre that no approval of [this type of consent agreement]/[of consent agreements by the name of constituent subdivision or agency is required].

164 Reed et al. (2004), 21.

165 ICSID [1997] ARB/95/2 – Award, 329 (351–352) paras 2.32–2.33; Schreuer (2009), 155 para 249.

166 An appropriate translation would be "public corporation".

167 Office national d'électricité (2010).

168 Schreuer (2009), 153 para 243.

169 Schreuer (2009), 153 para 243.

short comparison, with an ICSID case is necessary, which dealt with a comparable case.

In the Salini case, an Italian company concluded a contract with the Moroccan company ADM.[170] The Moroccan Treasury and other public entities held 89 % of all shares, the Minister was the President of the Board of Directors and according to the Articles of Association the Board of Directors had far-reaching powers. The contract dealt with the construction of a highway. The ICSID tribunal first assessed ADM from a structural perspective and concluded that ADM was an entity controlled by the Moroccan state.[171] In a second step, the tribunal analyzed ADM from a functional point of view and concluded that ADM performed tasks which were under state control, like: '(. . .) responding to structural needs of the Kingdom of Morocco (. . .).'[172] Then the tribunal highlighted that ADM represented the state of Morocco and performed Morocco's obligation and thus acts in the name of Morocco.[173] Comparing this to the situation of contracting with ONE, there are a lot of similarities.

As mentioned above, ONE is a public monopoly which includes a lot of state influence. Therefore, the structure of ONE and ADM are comparable. This is also illustrated by the purpose of both companies. On the one hand is ONE, which deals with the provision of energy, (a service of general interest), and on the other hand is ADM, which is concerned with infrastructure (also a service of general interest). Without a doubt, ONE fulfills the requirement of Art. 25(1) of the ICSID Convention as a state agency, but it is not listed as an agency under the ICSID Convention.[174] Thus, Morocco should designate ONE to ICSID jurisdiction if ONE contracts with foreign investors.[175] As mentioned above, this is the only way for ONE to be a party to ICSID jurisdiction.

3.5.1.1.2 The Host State's Scope of Responsibility

It is questionable, if Morocco is responsible for actions by ONE. In 2001, the ILC produced the ILC Draft Articles on Responsibility of States for Internationally Wrongful Acts (ILC Responsibility 2001). There is no doubt that according to customary international law the state is responsible for all its organs, provinces and municipalities.[176] However, the literature is dived on this issue of whether the ILC Responsibility is part of international customary law, although most seem to

---

170 ICSID [2001] ARB/00/4, 609 (617–618) para 32.

171 ICSID [2001] ARB/00/4, 609 (617–618) para 32.

172 ICSID [2001] ARB/00/4, 609 (618) para 33.

173 ICSID [2001] ARB/00/4, 609 (618) para 35.

174 ICSID (2014) List of contracting states and other signatories of the Convention.

175 Ecuador designated a company to ICSID jurisdiction, which allowed the investor to sue against Ecuador and its agency, in: ICSID [2008] ARB/05/12, 24–25 para 63.

176 Dolzer and Schreuer (2008), 195.

support the categorization as part of it.[177] Despite some not supporting this allocation, they also point out that the ILC Responsibility 2001 is the central international document concerning state responsibility.[178]

3.5.1.1.3 ILC Draft Articles on Responsibility of States for Internationally Wrongful Acts

Art. 4 of the ILC regulates the conduct of the state.[179] Since neither paragraph 1 or 2 of Art. 4 of the ILC Responsibility 2001 defines the term 'state organ', there is the proposal of a functional test.[180] Art. 5 of the ILC Responsibility 2001 stipulates the conduct of persons or entities exercising elements of governmental authority.[181] In both cases, there is also an imputation if the conduct is ultra vires, according to Art. 7 of the ILC Responsibility 2001.[182] Art. 8 of the ILC Responsibility 2001 regulates conduct directed or controlled by a state, (also known as *de facto*-entities).[183] Art. 33(2) of the ILC Responsibility 2001 only stipulates that all regulations of Part Two of the ILC Responsibility 2001 do not affect situations where the state faces any person or an entity other than the state.[184] However, Art. 5 to 8 of the ILC Responsibility 2001 are in Part One of Chapter 2, so they are not affected. As mentioned above, there is no doubt that ONE is a state entity. Art. 33(2) of the ILC

[177] Hobér (2008), 545 (545); Schreuer (2009), 150 para 233; Dolzer and Schreuer (2008), 195; Opposing opinion, in: Griebel (2008), 50; Opposing opinion, in: Gallus (2008), 157 (164).

[178] Griebel (2008), 50; Gallus (2008), 157 (164).

[179] Article 4(1): The conduct of any State organ shall be considered an act of that State under international law, whether the organ exercises legislative, executive, judicial or any other functions, whatever position it holds in the organization of the State, and whatever its character as an organ of the central Government or of a territorial unit of the State.

Article 4(2): An organ includes any person or entity which has that status in accordance with the internal law of the State.

[180] Only if domestic law defines it as state organ as well, functionality is irrelevant, in: Hobér (2008), 545 (550); The functionality test is also supported by Article 5 ILC Responsibility 2001, in: Hobér (2008), 545 (550–551).

[181] Article 5: The conduct of a person or entity which is not an organ of the State under article 4 but which is empowered by the law of that State to exercise elements of the governmental authority shall be considered an act of the State under international law, provided the person or entity is acting in that capacity in the particular instance.

[182] Article 7: The conduct of an organ of a State or of a person or entity empowered to exercise elements of the governmental authority shall be considered an act of the State under international law if the organ, person or entity acts in that capacity, even if it exceeds its authority or contravenes instructions.

[183] Article 8: The conduct of a person or group of persons shall be considered an act of a State under international law if the person or group of persons is in fact acting on the instructions of, or under the direction or control of, that State in carrying out the conduct.

[184] Article 33(2): This Part is without prejudice to any right, arising from the international responsibility of a State, which may accrue directly to any person or entity other than a State.

Responsibility 2001 also clearly points out that the articles can apply in the case of the Desertec Concept. Consequently, the ILC Responsibility 2001 might apply.

3.5.1.1.4 Object and Purpose of the ILC Responsibility 2001 and Problems

As mentioned in the General commentary of the ILC Responsibility 2001, the articles only deal with international responsibility (No. 1 and 4(c) of the General Commentary). The Commentary to Chapter III of the ILC Responsibility 2001 highlights this purpose again. According to the Commentary, attribution can only take place if the action of the state or the entity is within the field of international law and does not only take place on a national level. In the case of an attribution, it is not important, if the entity was able to represent the state based on the national laws of this state.[185] The ILC Responsibility 2001 is solely secondary norms and not primary norms.[186] As in the theory of treaties and the theory of responsibility, the theory of treaties comes first (question of whether there is a treaty obligation) and the theory of responsibility is second (question of who is responsible for the breach of the treaty obligation).[187] Hence, the ILC Responsibility 2001 regulations are secondary norms because they do not deal with the issue of treaty obligations.

The ILC Responsibility 2001 only applies to acts and not obligations, whereby 'an obligation is not an act and does not directly involve any breach of international law obligation.'[188] However, two tribunals solved this problem, whereby one focused on the attribution of the undertaking of an obligation instead of the attribution of an obligation,[189] and the second one mentioned that the entity was acting according to state authority and pursuant to domestic law.[190] There is also the problem that Art. 8 of the ILC Responsibility 2001 only stipulates entity actions because of governmental instructions, directions or control.[191] This problem arises in cases of a legally independent private company as the state is only able to express its will in the 'cooperation agreement' on how the private company should operate.[192]

---

[185] Dolzer and Schreuer (2008), 198; Crawford (2008), 351 (356); Hobér (2008), 545 (550); In addition it is irrelevant if the act committed by the state is wrongful according to national law, because characterization is according to international law, in: Hobér (2008), 545 (546).

[186] Binder (2009a), 608 (624).

[187] Binder (2009a), 608 (624–625).

[188] Gallus (2008), 157 (167).

[189] ICSID [2005] ARB/01/11—Award, 75–76 para 86; Gallus (2008), 157 (167).

[190] Ad hoc Arbitration [2005] Partial Award, 47 para 129; Gallus (2008), 157 (167); cf Dolzer and Schreuer (2008), 199.

[191] United Nations (2001), 31 (47) para Article 8 Commentary (1).

[192] Griebel (2008), 53; cf Hobér (2008), 545 (551–552); Further problems of Article 8 ILC Responsibility 2001, in: Gallus (2008), 157 (168–169).

3.5.1.1.4.1 International Arbitration and the ILC Responsibility 2001

There is also a lot of ICSID case law concerning the attribution of agencies or subdivisions. First of all, there are tribunals, which generally accept the ILC Responsibility as being applicable in ICSID arbitration processes.[193] Another ICSID tribunal did not assess the Articles of the ILC Responsibility 2001, but set up its own criteria for imputation. According to this tribunal, structure (e.g. owned by the state) and function (e.g. functions of governmental control) of the agency are relevant.[194] If the subdivision or agency manages the whole legal relationships instead of the state, there can be almost no doubt that the state is responsible for the actions of its agency.[195] In addition, other ad hoc tribunals and a SCC tribunal ruled in favor of an application.[196] One LCIA tribunal also supported the possibility of attributing the entity's action to the state.[197]

3.5.1.1.4.2 Attribution and Responsibility of Treaty Claim Breaches

As mentioned above, there is a difference between contract and treaty claims. In general, ICSID tribunals are only responsible for treaty claims. Treaty claims are based on the BIT, which is an international treaty between two subjects of international law. Therefore, international obligations exist and the ILC Responsibility applies. There is no question concerning the fact that a state cannot avoid its responsibility by "using" a state owned agency.[198] Nevertheless, there is the principle that state entities are separate from the state and its actions are not attributed to the state.[199] Overall, the attribution of the obligation is only useful if the obligation is that of the state.[200]

There are ICSID tribunals supporting the attribution in cases of an entity acting instead of the state.[201] Some authors point out that every time a subdivision or

[193] ICSID [2006] ARB/04/13, 26 para 84; ICSID [2005] ARB/01/11—Award, 69–70 paras 69–70; Highly persuasive statement on the attribution of responsibility, in: ICSID [2001] ARB(AF)/98/3, 20 para 70; ILC Responsibility 2001 is applicable in investment arbitration and U.S. Court of Appeal relied on ILC Responsibility 2001, in: Hobér (2008), 545 (548); Dolzer and Schreuer (2008), 200.

[194] ICSID [2000] ARB/97/7—Decision, 28–29 para 77; cf ICSID [2001] ARB/00/4, 609 (616–618) paras 31–33.

[195] ICSID [2005] ARB/01/11—Award, 73 para 79; cf ICSID [2001] ARB/00/4, 609 (616–618) paras 31–33; cf Loncle (2005), 3 (6); State cannot use domestic law to argue that it is not responsible under international law, in: Harten (2007), 65.

[196] State Treasury in Poland acted instead of Poland, in: Ad hoc Arbitration [2005] Partial Award, 48 para 134; Latvia holds 100 % of a company, in: SCC [2003], 29–30.

[197] LCIA [2006], 44 para 154.

[198] Loncle (2005), 3 (6); cf Walter (2006), 815 (823); Vinuesa (2005), 331 (337) and (354).

[199] Sinclair (2009), 92 (101); Dolzer and Schreuer (2008), 198.

[200] The state must only observe its rights and obligations, in: Gallus (2008), 157 (166).

[201] State owned company can be attributed to the state, in: ICSID [2005] ARB/01/11—Award, 75–76 para 86; ICSID [2003] ARB/01/13, 307 (362–363) para 166.

agency acts with public authority, there is an attribution to the state.[202] Others point out that it is irrelevant to categorize it (e.g. utilization of sovereign power) because the mere fact that the state breaches an existing obligation is sufficient.[203] Sometimes it is claimed that the state must designate its agency if this agency signs the investment contract.[204]

There is a consensus that treaty claims can be attributed to the agency because the ILC Responsibility 2001 applies to international obligations. Consequently, any investor—like Dii in the past—could be confident that any behavior on the side of ONE or Morocco which violates the BIT obligations will be included. These claims can also be filed against the state of Morocco as a respondent in ICSID arbitration. The fact that Morocco has not designated ONE is irrelevant because the BIT protection cannot be eroded by "using" a state agency.

3.5.1.1.4.3 Attribution of Contract Claim Breaches

Most tribunals deny attribution in cases of contract claims.[205] In one of the cases, the tribunal highlighted that there is no attribution, because the entity at issue is '(...) an autonomous corporate body, legally and financially distinct from (the host state) (...).'[206] The BIT does not extend to breaches of the contract to which the host state is not party to.[207] There must be a clear distinction between '(...) the responsibility of a State for the conduct of an entity that violates international law (e.g. a breach of Treaty), and the responsibility of a State for the conduct of an entity that breaches a municipal law contract (...).'[208] This is mainly due to the fact that the entity did not enter into the BIT because BITs are concluded by '(...) separate and distinct (entities).'[209] Similarly, another tribunal denied attribution, because '(...) (the host state) is not liable for the performance of contracts entered into by (an agency or subdivision) which possesses a separate legal personality under its own law and is responsible for the performance of its own contracts.'[210] This is also supported by another tribunal, which highlighted that:

---

[202] Walter (2006), 815 (822); Dolzer and Schreuer (2008), 198–199; There is always an attribution to the state, in: Hobér (2008), 545 (568).

[203] Crawford (2008), 351 (356). This is only different, if the primary obligation demands something differently, in: Crawford (2008), 351 (356).

[204] Loncle (2005), 3 (7); An investor should previously address the question, if the agency fulfills all ICSID requirements, in: Moses (2008), 223.

[205] ICSID [2006] ARB/01/12—Award, 138–139 para 384; ICSID [2005] ARB/03/3, 75 para 223.

[206] No attribution, if '(...) the entity involved has no link whatsoever with the (s)tate (...)', in: ICSID [2006] ARB/04/13, 26 para 84; ICSID [2005] ARB/03/3, 69–71 paras 199–209.

[207] ICSID [2005] ARB/03/3, 73 para 214; ICSID [2001] ARB/00/4, 609 (623–624) paras 60–61; cf ICSID [2001] ARB/00/6, 32 para 68.

[208] ICSID [2005] ARB/03/3, 71 para 210.

[209] cf ICSID [2006] ARB/01/12—Award, 138–139 para 384; ICSID [2005] ARB/03/3, 75 para 223.

[210] ICSID [2002] ARB/97/3—Annulment, 89 (128) para 96.

> (. . .) [T]o the extent that the obligations assumed by the State party are of a contractual nature, such obligations must originate in a contract between the State party to the Treaty and the foreign investor as (. . .).[211]

In addition, in a case at the SCC concerning the Czech-British BIT, the tribunal rejected attribution of any entity behavior to the state.[212] However, if the breach of the contract amounts to a violation of the BIT (and therewith can be connected to a treaty claim), ICSID tribunals have jurisdiction again.[213]

If contract claims are at issue, it depends on who is party to the respective contract. If the state is a direct party, an attribution might be possible. If an agency or subdivision is the contractual party, it depends on the legal personality. In cases, where the agency or subdivision is legally and financially independent, an attribution cannot take place according to the ILC Responsibility 2001. A mere contract between a legally independent agency and an investor is not part of international law and does not constitute any international obligations. An application of the ILC Responsibility 2001 is not possible. This case is treated in accordance to the applicable law of the contract (e.g. municipal law of the host state). The only exception is that a contract claim amounts to a treaty claim.

Consequently, there are certain risks for an investor if they are contracting solely with ONE. Such a cooperation excludes the jurisdiction of ICSID tribunals. Contractual claims could "only" be subject to domestic litigation/arbitration or international commercial arbitration. The only possibility to revive ICSID jurisdiction would be to conclude all contracts with ONE and Morocco.

3.5.1.1.4.4 Assessment of Attribution

As mentioned above, ONE is closely related to the state. Almost all its functions deal with the service of general interest and as it mostly acts for the state, discussion concerning the quality of it acts are irrelevant. Therefore it is clear that

> (. . .) if a State-owned corporation acts inconsistently with the international obligations of the State that owns and controls it, the State may be responsible provided that it has actually authorized the conduct in question as a result of its capacity to direct and control the actions of the corporation.[214]

There is no doubt that contracts between Dii or another investor and ONE could not have been claims in front of an ICSID tribunal, as long as they are contractual. Consequently, it would have been beneficial for Dii—or any other investor—to internationalize each contract and lift it to the level of international law. If international law applies to the contract, each breach of the contract would be a violation of international law. This would lead to the application of the ILC Responsibility 2001 and a possible attribution to Morocco. As mentioned above, the sole

[211] ICSID [2007] ARB/02/8—Award, 59 para 205.

[212] Case discussed, in: Gallus (2008), 157 (162).

[213] ICSID [2001] ARB/00/4, 609 (624) paras 62–64.

[214] Muchlinski (2014), para 46.

application of international law in contracts is not beneficial because of the limited number of laws on regulations in international law. Furthermore, it is doubtful that ONE would agree to a special provision mentioning that all contractual breaches are breaches of international law. The other option would be to conclude all contracts directly with the state of Morocco. In this case, the umbrella clause would lead to different results. It is unlikely that Morocco is willing to change the existing.

3.5.1.2 The Umbrella Clause

Should an investor conclude a state contract with Morocco, it is unclear whether the contract claims would be subject to ICSID arbitration, and covered by BIT protection. [215] The umbrella clause is generally '(. . .) a provision in (. . .) (a BIT) (. . .) that guarantees the observation of obligations assumed by the host state vis-à-vis the investor.'[216] As the term umbrella clause indicates, this provision acts as an umbrella and extends the application of international law by transferring any contractual issues into treaty claims.[217] One ICSID tribunal dealt with the issue of umbrella clauses in depth and stated that there are four different approaches towards the interpretation of such a clause.[218] The tribunal itself adopted the approach that '(. . .) umbrella clauses may form the basis for treaty claims, without transforming contractual claims into treaty claims (. . .).'[219]

First of all, there is the disintegrationist approach, mentioning that the BIT is self-contained and provides a closed list of rights and obligations.[220] Consequently, some tribunals point out that there is no extension of breaches of BIT due to the umbrella clause.[221] It is pointed out by one ICSID tribunal that the BIT contracting party did not express any intention to extend the scope of the BIT.[222] An extension of treaty claims would be against state sovereignty because of the broad consequences.[223] Furthermore, such an extension of the BIT would lead to a negation of other investment protection rules.[224] Other tribunals viewed the question

---

[215] Hunter (2007), 165 (170); Gallus (2008), 157 (158); Dolzer and Schreuer (2008), 153; Walter (2006), 815 (816); Hauschka (2005), 1550 (1555); Griebel (2008), 83; Moses (2008), 234–235.

[216] Dolzer and Schreuer (2008), 153.

[217] Griebel (2008), 85; Schöbener and Markert (2006), 65 (90); cf Walter (2006), 815 (815).

[218] ICSID [2009] ARB/07/12, 56–57 paras 197–200; Also in: Crawford (2008), 351 (367–368).

[219] ICSID [2009] ARB/07/12, 57 paras 200–201; This view is also called integrationist view and is not an internationalization of the respective contractual claim, because the tribunal applies the contractual law to it and not international law, in: Crawford (2008), 351 (370).

[220] Shany (2005), 835 (844).

[221] ICSID [2004] ARB/03/11—Award, 20 paras 81–82; ICSID [2003] ARB/01/13, 307 (360–361) para 161 and (365–366) para 171.

[222] ICSID [2003] ARB/01/13, 307 (367) para 173.

[223] ICSID [2007] ARB/02/16—Award, 92 para 310; ICSID [2003] ARB/01/13, 307 (363–364) paras 167–168.

[224] ICSID [2003] ARB/01/13, 307 (364) para 168.

restrictively as well, but highlighted that the application of the umbrella clause depends on the terms of the respective BIT and its interpretation.[225]

The idea that ICSID tribunals can deal with contract claims as well for a better harmonization of '(. . .) overlapping norms and procedures (. . .)' is also called the integrationist approach.[226] Therefore, some tribunals accepted the extension of jurisdiction according to application of the umbrella clause.[227] Hence, the investment contract between the investor and the state is within the scope of the BIT.[228] If the BIT does not cover non-BIT related 'legal positions' of an umbrella clause, some view this as an 'internationalization' of contract claims.[229] However, an umbrella clause does not transfer the question of contracts into the question of treaty law, which means the contract is examined according to the applicable law concluded in the contract by the ICSID tribunal.[230] It is essential to judge each umbrella clause individually and hence there is an obligation to interpret each umbrella clause.[231] In recent years, more ICSID tribunals have allowed the application of an umbrella clause to put contract claims on the same level as treaty claims.[232]

3.5.1.2.1 Application of the Umbrella Clause and Its Implications

The application of the umbrella clause can be quite difficult as ICSID tribunals also differ in definition and setup of an umbrella clause application. In most cases, the umbrella clause comes into play if the dispute is in front of the national courts of the host state.[233] In these cases, the investor often tries to justify ICSID tribunal's jurisdiction because of application of the umbrella clause.[234]

[225] Reviewing the BIT and interpretation of respective regulations, in: ICSID [2009] ARB/07/9, 55–56 paras 141–142; ICSID [2004] ARB/02/13, 42 paras 127–128.

[226] Shany (2005), 835 (844).

[227] ICSID [2007] ARB/02/8—Award, 59–60 paras 204–206; ICSID [2007] ARB/01/3—Award, 87 para 274; ICSID [2005] ARB/01/11—Award, 66 paras 60–62; ICSID [2004] ARB/02/6, 43–49 paras 113–129; To a limited extend, in: ICSID [2002] ARB/97/3—Annulment, 89 (128) para 96; Peterson (2010) In parallel decisions, tribunals diverge on admissibility of umbrella clause claims, 1 (3); *Ad Hoc* Arbitration under the Agreement between the Kingdom of the Netherlands and the Republic of Poland on Encouragement and Reciprocal Protection of Investment [2005] Partial Award, 85 para 260; In general supporting umbrella clauses, in: Hunter (2007), 165 (168).

[228] ICSID [2004] ARB/02/6, 45 para 119.

[229] Griebel (2008), 85; cf Walter (2006), 815 (816).

[230] ICSID [2004] ARB/02/6, 48 para 126.

[231] ICSID [2005] ARB/01/11—Award, 61 para 50; *Ad Hoc* Arbitration under the Agreement between the Kingdom of the Netherlands and the Republic of Poland on Encouragement and Reciprocal Protection of Investment [2005] Partial Award, 78–79 para 247.

[232] Fiebiger (2010), 269 (270).

[233] Walter (2006), 815 (821).

[234] Walter (2006), 815 (821).

However, ICSID tribunals applying the umbrella clause do not even agree on the prerequisites for utilizing the umbrella clause. Some tribunals point out that not every breach of contract is a violation of the umbrella clause, as the umbrella clause is only affected if the breach relates to a sovereign state function or power.[235] In another case, the ICSID tribunal concluded that the wording of the BIT was clear and thus there was coverage of 'any obligation.'[236]

It is obvious that the ICSID tribunal's decisions do not provide legal certainty. Reviewing the literature, most authors are in favor of an investor-friendly interpretation of umbrella clauses. In one case, the ICSID tribunal pointed out that laws and obligations can become obligations within the BIT if these regulations target foreign investors.[237] Hence, an abstract law, which applies in general, can be subject to the umbrella clause if it stimulates investments and creates a 'legitimate expectation.'[238] There are concerns that the umbrella clause might "get out of hand."[239] Others point out that the umbrella clause would be meaningless if it is restricted contrary to its exact wording and meaning.[240] One example is a German formulation of an umbrella clause. If there is a formulation like, the jurisdiction of an ICSID tribunal encompasses 'alle Streitigkeiten zwischen Staat und Investor',[241] every claim, even contract claims, is within the scope of the ICSID Convention.[242] Consequently, it is important to analyze and interpret each BIT umbrella clause individually.

3.5.1.2.2 Interpretation of Art. 11 of the G/M-BIT

The purpose of interpretation is to find the exact meaning of the provision without changing it.[243] As mentioned above, most ICSID tribunals highlighted the importance of analyzing each umbrella clause individually. In the case of the G/M-BIT, Art. 11 deals with the issue. Due to the different interests of developed and developing countries, treaty interpretation is nowadays an important tool.[244] International law recognizes three different types of treaty interpretation. The first emphasizes the wording of the treaty, the second examines the intention of the parties, and a third approach asserts the object and purpose of the treaty.[245]

---

[235] ICSID [2007] ARB/02/16—Award, 92 paras 310–311; ICSID [2005] ARB/03/3, 85 para 260.

[236] ICSID [2007] ARB/02/8—Award, 59–60 paras 205–206.

[237] ICSID [2006] ARB/02/1—Decision on Liability, 53 para 175.

[238] Walter (2006), 815 (822); Griebel (2008), 87.

[239] Hauschka (2005), 1550 (1555–1556).

[240] Griebel (2008), 86.

[241] All disputes between the investor and the Contracting State.

[242] Gill et al. (2004), 397 (409); Walter (2006), 815 (821); cf Griebel (2008), 83 and 85; ICSID [2004] ARB/02/6, 49–52 paras 130–135; ICSID [2002] ARB/97/3—Annulment, 89 (115) para 55.

[243] Shaw (2008), 934.

[244] Matz (2005), 276–277.

[245] Shaw (2008), 932–933.

However, the last approach is subject to criticism, since it leaves room for judicial law-making, because judges or arbitrators will be called upon to define the object and purpose.[246] For a comprehensive interpretation it is necessary to take all abovementioned aspects into account. The ICJ already pointed out in 1950 that:

> (...) [T]he first duty of a tribunal which is called upon to interpret and apply the provisions of a treaty, is to endeavor to give effect to them in their natural and ordinary meaning in the context in which they occur.[247]

An interpretation shall take all circumstances into account which prevailed at time of treaty conclusion.[248] That does not exempt it from considering present-day state of scientific knowledge, as far as this is available to the respective court.[249] Above all, the interpretation must be based on the text of the treaty.[250] The principle of effectiveness gives effect to provisions relating to the party's intentions and rules of international law.[251] A combination of a broader purpose approach and the principle of effectiveness have been used in the case of constituting international organization treaties.[252]

3.5.1.2.3 Interpretation According to the Vienna Convention on the Law of Treaties

In most cases, interpretation of international treaties is done according to the 1980 VCLT. There are also several ICSID tribunals using the VCLT for interpretation.[253] This does not create a problem as '(...) every person or organ concerned with a treaty' is competent to interpret the respective treaty based on the VCLT.[254] The VCLT was a product of the ILC.[255] Art. 1 to 4 of the VCLT define the scope of the

---

[246] Shaw (2008), 933.

[247] ICJ [1950] Competence of the General Assembly for the Admission of a State to the United Nations, 4 (8).

[248] '(...) the Court must seek to ascertain the intention of the parties at the time', in: ICJ [2002] Land and Maritime Boundary between Cameroon and Nigeria—Judgement, 303 (346).

[249] ICJ [1999] Case concerning Kasikili/Sedudu Island—Judgment, 1045 (1060).

[250] ICJ [1994] The Territorial Dispute—Judgment, 6 (22); Binder (2009b), 187 (192).

[251] Principle of effectiveness is an important role of international law, in: ICJ [1998] Fisheries Jurisdiction—Judgement, 432 (455) para 52; In accordance with intention of the parties, in: ICJ [1949] The Corfu Channel—Judgment, 4 (24); In accordance with international law, in: ICJ [1998] Fisheries Jurisdiction—Judgement, 432 (455) para 54.

[252] The main idea is to gain flexibility, which is essential to accomplish the stated aims of that organization, in: Shaw (2008), 936–937; Concerning programmatic approach, in: Shaw (2008), 937.

[253] ICSID [2005] ARB/01/11—Award, 61 para 50; ICSID [2004] ARB/02/8—Decision, 30 para 80.

[254] Dörr in Dörr and Schmalenbach (2012), Art. 31 para 18.

[255] Aust (2007), 6–7.

VCLT. Art. 1 of the VCLT stipulates that the convention applies on written treaties between states. Then, Art. 3 of the VCLT regulates that the Convention only applies to treaties of states, excluding other subjects of international law. Art. 3(b) of the VCLT expressively mentions that if the VCLT does not apply on a contract, this does not exclude other obligations under international law (e.g. customary international law). Finally, Art. 4 of the VCLT mentions that the Convention only applies to treaties which entered into force after the VCLT.

Within the VCLT, section 3 deals with interpretation of treaties; this encompasses Art. 31–33 of the Convention.[256] Art. 31 of the VCLT is the general rule of interpretation. Its main idea is that the contractual text correctly reflects the contract party's intentions and thus an objective interpretation is sufficient.[257] According to Art. 31(1) of the VCLT, interpretation 'shall result in good faith and according to the ordinary meaning' and hence pursuant to the Convention's 'object and purpose.' Ordinary meaning means that the interpretation must be according a linguistic and grammatical analysis of the treaty based on what is '(...) regular, normal or customary'.[258] The word 'context' expresses that preamble, annexes and all agreements or instruments used by the states in connection with the conclusion of the treaty are relevant, pursuant to Art. 31(2)(a) and (b). This also means that the '(...) systematic structure of a treaty is thus of equal importance to the ordinary linguistic meaning (...)'.[259] The criterion "objective and purpose" introduces a 'teleological or functional element' and therewith includes 'the principle of effectiveness' into the general rule of Art. 31(1) of the VCLT.[260] As some treaties do not state their objective or purpose clearly, it can be difficult to identify these features, but the limit of interpretation is set by the text of the treaty.[261] During the whole process the interpretation must be "in good faith", as this is a '(...) fundamental requirement of reasonableness (...)'.[262]

Paragraph 3 goes on, stipulating that subsequent agreements (a), subsequent state practice (b) and all rules of international law (c) shall be taken into account together with the context. These "tools of interpretation" have the same value as the tools mentioned in Art. 31(2) VCLT.[263] Subsequent agreements is comparable to agreements mentioned in Art. 31(2) VCLT.[264] Very important is the interpretation

[256] Some differ between the interpretation of contractual agreements and contracts establishing law, whereby the subjective will of the party is more relevant in the case of contractual agreements. The VCLT however does not expressively mention the will of party as decisive, in: Matz (2005), 282–283. In the case of the G/M-BIT, this does not matter, because the BIT is a contract establishing a legal framework and thus the will of the party can be found in the contractual text.

[257] Herdegen (2010), § 15 para 28; Matz (2005), 288–289.

[258] Dörr in Dörr and Schmalenbach (2012), Art. 31 para 41.

[259] Dörr in Dörr and Schmalenbach (2012), Art. 31 para 44.

[260] Dörr in Dörr and Schmalenbach (2012), Art. 31 para 53.

[261] Dörr in Dörr and Schmalenbach (2012), Art. 31 paras 56 and 58.

[262] Dörr in Dörr and Schmalenbach (2012), Art. 31 paras 60–61.

[263] Dörr in Dörr and Schmalenbach (2012), Art. 31 para 71.

[264] Dörr in Dörr and Schmalenbach (2012), Art. 31 para 72.

according to subsequent state practice as only this guarantees an authentic interpretation.[265] Lastly relevant rules of international law introduce 'the international legal system as a whole' as a tool to interpret a treaty, although this is handled restrictively by many courts and tribunals.[266]

Nevertheless, Art. 31(3)(a) of the VCLT does not require that the used document is an international contract or has any specific legal quality.[267] This illustrates that the text of the treaty '(...) as it stands since the time of its conclusion is not all that matters for an interpretation *lege artis*'.[268] It is noteworthy that

> (...) consensus, which exists at the time of interpretation, may in some cases even override the original understanding of the text of the treaty, which prior to the subsequent developments may have appeared perfectly clear.[269]

Art. 31(4) of the VCLT regulates that 'a special meaning shall be given to a term if it is established that the parties so intended.' Art. 31(4) of the VCLT is an exception to Art.31(1) VCLT '(...) for cases where the parties have agreed, even implicitly, to replace the ordinary meaning of a term contained in a treaty provision by a special meaning'.[270] It must be clear that the burden of proof is very strict and '(...) lies on the party invoking the special meaning of the term (...)'.[271] However Art. 31(4) of the VCLT does not mention what kind of evidence may be used.[272] The three paragraphs of Art. 31 of the VCLT are not a legal hierarchy, but represent the logical order of application.[273] Overall, Art. 31 of the VCLT encompasses four elements, (1) recognition of good faith, (2) ordinary meaning rule, (3) consideration of objective and intention, and (4) context.

Art. 32 of the VCLT lists supplementary means of interpretation, if the tools in Art. 31 of the VCLT are unable to deliver a proper interpretation. Art. 32 of the VCLT regulates that the preparatory work of the treaty and the circumstances of its conclusion are supplementary tools. The preparatory work can be incomplete or misleading, which supports the supplementary character as an interpretation tool.[274] Finally, Art. 33 of the VCLT deals with the issue that a treaty has more than one official language. According to Art. 33(1) of the VCLT, all official languages are equally important for interpretation. In cases where a treaty has several official languages, Art. 33 of the VCLT addresses this problem. In addition, paragraph 3 stipulates the assumption that terms of the treaty always have the same

---

[265] Dörr in Dörr and Schmalenbach (2012), Art. 31 para 76.

[266] Dörr in Dörr and Schmalenbach (2012), Art. 31 paras 89–104.

[267] The parties should be bound by the contract, in: Herdegen (2010), §15 para 30; Matz (2005), 294.

[268] Dörr in Dörr and Schmalenbach (2012), Art. 31 para 4.

[269] Dörr in Dörr and Schmalenbach (2012), Art. 31 para 4.

[270] Dörr in Dörr and Schmalenbach (2012), Art. 31 paras 5 and 105.

[271] Dörr in Dörr and Schmalenbach (2012), Art. 31 para 106.

[272] Dörr in Dörr and Schmalenbach (2012), Art. 31 para 107.

[273] Aust (2007), 234.

[274] Aust (2007), 244–245; cf Matz (2005), 304.

meaning, despite the official language. If the interpretations of the different languages result in different interpretations, the meaning which best represents the object and purpose of the treaty should be adopted.[275]

Finally, the VCLT provisions on treaty interpretation are seen as part of customary international law.[276] The application of the VCLT in ICSID case law is slowly increasing since the year 2000.[277] This means they apply to all international treaties, regardless of Art. 1 to 5 of the VCLT and it adds an erga omnes effect to certain norms of the VCLT. Until today it is not clear to what moment in time the process of interpretation refers, as the meaning of the treaty can be interpreted according to the circumstances at the time of the conclusion of the treaty (static approach) or according to today's perception (principle of contemporaneity).[278]

3.5.1.2.4 Art. 11 of the G/M-BIT

The G/M-Bit entered into force in 2004, which was after the VCLT. Both contractual partners are states which signed and ratified the VCLT. As mentioned above, Art. 11(1) of the G/M-BIT includes the umbrella clause. The article addresses "disagreements with regard to investments" in paragraphs one and two.

3.5.1.2.4.1 Wording

The wording of the German and the French version of Art. 11(1) of the G/M-BIT is identical. As mentioned above, this is a broad notion of a dispute settlement agreement, which also means that it is an umbrella clause. This is also supported by the broad definition of the term investment in Art. 1 of the G/M-BIT. The protection of the investment should be comprehensive. The wording of the article indicates that all disagreements will be covered by the BIT. There is no limitation of the BIT's scope to treaty claims, as disagreements can also imply contractual dissents. Hence, the wording illustrates that contract claims can be treaty claims if a relationship to the investment exists.

---

[275] Art 33(4) VCLT as customary international law, in: ICJ [2001] LaGrand case—Judgement, 466 (502) para 101; Matz (2005), 306.

[276] ICJ [2007] Case concerning the application of the Convention on the Prevention and Punishment of the Crime of Genocide—Judgement, 60 para 160; ICJ [2002] Case concerning Sovereignty over Pulau Ligitan und Pulau Sipadan—Judgement, 625 (645–646) paras 37–38; ICJ [1999] Case concerning Kasikili/Sedudu Island—Judgment, 1045 (1059) para 18; Dörr in Dörr and Schmalenbach (2012), Art. 31 paras 6–7; Article 31 para 4.Article 31 VCLT is customary international law, in: Shaw (2008), 933; Matz (2005), 277; Mortenson (2010), 257 (309); Herdegen (2010), § 15 para 4; Wälde (2009), 724 (745).

[277] Wälde (2009), 724 (746).

[278] Dörr in Dörr and Schmalenbach (2012), Art. 31 para 23. Further references, in: Dörr in Dörr and Schmalenbach (2012), Article 31 paras 24–28.

3.5.1.2.4.2 Context, Agreements, State Practice, International Law and Historical Background

Besides the wording of the BIT, other facts support the assumption of a "broad" umbrella clause as well. As mentioned above, ICSID tribunals have dealt with the question of umbrella clauses for quite some time, and states repeatedly include them in their BITs. Therefore, there is state practice concerning umbrella clauses. Since there are different notions of umbrella clauses, which also been subject to ICSID tribunals, state practice also involves knowledge of the effect of the clauses' setup. International law does not forbid the use of an umbrella clause. The ICSID Convention itself is an example of "internationalizing" certain behaviors to secure them by international law. Germany and Morocco have been subject to ICSID tribunals before, so they have experience with the application of the umbrella clause. Some cases mentioned above involved German companies like Siemens and expressively dealt with the question and scope of an umbrella clause. The fact that the most controversial decisions in *SGS v Pakistan*[279] and *SGS v Philippines*[280] took place in 2003 and 2004 indicates that Morocco and Germany were aware of the effect of the umbrella clause and correct choice of words to conclude such a clause prior to the entry into force of the G/M-BIT. In both cases the application and possible consequences were subject to extensive discussion. Hence, it could be argued that the intentions of both parties must be that Art. 11(1) of the G/M-BIT includes a broad approach to the umbrella clause, which also encompasses contract claims.

Overall, the G/M-BIT umbrella clause is broadly defined and covers the question of contracts too, as long as they are with regard to the investment. Hence, contract claims can be treated as treaty claims. In the case of the implementation of the Desertec Concept, both contract parties should be aware of this fact. A possible investment contract can be subject to the ICSID tribunals. If they do not want this broad interpretation, they must include a special section within the state contract which states that ICSID tribunals only have jurisdiction concerning BIT related claims. However, this only serves as an indicator, as only Germany and Morocco can conclude how to interpret this section. A unilateral expression is not sufficient as both contracting parties previously agreed on the broad clause. In respect of the Desertec Concept, this might have been beneficial for Dii as long as Morocco is part of the contract at issue.

3.5.1.2.5 Evaluation of the Umbrella Clause

It is noteworthy that the umbrella clause is complicated. The main question is "how far it reaches". In respect of the Desertec Concept, it is irrelevant to discuss

[279] ICSID [2003] ARB/01/13, 307.

[280] ICSID [2004] ARB/02/6.

legitimacy of the umbrella clause. The scope is only relevant. The above analysis shows that there is a restrictive interpretation and a broad interpretation. Due to the fact that there is no stare decisis within ICSID case law, only the restrictive approach can be considered—despite the fact that more ICSID tribunals are now supporting the broad notion of the umbrella clause. If an ICSID tribunal decides to apply the broad notion, this would only be beneficial for the investor. However, taking the wording and the intention of Art. 11(1) of the G/M-BIT into account, even the restrictive approach can lead to the acceptance of the umbrella clause. First of all, both BIT parties desired an extension. The G/M-BIT is the second BIT between Germany and Morocco. The BIT entered into force on the 12th of April 2008. The two major ICSID decisions concerning the treatment of the umbrella clause took place between 2003 and 2004. Therefore, both parties knew about the issue of interpretation of the umbrella clause approximately 4 years prior to its ratification and decided not to change the text of the BIT and did not include any commentary or limitations on Art. 11(1) of the G/M-BIT. The interpretation of the BIT clause revealed a broad formulation, which indicates that both parties are willing to limit their sovereignty. As a result, Art. 11(1) of the G/M-BIT is an umbrella clause which transforms all contract claims into investment claims.

As mentioned above, there is only an attribution if the state is part of the contract. The same must apply in the case of the umbrella clause because mere contractual violation by a third party cannot lead to liability on the side of the Moroccan state. This is different in the case of ONE as it is a state monopoly. This does not mean, however, that the investor should rely on the application of the umbrella clause. Therefore, it is very important that the PPP includes specific rules on the handling of contract claims.

3.5.2 Ratione Materiae *Under the ICSID Convention*

After clarifying who can be party to the dispute (*ratione personae*), the following section deals with the question of *ratione materiae*. The ratione materiae deals with the question of what can be subject to Art. 25(1) of the ICSID Convention. In the case of the Desertec Concept, the question of "investment" is particularly relevant. Since the project is still in the planning stage, only selected topics will be discussed.

3.5.2.1 Requirements of Art. 25 of the ICSID Convention

The basic requirements of Art. 25(1) of the ICSID Convention are consent and a '(. . .) legal dispute arising directly out of an investment (. . .).' The basic requirement for any form of international arbitration is consent. Then, the question of what is a legal dispute is the subject of review. Finally, there is an analysis of investment and its prerequisites.

3.5.2.1.1 Consent

Since consent is the requirement for any ICSID arbitration, it is important to clarify its scope. To have rights or obligations according to the ICSID convention, both parties must consent to the proceeding.[281] Consent relates to the respective dispute.[282] Ratification of the ICSID Convention does not automatically lead to acceptance ICSID arbitration.[283] However, it is a well-known practice that a state gives consent by accepting ICSID jurisdiction in a BIT.[284] In this case, it depends on whether the consent of the state is conditional or not.[285] Parties have great freedom to express their consent to ICSID application (e.g. direct agreement within an investment agreement).[286] There are several ways to reach consent between the parties.[287] Consent through BIT is known as 'arbitration without privity', which does not replace a consent in writing by both parties.[288] In addition, No. 24 of the 1965 Report of the Executive Directors of the International Bank for Reconstruction and Development on the Convention on the Settlement of Investment Disputes between States and Nationals of Other States (Report of Executive Directors) also points out that the consent must rise out of an agreement or in a *compromis*, but the expression must not be within a single instrument. In practice, the request for initiation of proceedings by the investor is frequently the acceptance of 'the offer' and hence the investor's consent.[289] Thus, the state cannot bring arbitration against the investor itself. [290] There have been cases of the host state suing the investor according to the ICSID Convention.[291] In a rare case, a company which was wholly owned by the state sued the investor, who was a part of a joint venture.[292]

The acceptance of the investor is limited to the scope of the offer by the contracting state.[293] Both parties must consent to the ICSID jurisdiction, which means that two states cannot enforce ICSID jurisdiction on their nationals by

[281] Reed et al. (2004), 21; Egonu (2007), 479 (481); Dolzer and Schreuer (2008), 238; Hobe and Müller (2009), 65 (67); Schreuer (2010), 353 (357); Loncle (2005), 3 (5).

[282] Harten (2007), 67–68.

[283] Moses (2008), 222; Hobe and Müller (2009), 65 (67); Lörcher (2005), 11 (13–14).

[284] ICSID [2009] ARB/07/9, 25–26 para 65 and 27 para 73.

[285] ICSID [2009] ARB/07/9, 26 para 68.

[286] Schreuer (2010), 353 (357); cf Loncle (2005), 3 (4–5); Reed et al. (2004), 22; Herdegen (2003), 13 (32–33).

[287] Tietje (2003), 5 (10).

[288] Reed et al. (2004), 22; However the agreement is achieved indirectly, without direct contact between the parties, in: Schreuer (2010), 353 (357) ; Loncle (2005), 3 (5); Harten (2007), 63.

[289] Hobe and Müller (2009), 65 (68).

[290] Loncle (2005), 3 (5).

[291] Schreuer (2010), 353 (360).

[292] ICSID [2001] ARB/98/8—Award, 1 paras 1–2.

[293] Schreuer (2010), 353 (358).

simply including it in their BIT.[294] According to Rule 2 of the Institution Rules, the date of consent is when the parties '(. . .) consented in writing to submit it to the Centre (. . .)' or the day when the last of the two parties acted. The date of consent is important because determination of the nationality of the investor is according to this day.[295] Finally, the last sentence of Art. 25(1) of the ICSID Convention stipulates that once a party gives consent, it is not irrevocable unilaterally. No. 23 of the Report of Executive Directors also highlights this. The benefit of the previous state consent lies in the fact that it is predictable and manageable for the investor, should a dispute arise.[296]

Consequently, the fact that Morocco ratified the G/M-BIT is an expression of consent. Art. 11(2) of the G/M-BIT mentions that the ICSID Convention applies, despite the existence of a different agreement between the parties. The only condition is that there should be a 6-month period after the claim of the breach before it is submitted to ICSID. Within this period, both parties should try to settle the dispute. It is advisable for an investor and Morocco/ONE to conclude a separate section in the state contract which deals with the issue of consent and express their will to use the ICSID Convention to solve investment disputes.

3.5.2.1.2 Legal Disputes

ICSID tribunals have a similar description and definition of the term dispute like the ICJ and the PCIJ.[297] The PCIJ mentioned that a dispute is '(. . .) a disagreement on a point of law or fact, a conflict of legal views or of interest between two parties (. . .).'[298] Similarly, the ICJ concluded that a dispute is '(. . .) a situation in which the two sides hold clearly opposite views concerning the question of the performance or non-performance of certain treaty obligations.'[299] These definitions are subject to critics because of their broad or narrow nature.[300] Concerning ICSID jurisdiction, Art. 25(1) of the ICSID Convention stipulates that the jurisdiction of ICSID extends '(. . .) to any legal dispute arising directly out of an investment (. . .).' In Section V (Jurisdiction of the Centre) of the Report of Executive Directors, No 26 defines the term more narrowly. Within No 26, it is written that:

> [T]he expression "legal dispute" has been used to make clear that while conflicts of rights are within the jurisdiction of the Centre, mere conflicts of interests are not. The dispute must concern the existence or scope of a legal right or obligation, or the nature or extent of the reparation to be made for breach of a legal obligation.

---

[294] ICSID [1997] ARB/93/1—Award, 21 para 5.18.

[295] Hobe and Müller (2009), 65 (70); Schreuer (2010), 353 (361).

[296] Harten and Loughlin (2006), 121 (144).

[297] Schreuer (2008), 959 (960).

[298] PCIJ [1924] The Mavrommatis Palestine Concessions, 7 (11).

[299] ICJ [1950] Interpretation of Peace Treaties with Bulgaria, Hungary and Romania, 65 (74).

[300] Dispute is a complex term and difficult of define, in: Schreuer (2008), 959 (960).

Furthermore, some ICSID tribunals dealt with the issue of legal dispute according to Art. 25(1) of the ICSID Convention. One tribunal mentioned that a '(...) legal dispute, in the ordinary meaning of the term, is a disagreement about legal rights or obligations.'[301] In another decision, an ICSID tribunal highlighted that a legal dispute exists, if a state measure affects the legal rights of the investor and if the BIT offers legal protection.[302]

3.5.2.1.3 The Meaning of "Directly"

The first drafts of the ICSID Convention only included jurisdiction for all legal disputes '(...) arising out of or in connection with any investment (...).'[303] However, this definition was subject to a lot of criticism and the term "directly" was added.[304] This change was done without providing any definition of the word direct.[305] Directness is an objective criterion of Art. 25(1) ICSID Convention, which means it cannot be differently agreed upon by the parties.[306] One ICSID tribunal mentioned that:

> (...) [I]t is apparent that the term "directly" relates (...) (Article 25(1) ICSID Convention) (...) to the "dispute" and not to the "investment". It follows that jurisdiction can exist even in respect of investments that are not direct, so long as the dispute arises directly from such transactions.[307]

Other tribunals support this decision.[308] Due to the complicated task to separate between direct or indirect claims, a case by case analysis is necessary.[309] Overall, requirements of legal dispute and directness are subject to a case by case evaluation. Not fulfilling these criteria would result in a decline of jurisdiction and lead to unnecessary costs.

3.5.2.2 Unity of Investment

The Desertec Concept encompasses several contracts, e.g. concession contracts, construction contracts, and state contracts. Therewith, the question arises if each

---

[301] ICSID [2006] ARB/03/17, 18–19 para 34.

[302] ICSID [2006] ARB/03/9, 24 para 67.

[303] ICSID (1970) vol I, 116.

[304] ICSID (1968) vol II(2), 700 and 826.

[305] Schreuer (2009), 106 para 83.

[306] Schreuer (2009), 106–107 para 85.

[307] ICSID [1997] ARB/96/3, 1378 (1384) para 24.

[308] Ambivalent concerning the requirement, in: ICSID [2006] ARB/03/16—Award, 62–63 para 331; ICSID [2003] ARB/01/8, 788 (796) para 52; ICSID [1999] ARB/97/4—Decision, 251 (275) paras 71 and 72.

[309] Further indicators for a possible distinction, in: Schreuer (2009), 112 para 104.

contract must be or can be an investment according to the ICSID Convention. If ICSID tribunals only regard one contract as an investment, the others might be excluded. This might argue against the use of PPPs, as it is impossible to predict which part might be protected by the ICSID Convention. However, ICSID tribunals handled comparable contractual setups not too strict.[310] The tribunals established the idea of 'unity of investment', also qualifying contracts as investments, which individually would not qualify.[311] One ICSID tribunal mentioned that it would be contrary to the economic reality to consider each contract separately (e.g. in the case of a PPP).[312] Other ICSID tribunals decided similarly.[313] Consequently, the Desertec Concept, set up as a PPP (see Figs. 1.1, 1.2 and 1.3), is going to be treated as one.

3.6 Definition of the Investment According to the ICSID Convention

Art. 25(1) sentence 1 of the ICSID Convention addresses the term "investment". The article deals with the investment in the context of the jurisdiction of the centre, stipulating all the necessary requirements to invoke an ICSID tribunal.

3.6.1 *Interpretation of Art. 25(1) of the ICSID Convention*

First of all, Art. 25(1) of the ICSID Convention requires interpretation. As mentioned above, the general rules of interpretation in international law are within the VCLT. However, the ICSID Convention was ratified before the VCLT and therefore it does not apply directly to the ICSID Convention. Due to VCLT's status as customary international law, the statutes concerning interpretation also apply to the ICSID Convention.[314]

---

[310] Grubenmann (2010), 60.

[311] Grubenmann (2010), 60; cf ICSID ICSID [2013] ARB/08/9—Decision on Jurisdiction and Admissibility 139 paras 424–425; ICSID [2011] ARB/07/5—Decision on Jurisdiction and Admissibility, 141–142 para 364.

[312] Grubenmann (2010), 60.

[313] ICSID [2004] ARB/01/3—Decision on Jurisdiction, 29 para 70; ICSID [1999] 97/4—Decision, 251 (275) para 72; ICSID [1997] ARB/96/3, 1378 (1383) para 26; ICSID [1988] ARB/82/1—Award, 125 (144) para 4.13 and (145–146) para 4.17.

[314] ICSID [2007] ARB/05/10—Award, 19–20 para 65.

3.6.1.1 Wording of Art. 25(1) of the ICSID Convention

According to Art. 31(1) of the VCLT, the wording of the ICSID Convention is the first important part concerning an interpretation. Taking a closer look at Art. 25 (1) of the ICSID Convention, several prerequisites for jurisdiction of ICSID need to be present. These encompass: (1) a legal dispute arising directly out of an investment between (2) a contracting state and a national of another contracting state, and (3) written dispute consent by both parties. However, neither Art. 25(1) of the ICSID nor its other paragraphs include any definition of the term investment. Paragraph 1 just mentions the precondition of an investment, but does not offer any explanation. It is only highlighted that the jurisdiction of ICSID implies a legal dispute arising directly out of an investment. In addition, no other article within the ICSID Convention offers a definition of investment. Therefore, Art. 25(1) of the ICSID Convention is silent on the notion of investment.

The basic understanding of the term investment includes some "logical" requirements. First of all, it involves two parties, one as the investor and a one as the person (e.g. host state) receiving the money. The main purpose of the investor is gaining benefit from its investments.[315] On the other hand, the host state desires some benefit as well for allowing an investment within its territory.[316] Both parties know each other's interests and try to balance them.[317] Thus, each investment encompasses benefits for the host state, e.g. economic development.

3.6.1.2 Context, Subsequent Agreements to the ICSID Convention and State Practice

As mentioned above, interpretation of the wording includes: (1) recognition of good faith, (2) ordinary meaning rule, (3) consideration of objective and intention, and (4) context. First of all, there is the Preamble of the ICSID Convention which can help to interpret the term. The Preamble highlights the '(. . .) need for international cooperation for economic development (. . .).' Hence there is a need for the cooperation concerning economic development, possibly making one requirement of investments the economic development within the host state.

There are also subsequent agreements concerning the ICSID Convention. One of the main documents is the Report of Executive Directors. There is section III No. 9 of the 1965 Report of Executive Directors. It states that one of the Executive Directors wanted to strengthen the partnership between the participating countries 'in the cause of economic development.' However, the Executive Directors pointed out in section V No. 27 of the Report of Executive Directors that no definition of investments offers the possibility to accommodate new forms of investment. This

[315] European Commission (2010), 2–3.

[316] European Commission (2010), 2–3; Universität Frankfurt (2014).

[317] Universität Frankfurt (2014).

results in the fact that the Report of Executive Directors supports both interpretations.

Concerning the question of state practice, a closer look at the different BITs is necessary. After the ratification of the ICSID Convention, states concluded several BITs to protect their investors. Almost all BITs include a definition of investment. They are the perfect example of what state practice is. Examining BITs, most of them define investment very broadly.[318] Consequently, states do not view investment in an overly restrictive way.

Overall, the wording of Art. 25(1) of the ICSID Convention does not deliver a completely clear picture, leaving both forms of interpretation. On the one hand, there might be some requirements within the Preamble; on the other hand investment remains rather "undefined". Therefore, a closer look at the preparatory work of the ICSID Convention is necessary.

3.6.1.2.1 Historical Background of Art. 25(1) of the ICSID Convention

The start of the ICSID Convention is dated back to 1962 and a World Bank working group completing the first draft of the ICSID Convention.[319] The first draft did not include any restrictions on the notion of investment.[320] In the years 1963[321] and 1963–1964,[322] further meetings followed and introduced the first restriction of investment. The 1964 ICSID Convention draft included a notion[323] or restrictions of the term investment, e.g. a duration of at least 5 years.[324] However, these ideas and definitions were not supported as many desired that the parties can decide what falls under ICSID jurisdiction.[325] The word "contribution" and the time element were subject to particular opposition.[326] The main argument concerning the definition of investment was between the developed and developing states.[327] At the end of the negotiation, the contracting states were not able to agree on a definition

[318] Krishan (2009), 1 (14); Mortenson (2010), 257 (311).

[319] Working paper in the form of a draft convention (June 5, 1962), in: ICSID (1968) vol II(1), 19.

[320] Working paper in the form of a draft convention, art. IV, § 1 (June 5, 1962), in: ICSID (1968) vol II(1), 33.

[321] Preliminary draft of a convention on the settlement of investment disputes between a state and nationals of other states, art. II, § 1 (October 15, 1963), in: ICSID (1968a) vol II(1), 202.

[322] Consultative meeting of legal experts, addis ababa (April 30, 1964), in: ICSID (1968a) vol II (1), 236 (et seq.).

[323] ICSID (1968b) vol II(2), 844; ICSID (1970) vol I, 116.

[324] ICSID (1968a) vol II(1), 623.

[325] Johannsen (2009), 5 (6); Mortenson (2010), 257 (287); ICSID (1968b) vol II(2), 652, 661, 668, 669, 700, 703 and 707.

[326] ICSID (1968b) vol II(2), 702, 703, 708, 709 and 710.

[327] Mortenson (2010), 257 (280) and (284–286); Abraham (2009), 206 (209).

of the term investment and thus left the term 'open.'[328] This was due to a British proposal which omitted any definition of the term investment.[329] This proposal was adopted unmodified in the final ICSID version of 1965.[330] Hence, considerable importance should be attached to the consent of the parties in respect of the assessment of ICSID jurisdiction.[331]

Chairman Broches mentioned that the failure of the proposed definition was closely connected to the individual fitting of restrictions by the proposing country and was not an overall notion for the legal framework of international investment.[332] Some say that the wide definition of investment was an intentional deal of the developed states in exchange for opt-out possibilities for developing states.[333] The main idea of the ICSID Convention was to offer states a new way to promote economic development and to protect transnational investors.[334] Legal author often argue that the *ex ante* perception of states was to set up an umbrella for foreign investors.[335] It is also frequently proposed that one of major benefits is that the ICSID Convention is more flexible and can be applied appropriately to every individual case.[336] Furthermore, an overly restrict approach could impede transnational investments and therefore would be contrary towards ICSID founding idea.[337] The preparatory work supports the assumption that investment is deliberately undefined within the ICSID Convention. The definition is mostly up to the parties, which have chosen, so far, to use a broad understanding. Even countries which opposed the broad definition include wide investments definitions in their BITs today.

3.6.1.2.2 The Different Languages of Art. 25(1) of the ICSID Convention

To summarize the interpretation, a brief overview of the official languages of the ICSID Convention is necessary, due to the ambivalent results. The ICSID Convention contains several arbitrational languages, namely English, French and Spanish.[338] Assessing all the documents, there is no difference in the meaning of Art. 25

[328] Gaillard (2009), 403 (403); Yala (2005), 105 (106); Hamida (2007), 287 (288–289); Andreeva (2008), 161 (168); cf Hobe and Müller (2009), 65 (68); Johannsen (2009), 5 (6); cf Delaume (1966), 64 (70); Broches (1966), 263 (268).

[329] ICSID (1968b) vol II(2), 821 and 826.

[330] Mortenson (2010), 257 (292).

[331] Mortenson (2010), 257 (290); Broches (1966), 263 (268).

[332] Summary of proceedings of the legal committee meeting (November 27, 1964), in: ICSID (1968b) vol II(2), 707.

[333] Mortenson (2010), 257 (280); Most of the Salini requirements were considered and rejected, in: Mortenson (2010), 257 (281).

[334] Nmehielle (2001), 21 (23); Mortenson (2010), 257 (304) and (311).

[335] Mortenson (2010), 257 (311).

[336] Nmehielle (2001), 21 (26); Mortenson (2010), 257 (311).

[337] Mortenson (2010), 257 (311).

[338] Reed et al. (2004), 21.

(1) of the ICSID Convention. All three versions do not include any restrictions on investment.

Unlike the ICSID, NAFTA and the ECT have a definition of investment.[339] In both cases, the definition of investment is broad and thus comparable to most BIT definitions.[340] Taking all the above-mentioned into account, most parties seem to favor an undefined term. Although there are several indications that an investment includes certain requirements, the opposing arguments outweigh these. The fact that Art. 25(1) of the ICSID Convention does not include a definition combined with the preparatory work and the different language versions support the interpretation of a broad investment term. Hence, only the parties to the respective BIT can limit the notion.

3.6.2 The Desertec Concept as an Investment According to the G/M-BIT

Taking a closer look at the G/M-BIT, Art. 1(1)(a-e) defines investments. Similarly to most BITs, the G/M-BIT uses a broad definition of investments. There are almost no exclusions of investment types. Art. 1(1)(e) of the G/M-BIT is very relevant for the Desertec Concept. There it is mentioned that 'Öffentlich-rechtliche und vertragliche Konzessionen einschließlich Aufsuchungs- und Gewinnkonzessionen für natürliche Ressourcen'[341] are investments according to the BIT. The French version is comparable to the German version of the BIT.[342] There could be some doubts as to whether the protection only includes concessions concerning natural resources. As mentioned above, implementation of the Desertec Concept is similar to typical natural resource concessions. However, Art. 1(1)(e) of the G/M-BIT might restrict the protection for concessions only for cases of exploration and profit. The word 'einschließlich'[343] indicates that these concession types are specifically included. It does not mean that all concessions must relate to natural resources in general. Hence, Art. 1(1)(e) of the BIT applies to the above-proposed form of cooperation for the implementation of the Desertec Concept. The burden of proof, i.e. if it is an investment according to the BIT, lies on the side of the claimant.[344]

---

[339] Karl (1994), 809 (811).

[340] Karl (1994), 809 (811).

[341] Public and private concessions, including exploration and profit concessions concerning natural resources.

[342] Article 1(1)(e) G/M-BIT (French version): 'les concessions de droit public ou contractuelles, y compris celles relatives à la prospection et l'exploitation de resources naturelles'.

[343] Including.

[344] *Sergei Paushok, CJSC Golden East Company and CJSC Vostokneftegaz Company v Mongolia* [2011] Award—Proceeding pursuant to UNCITRAL Arbitration Rules, 32 para 200.

Also, as there is no specific case law concerning the requirements of Art. 25(1) of the ICSID Convention, it is possible that the same rule applies.

3.6.3 *ICSID Case Law and Art. 25(1) of the ICSID Convention*

ICSID tribunal's case law is also important concerning the application of investment. As Art. 41(1) of the ICSID Convention mentions, an ICSID tribunal can decide on its own competence. Furthermore, Art. 42(2) of the ICSID Convention offers ICSID tribunals the possibility to "close" legal loopholes within substantive rules. According to Art. 44 sentence 2 of the ICSID Convention, ICSID tribunals can also decide subsidiary procedural questions. As a result, ICSID tribunals remain influential concerning admittance to ICSID arbitration. As mentioned above, there is no stare decisis within ICSID case law, but a de facto stare decisis leads to a certain equal treatment, which makes it necessary to examine ICSID case law in detail.

3.6.3.1 Salini Costruttori S.P.A. and Italstrade S.P.A. v. Kingdom of Morocco

In 2001, the investment arbitration world was shaken up when one ICSID tribunal defined the term investment. This happened in the case *Salini Costruttori S.P.A. and Italstrade S.P.A. v Kingdom of Morocco*. Until today, the Salini test remains one of the most diversely discussed topics of international investment arbitration. The Salini decision established four criteria to meet the investment notion of Art. 25 (1) of the ICSID Convention, namely (1) a contribution (in cash, in kind or in labor), (2) certain duration of performance, (3) participation in the risks of the transaction, and (4) a contribution to the economic development of the host state.[345] These elements of investments may be independent, but they normally require a global assessment.[346] Furthermore, the tribunal highlighted that these elements are characteristics of an investment.[347] The tribunal noted that there is no definition within the ICSID Convention.[348] However, there are no doubts that the notion of investment is not solely up to the discretion of the parties.[349] The criteria stem from the different opinions in the literature, from different ICSID cases, the registration

[345] ICSID [2001] ARB/00/4, 609 (622) para 52.
[346] ICSID [2001] ARB/00/4, 609 (622) para 52.
[347] ICSID [2001] ARB/00/4, 609 (622) para 52.
[348] ICSID [2001] ARB/00/4, 609 (622) para 51.
[349] ICSID [2001] ARB/00/4, 609 (622) para 52.

requirement by the Secretary General and the Preamble of the ICSID Convention.[350]

The Salini tribunal was the first to assess the BIT and its scope concerning investment.[351] Afterwards, it proceeded to analyze the requirements originating from Art. 25(1) of the ICSID Convention.[352] The Salini test was later on also known as 'double-barrelled test', which stems from the establishment of a limitation of ICSID jurisdiction, which is completely distinct from the party's consent.[353] The double-barrelled test is performed by primarily assessing the BIT and its definition and requirements of investment, and afterwards requirements of the ICSID Convention.[354]

The decision was the starting point for the discussion whether there is a notion of investment within Art. 25(1) of the ICSID Convention. Down to the present day, three schools of thought have developed. These are the objective, subjective and liberal approach, which all handle the Salini test differently. In respect of the Desertec Concept, there might be two difficulties related to the Salini test. First, there could be problems concerning the risk element, and second concerning the contribution to the economic development of Morocco. Consequently, a closer assessment of the Salini test is necessary. In the following paragraphs, the three approaches are discussed, and this will be followed by an evaluation of the literature.

3.6.3.2 The Objective Approach

First of all, there is the objective approach towards the question of the Salini test. This is also known as the deductive approach, which tries to give a "true definition".[355] Based on the objective theory, there is an objective definition of the term investment within Art. 25(1) of the ICSID Convention, which does not require the consent of the parties.[356] The term investment is not found within the scope of the ICSID Convention, but is an objective requirement.[357] The prevailing idea is that the parties cannot define anything as an investment for the purpose of ICSID jurisdiction, because this would turn Art. 25(1) of the ICSID Convention into a

[350] ICSID [2001] ARB/00/4, 609 (622) para 52.

[351] ICSID [2001] ARB/00/4, 609 (619–621) paras 43–49

[352] ICSID [2001] ARB/00/4, 609 (621–623) paras 50–58.

[353] ICSID [2007] ARB/05/10—Award, 16 para 55; ICSID [2004] ARB/03/11—Award, 10 paras 42–43 and 11 para 48; Andreeva (2008), 161 (165).

[354] ICSID [2004] ARB/02/18—Dissenting Opinion (Weil), 245 (250) para 14; ICSID [2001] ARB/00/4, 609 (619–623) paras 43–58; Andreeva (2008), 161 (173); cf Loncle (2006), 319 (329–330); 'Double-barrelled' test is also known as 'double keyhole' approach, in: Schreuer (2009), 117 para 124.

[355] Gaillard (2009), 403 (410).

[356] Hamida (2007), 287 (290).

[357] ICSID [2002] ARB/00/2—Award, 142 (157) para 52.

meaningless provision.[358] There were also cases where the decision was later on annulled because of the missing application of the objective approach.[359] Furthermore, one tribunal mentioned that all Salini criteria must be satisfied individually and not in an overall perspective.[360] If the project does not meet the ICSID requirement of investment, the fact that it is an investment concerning the respective BIT does not matter.[361] Overall, there are several cases supporting the objective approach.[362] In addition, the Permanent Court of Arbitration (PCA) mentioned that investment always has an inherent meaning, regardless of the applicable arbitration institution.[363] This encompasses a contribution, which extends over a certain amount of time and involves risk.[364]

3.6.3.3 The Subjective Approach

The main idea of the subjective approach is that the parties subjectively define investment based on their consent to ICSID arbitration.[365] Consequently, there are no grounds in Art. 25(1) of the ICSID Convention to apply the Salini test because it is against the will of the drafters and signatories.[366] As a result, the scope of the BIT and hence its definition of investment is crucial for ICSID jurisdiction.[367] This approach is also known as the intuitive approach.[368] A lot of tribunals champion the subjective approach.[369]

---

[358] ICSID [2004] ARB/03/11—Award, 11 paras 49–50.

[359] ICSID [2006] ARB/99/7—Annulment, 12 para 27.

[360] ICSID [2006] ARB/05/19, 23–24 para 77.

[361] ICSID [2007] ARB/05/10—Award, 49 para 148.

[362] ICSID [2009] ARB/06/5—Award, 29 para 74 and 45 para 114; ICSID [2009] ARB/05/10—Annulment—Dissenting Opinion (Shahabuddeen), 38 (50–57) paras 40–61; ICSID [2008] ARB/05/19—Award, paras 35–36 117–120 ICSID [2007] ARB/05/18, 32 para 116; ICSID [2007] ARB/05/07, 28 para 99; cf ICSID [2007] ARB/05/10—Award 48–49 paras 144–146; ICSID [2006] ARB/05/19, 23–24 para 77; ICSID [2006] ARB/99/7—Annulment, 8–16 paras 23–41; ICSID [2006] ARB/04/13, 28–29 paras 91–92; ICSID [2005] ARB/03/29, 34 para 130; ICSID [2004] ARB/03/11—Award, 12 para 53; Also not directly, the tribunal supports the idea of objective criteria, in: ICSID [1997] ARB/96/3, 1378 (1387) para 43.

[363] Permanent Court of Arbitration [2009] PCA Case No. AA280—Award, 53–54 para 207.

[364] Permanent Court of Arbitration [2009] PCA Case No. AA280—Award, 53–54 para 207.

[365] Hamida (2007), 287 (289); Johannsen (2009), 5 (6).

[366] ICSID [2010] ARB/07/16—Award, 109–110 para 311.

[367] ICSID [2009] ARB/07/9, 28–29 para 78; ICSID [1999] ARB/95/3—Award, 458 (494–495) para 83.

[368] Gaillard (2009), 403 (410).

[369] ICSID [2014] ARB/07/8—Decision, 143–144 para 296; ICSID [2014] ARB/07/27—Award, 67 para 185 ICSID [2014] ARB/12/2—Award, 46–47 para 148; ICSID [2013] ARB/11/23—Award, 89–90 para 369; ICSID [2013] ARB/10/5—Decision on Jurisdiction, 47–48 para 142; ICSID [2012] ARB/07/29—Award, 20–22 paras 71–78; Only considering the BIT, in: ICSID [2012] ARB/09/1—Decision on Jurisdiction, 47–49 paras 208–214; ICSID [2012] ARB/08/11—

In some cases, Art. 25(1) of the ICSID Convention was not even considered at all because the BIT's notion of investment was met.[370] The most popular example is that a commercial contract cannot be an investment due to a BIT definition of investment.[371] Another example is a decision by the PCA, which argued that an investment which is undefined in a BIT cannot be interpreted 'in the sense of making a substantial contribution to the local economy.'[372] Hence, only the definition of the parties is relevant. There were cases where tribunals explicitly annulled previous decisions which applied the Salini test[373] or the Salini test was a vital part of the prior decision.[374] Hence, the term investment is deliberately undefined.[375] The only requirement for an investment is the consent of the parties; any further definition would be a gross error and lead to annulment.[376]

Award, 36 para 132; Only reviewing the BIT, in: ICSID [2012] ARB/10/12—Award, 53 paras 230–232; ICSID [2012] ARB/09/5—Award, 68–69 paras 280–286; ICSID [2012] ARB/09/12—Decision on Jurisdiction, 5 paras 5.27–5.28; ICSID [2011] ARB/08/16—Award, 34 para 127 and 38 para 137; ICSID [2011] ARB/07/17—Award, 24 para 94; *Sergei Paushok, CJSC Golden East Company and CJSC Vostokneftegaz Company v Mongolia* [2011] Award—Proceeding pursuant to UNCITRAL Arbitration Rules, 32–34 paras 200–207; By rejecting Argentina's annulment request concerning the excess of power related to the definition of investment, the tribunal followed its precedent tribunals and there assessment of the BIT, in: ICSID [2010] ARB/01/3—Annulment, 26 para 81, 37–38 paras 93–94, 40–41 paras 104–106 and 43 para 111; ICSID [2010] ARB/07/23, 53–55 paras 139–147; ICSID [2010] ARB/06/18, 17–18 paras 52–55; ICSID [2010] ARB/07/16—Award, 89–109 paras 254–309 and summary on (111) paras 314–316, ICSID [2010] ARB/08/2—Award, 58 para 111; ICSID [2010] ARB/07/27, 44–45 paras 162–166; ICSID [2009] ARB/01/12—Annulment, 30 para 59; ICSID [2009] ARB/07/9, 34 para 96; Mentioning that the ICSID Convention does not have a definition and relying on the BIT, in: ICSID [2007] ARB/05/8—Award, 55 paras 249–250; Considering the BIT definition of investment, in: ICSID [2007] ARB/03/25—Award, 160 paras 334–335; Considers "risk", but not as part of an investment requirement, in: ICSID [2006] ARB/03/16—Award, 59–60 para 317 and 61 para 325; ICSID [2006] ARB/04/15, 28–30 paras 59–62; ICSID [2005] ARB/03/2, 16–17 paras 54–58; Only considering the BIT definition, in: ICSID [2004] ARB/01/3—Decision, 9–10 para 29; ICSID [2004] ARB/02/6, 40 para 103 and 43 para 112; ICSID [2004] ARB/02/5—Decision, 35 para 124; ICSID [2003] ARB/01/13, 307 (350) para 140; Rejecting the claim to apply the objective approach, in: ICSID [2003] ARB/02/10, 9 para 39 and 12 para 49; ICSID [2003] ARB/00/9—Award, 33 para 8.2, 34 para 8.5 and 37 para 8.14; Only considering the BIT definition, in: ICSID [2003] ARB/01/12—Decision, 33–34 para 62; ICSID [2002] ARB/99/6—Award, 33 n. 136; Pointing out that ICSID does not define investment, in: ICSID [2000] ARB/99/3, 484 (492) para 13.6; ICSID [1998] ARB/97/6—Preliminary Decision, 457 (470) para 4; Permanent Court of Arbitration [2009] PCA Case No. AA280, 54 para 208 and 62 paras 242–243.

[370] ICSID [2002] ARB/99/6—Award, 20 para 86 and 33 para 136.

[371] Permanent Court of Arbitration [2009] PCA Case No. AA280, 46 para 185.

[372] Permanent Court of Arbitration [2006] Partial Award (under the UNICTRAL Arbitration Rules 1976), 43 para 211.

[373] ICSID [2009] ARB/05/10—Annulment, 22–23 para 56.

[374] ICSID [2014] ARB/09/4—Decision on Elsamex S.A.'s Preliminary Objections, 39–40 para 136; ICSID [2014] ARB/09/4—Decision on the Termination of the Stay of Enforcement of the Award, 11 para 35.

[375] ICSID [2009] ARB/05/10—Annulment, 26 paras 62 and 63.

[376] ICSID [2009] ARB/05/10—Annulment, 30 para 71, 31 para 74 and 36 para 83.

3.6.3.4 The Liberal Approach

The main idea of the liberal approach is that the tribunal needs to clarify if the investment is according to the BIT and afterwards if the investment fulfills all requirements of the ICSID convention.[377] By taking the BIT as well as the Art. 25 ICSID Convention into account, this approach is in between the objective and subjective approach, as it checks both sources. Within the liberal approach, there are two different streams of how to handle criteria of the Salini test.

First of all, there is the 'Jurisdictional Approach', which claims that all Salini criteria must be present cumulatively.[378] On the other hand, there is the 'Typical Characteristics Approach', which points out that it depends on the Salini test criteria in its entity and not on every criterion being present cumulatively.[379] After reviewing the ICSID cases using the liberal approach, it is clear that the 'Typical Characteristic Approach' dominates. Most tribunals refuse to apply the Salini too strictly because of the missing definition within the ICSID Convention.[380] The Salini test can only introduce subjective elements of judgment, but not objective requirements.[381] They are proposing that the Salini test must be a flexible test, whereby the Salini test is only one feature to identify an investment.[382] As a review of the ICSID case law reveals, the distinction between the liberal and objective approach can be quite difficult. The prime example of the liberal approach is the Salini decision mentioned above and the 'double-barrelled test.' This tribunal highlighted that the elements of the Salini test are characteristics of an investment which must be assessed in total, arguing against a strict definition according to Salini, but rather a guideline. A not fully represented element can be outbalanced by another very strong existing Salini element. There are several tribunals supporting

[377] ICSID [2002] ARB/00/2—Award, 142 (158) para 56; Andreeva (2008), 161 (165).

[378] On the verge of supporting the objective approach, in: ICSID [2013] ARB/08/9—Dissenting Opinion of Santiago Torres Bernárdez 65–68 paras 199–209; ICSID [2007] ARB/05/10—Award, 21–22 para 71; ICSID [2004] ARB/03/11—Award, 12 para 53; cf ICSID [2006] ARB/99/7—Annulment, 12 para 27.

[379] ICSID [2013] ARB/08/9—Decision on Jurisdiction and Admissibility, 142–174 paras 435–520; Discussing Salini and mentioning that the investment is within the scope of Article 25 ICSID Convention and the BIT, in: ICSID [2013] ARB/10/7—Decision, 63–68 paras 193–210; Using the Salini criteria but not in a too strict way, in: ICSID [2012] ARB/09/2—Award, 59–63 paras 293–312; ICSID [2011] ARB/07/5 Decision on Jurisdiction and Admissibility, 387 para 149; ICSID [2007] ARB/05/10—Award, 21–22 para 71; ICSID [2006] ARB/05/3, 16–17 para 72; ICSID [2005] ARB/03/08—Award, 426 (450) para 13; ICSID [2001] ARB/00/4, 609 (622) para 54; ICSID [1999] ARB/97/4—Decision, 251 (275) para 72.

[380] ICSID [2012] ARB/09/4—Award, 66–74 paras 253–283; Criterions might be of some use, but not in the respective case, in: ICSID [2010] ARB/07/16—Award, 110 para 313; ICSID [2009] ARB/05/10—Annulment, 33–35 paras 77 and 79; ICSID [2009] ARB/07/12, 26 para 81; ICSID [2008] ARB/05/22—Award, 86 para 312; ICSID [2007] ARB/03/6—Award, 38 para 165.

[381] ICSID [2009] ARB/07/21—Award, 10 para 43.

[382] ICSID [2009] ARB/05/10—Annulment, 33–35 para 79; ICSID [2008] ARB/05/22—Award, 87 paras 314–316; ICSID [2007] ARB/03/6—Award, 38 para 165.

the liberal approach.[383] One tribunal even used the liberal approach (therewith the Salini test) in a case which was according to UNCITRAL arbitration rules and not the ICSID Convention.[384] The tribunal highlighted that there is no difference in the term investment under UNCITRAL or ICSID rules, and that a BIT cannot turn everything into an investment.[385]

One ICSID tribunal dealt with issue in depth and mentioned that this discussion is purely academic, because it often leads to the same result.[386] Different cases may require a different emphasis on the Salini criteria.[387] All the Salini criteria are interrelated and depend on each other, and it does not depend on the quantity, but rather on the quality of each criteria.[388] In another case, the tribunal highlighted the importance of the agreements of the parties within the BIT and mentions that the Salini test should only be performed in cases where the parties did not include an investment definition.[389] In 2013 one tribunal checked all Salini criteria and mentioned that it needs no further discussion if they are a 'jurisdictional requirement' as long as the criteria are met in the respective case.[390]

In 2009, one ICSID tribunal decision dealt intensively with the liberal approach. Besides the normal Salini criteria, it moved on and established an even wider test. ICSID case law does not provide any proof for a definition, which completely depends on party's intentions, as there are always same basic criteria concerning the presence of an investment.[391] The most frequently referred definition of investment is the Salini test, if an ICSID tribunal decides to define the term.[392] The tribunal highlights that the main purpose of the ICSID Convention is to encourage and

---

[383] ICSID [2010] ARB/07/20—Award, 35–36 paras 110–111; ICSID [2010] ARB/08/20, 28–29 paras 79–80; ICSID [2010] ARB/08/8, 58–59 paras 128–129 and 61 para 133; cf ICSID [2009] ARB/07/21—Award, 8–9 para 36 and 10 para 39; ICSID [2009] ARB/05/6—Award, 28 para 95; ICSID [2009] ARB/06/5—Award, 29 para 74; ICSID [2009] ARB/07/12, 19–21 paras 61–65 and 25–28 paras 77–87; ICSID [2008] ARB/05/22—Award, 86–87 paras 312–316; ICSID [2008] ARB/05/12, 39 para 128; ICSID [2007] ARB/03/6—Award, 38 para 165; cf ICSID [2007] ARB/01/8—Annulment, 17–18 paras 71–72; cf ICSID [2006] ARB/03/9, 29 para 81; Seems to be subjective approach, but assess all requirements of Salini, in: ICSID [2005] ARB/02/17, 29 para 88; cf ICSID [2004] ARB/02/18, 205 (213) para 19 and (220) para 39; Article 25(1) ICSID Convention includes objective criteria, in: ICSID [2002] ARB/00/2—Award, 142 (157) para 52; ICSID [2001] ARB/00/6, 28–31 paras 58–66; ICSID [2001] ARB/00/4, 609 (622) para 54; Applied in the *RSM Production Corporation v Grenada* case (ICSID [2009], ARB/05/14—Award, para 241), in: Peterson (2009a).

[384] Peterson (2009b).

[385] Peterson (2009b).

[386] ICSID [2007] ARB/05/10—Award, 33 para 105.

[387] ICSID [2007] ARB/05/10—Award, 21–22 para 71 and 33 para 105.

[388] ICSID [2007] ARB/05/10—Award, 33–35 para 106.

[389] ICSID [2010] ARB/08/8, 58–59 paras 128–129.

[390] ICSID [2013] ARB/10/11 and ARB/10/18—Decision on Jurisdiction, 98 para 352.

[391] ICSID [2009] ARB/06/5—Award, 32 para 82.

[392] ICSID [2009] ARB/06/5—Award, 33 para 83.

protect the contribution to the economy of the host state.[393] No national investment is allowed to be subject to the ICSID convention.[394] Furthermore, the tribunal emphasized the need for economic activity, in order to prevent an abuse of investment protection rights by the investor.[395] The BIT must be within the scope of the ICSID Convention.[396] The tribunal created a new catalog for investment, which includes (1) a contribution in money or other assets, (2) a certain duration, (3) an element of risk, (4) an operation made in order to develop an economic activity in the host State, (5) assets invested in accordance with the laws of the host State, and (6) assets invested *bona fide*.[397] Nevertheless, the tribunal highlights that an extensive assessment of these requirements is not always necessary (e.g. some may overlap) and that all requirements must be analyzed '(. . .) with due consideration of all circumstances.'[398]

In 2013 another ICSID tribunal dealt with the liberal approach intensively, by assessing the 'double-barrelled test' and the respective BIT.[399] For the first time an ICSID tribunal interpreted the notion of investment according to the drafting process of the ICSID Convention and the VCLT.[400] Afterwards the concept of investment was checked according to the VCLT,[401] pertinent case law and doctrine.[402] Lastly it also reviewed the Salini test within ICSID case law and accepted the liberal approach, but only restricted it to the Salini criteria.[403]

Down to the present day, a lot of other problems remain within the liberal approach. One major issue relates to the consideration of the "relationship" of BIT regulations and Art. 25(1) ICSID Convention within the 'double-barrelled test.'[404] There is no real decision; however it seems that the emphasis is more on the ICSID Convention. This might be due to the fact that most BITs include such a wide definition that there is almost never a problem. The second issue revolves around the question of which Salini elements should be considered as characteristics of an investment.[405] The element of 'contribution to the local economy' is subject to a lot of criticism. There are some tribunals highlighting that the

[393] ICSID [2009] ARB/06/5—Award, 35 para 87.

[394] ICSID [2009] ARB/06/5—Award, 35–36 para 89.

[395] ICSID [2009] ARB/06/5—Award, 37 para 93.

[396] ICSID [2009] ARB/06/5—Award, 38 para 96.

[397] ICSID [2009] ARB/06/5—Award, 45 para 114.

[398] ICSID [2009] ARB/06/5—Award, 45 para 115.

[399] ICSID [2013] ARB/08/9—Decision on Jurisdiction and Admissibility, 142–174 paras 435–520.

[400] ICSID [2013] ARB/08/9—Decision on Jurisdiction and Admissibility, 144–149 paras 441–454.

[401] ICSID [2013] ARB/08/9—Decision on Jurisdiction and Admissibility, 150–155 paras 455–465.

[402] ICSID [2013] ARB/08/9—Decision on Jurisdiction and Admissibility, 155–159 paras 466–474.

[403] ICSID [2013] ARB/08/9—Decision on Jurisdiction and Admissibility, 159–162 paras 475–482.

[404] ICSID [2008] ARB/06/3, 24 para 82.

[405] ICSID [2010] ARB/07/20—Award, 35–36 para 110; cf ICSID [2008] ARB/98/2—Award, 76–77 para 232.

contribution to the country's economic development is difficult to assess and mostly covered by the other Salini requirements.[406]

3.6.4 *Evaluation on the Basis of the Literature*

The different approaches of ICSID tribunals lead to a great response within the literature. Comparable to the ICSID case law, the literature is divided on the issue of how to handle the Salini test. There is no consent on the question of which approach is prevailing at the moment. Some claim that most ICSID decisions today do not apply the sole subjective approach as they find this insufficient in regard to Art. 25(1) of the ICSID Convention.[407] Thus, there is a strict handling of investment today because of the acceptance of the Salini criteria.[408] Others claim that in practice most ICSID tribunals have adopted a broad definition of investment and only a minority a narrow approach.[409] Some authors mention that some ICSID tribunals try to 'beat' each other to see who has the better test, thus forgetting that they should observe each case separately.[410]

3.6.4.1 The Validity of the Subjective Approach

First of all, there some authors who support the subjective approach of ICSID tribunals.[411] Most of them claim that the Salini test has led to a crisis in the international investment protection.[412] Their main claim is that the missing definition of investment within the ICSID Convention was intentional to prevent a possible limitation of the Convention's scope.[413] During the negotiation of the ICSID Convention, the objective approach was rejected.[414] Hence, only parties can limit ICSID jurisdiction.[415]

---

[406] ICSID [2006] ARB/05/3, 16–17 para 72; ICSID [2005] ARB/03/08—Award, 426 (450) para 13 (iv); Arbitration under UNCITRAL Rules, in: Peterson (2009b).

[407] Gaillard (2009), 403 (411); cf Harten (2007), 78.

[408] Dolzer and Schreuer (2008), 69; Schreuer (2009), 118 para 125; Mortenson (2010), 257 (259) and (277).

[409] Schöbener et al. (2010), 262, Kap. 4 § 18 para 156; Loncle (2006), 319 (321).

[410] Ho (2010), 633 (641).

[411] Nmehielle (2001), 21 (26); cf Tietje (2005), 47 (54); McIlwrath and Savage (2010), 375 para 7–019; Happ (2008), 19 (24); Nariman (2004), 123 (134); cf Moses (2008), 225; Mortenson (2010), 257 (300–301); Tietje (2003), 5 (10); Broches (1966), 263 (268).

[412] Reinisch (2014), 111 (145–146); Diehl (2007), 268 (269); Krishan (2009), 1 (2–3).

[413] Nathan (1995), 27 (30); Hobe and Müller (2009), 65 (68); Mortenson (2010), 257 (259), (304) and (311); cf Ho (2010), 633 (646); Delaume (1966), 64 (70) 64 (70); Abraham (2009), 206 (207); Andreeva (2008), 161 (168); Krishan (2009), 1 (17); Loncle (2006), 319 (319).

[414] Mortenson (2010), 257 (299).

[415] Hamida (2007), 287 (289); Hobe and Müller (2009), 65 (68); Mortenson (2010), 257 (301).

Nevertheless, some authors favor a so-called 'outer limit' of investment. If a BIT definition goes beyond the ICSID's concept, there will be no ICSID jurisdiction.[416] Even in cases where ICSID tribunals expressively declined the application of the Salini test, tribunals pointed out that there are limits concerning the definition of investments.[417] The drafting history of the ICSID Convention supports these 'outer limits of definition' because the missing notion of investments does not mean that ICSID parties were incapable to define it, but unable.[418] The fact that the ICSID Secretary-General does not register an arbitration request is sometimes understood as proof that Art. 25(1) of the ICSID Convention has an unchangeable core, which always applies.[419] Consequently, these "core" elements of investment imply that there is an outer limit of the investment, which must be met at all times.

There is also the risk that investments which are in accordance with the ICSID Convention are excluded due to the Salini test.[420] Thus, application of the Salini test defines investment and makes it less interesting for states, to seek ICSID jurisdiction, if they disagree with the definition.[421] Investment is a fluid concept which should not be frozen in a certain notion.[422] Due to the application of the Salini test, the requirement of investment almost merges into the requirement of consent of jurisdiction.[423] Aaron Broches said 'it was always up to the parties to give or withhold consent (. . .) in other words, it should be the terms of consent that ultimately define the Centre's jurisdiction.'[424]

Therefore, ICSID tribunals should not limit the convention's scope because this is against the objective and purpose' of the ICSID Convention.[425] By ignoring BITs and just taking the ICSID Convention into account concerning investment, the tribunals 'raise(s) concerns about the completeness of the (ICSID tribunals) analysis.'[426] Imposing an objective definition of investment violates the state

[416] Schreuer (2009), 124 para 144; However most BIT definitions are in accordance with ICSID in: Schreuer (2009), 124 para 145; No real definition of commercial dispute, which makes it into a "slippery slope" for the exclusion of investments, in: Mortenson (2010), 257 (298–299); Hobe and Müller (2009), 65 (69); cf Yala (2005), 105 (117).

[417] ICSID [2009] ARB/05/10—Annulment—Dissenting Opinion (Shahabuddeen), 38 (40–42) paras 9–13.

[418] ICSID [2009] ARB/05/10—Annulment—Dissenting Opinion (Shahabuddeen), 38 (41) paras 10–11; cf Schöbener and Markert (2006), 65 (82–83).

[419] Ho (2010), 633 (642).

[420] Krishan (2009), 1 (11).

[421] Krishan (2009), 1 (8).

[422] Krishan (2009), 1 (9).

[423] Hobe and Müller (2009), 65 (68); Broches (1966), 263 (268).

[424] Grubenmann (2010), 72.

[425] Lörcher (2005), 11 (13); Hobe and Müller (2009), 65 (69); This also includes a prohibition to enlarge ICSID jurisdiction to every possible case, in: Nathan (1995), 413, 27 (50); Loncle (2006), 319 (319–320); Wrong belief that there are objective criteria, in: Ho (2010), 633 (646); Mortenson (2010), 257 (311–312).

[426] Andreeva (2008), 161 (169).

sovereignty of all the ICSID Convention member states.[427] ICSID tribunals are not empowered to create a general applicable objective requirement of investment.[428] They can only decide on a case by case evaluation, if an investment exists.[429] The great variety of different BITs and state opinions highlights that there is no state practice to rely on.[430] The problem of comparing investment cases lies in the fact that all cases do have a different factual background, e.g. different BITs.[431] The way the ICSID tribunal proceeded in the Salini case might be a departure from an interpretation according to Art. 31 of the VCLT.[432]

3.6.4.2 The Validity of the Objective Approach

There are also some authors in favor of the objective approach. They point out that today an objective restriction of the investment term is recognized.[433] The ICSID Convention is a special legal framework within the World Bank which deals with issues of the investment. The definition of investment encompasses the basic ideas of the World Bank, which includes a development aspect for less developed countries.[434] Furthermore, the argument of Art. 41(1) of the ICSID Convention (stipulating that ICSID tribunals decide *ex officio* concerning their jurisdiction) is put forward.[435]

Today, BITs have a wide definition of investment.[436] Although most BITs have similar definitions of the term investment, this does not necessarily reflect the ICSID Convention's concept of investment.[437] Customary international law has a more narrow definition, "which historically limited the protection of foreign-owned property to assets that are tangible and directly owned".[438] Thus, limitations of investment are nothing new within the history of investment protection. It is

[427] Krishan (2009), 1 (9–10); cf Mortenson (2010), 257 (305); cf Ho (2010), 633 (646–647); Violation of party autonomy, in: Krishan (2009), 1 (11) and (14).

[428] Krishan (2009), 1 (10); Mortenson (2010), 257 (306); cf Ho (2010), 633 (646–647).

[429] Krishan (2009), 1 (10).

[430] Mortenson (2010), 257 (312).

[431] Andreeva (2008), 161 (171).

[432] Andreeva (2008), 161 (170).

[433] Hunter (2007), 165 (169); Lörcher (2005), 11 (12); Diehl (2007), 268 (269); cf Johannsen (2009), 5 (8); cf Grubenmann (2007), 3 and 4–5; Metje (2008), 162; Highlighting that the mere party consent is not sufficient, in: Hobe and Müller (2009), 65 (69).

[434] Johannsen (2009), 5 (9).

[435] Johannsen (2009), 5 (9); Nathan (1995), 413, 27 (39–41).

[436] States define investment to their own advantage, in: Mortenson (2010), 257 (261); Zampetti and Sauvé (2007), 211 (217); Harten (2007), 75; Including FDIs and portfolio investments, in: Zampetti and Sauvé (2007), 211 (217); Including FDIs and portfolio investments, in: Harten (2007), 78.

[437] Schreuer (2009), 124 para 144.

[438] Harten (2007), 77.

important to strengthen investments by making ICSID tribunal decisions more predictable and defining the term investment.[439] The mere definition according to a BIT would set the ICSID framework and its credibility at risk.[440]

Rule 2 of the Institution Rules also supports an objective meaning of investment as it mentions in Rule 2(1)(e) of the Institution Rules that a request for arbitration or conciliation must '(. . .) contain information concerning the issues in dispute indicating that there is, between the parties, a legal dispute arising directly out of an investment (. . .)', beside the consent of the parties (Rule 2(1)(c) of the Institution Rule).[441] Another argument against the subjective view is that there is no definition of investment. The contracting states were not able to agree on a common notion, so not defining the term was an expedient.[442] In addition, the option of Art. 25(4) of the ICSID Convention to previously exclude subjects from ICSID jurisdiction does not support the subjective approach, as it would be meaningless if state parties determined ICSID jurisdiction solely through BITs.[443] Art. 2(b) of the Rules Governing the Additional Facility for the Administration of Proceedings by the Secretariat of the International Centre for the Settlement of Investment Disputes (Additional Facility Rules) stipulates that the Secretariat of the Centre is able to administer disputes between a state and a national in cases of arbitration which lacks the requirement of 'arise directly out of an investment.' A tribunal also explicitly mentioned that this provision encompasses the question of the notion of investment and its coverage.[444] Thus, protection for investors exists, even if the term investment is defined in a narrow way, thus arguing against the subjective theory.[445]

3.6.4.3 The Validity of the Liberal Approach

Some authors argue that the Salini criteria (as an expression of the ICSID Convention) should be used as guiding factors in addition to the BIT.[446] They highlight that ICSID tribunals champion the liberal approach.[447] ICSID tribunals frequently misunderstand the Salini test as an objective requirement; however, the criteria of

[439] Diehl (2007), 268 (269).

[440] Yala (2005), 105 (117).

[441] Schreuer (2009), 117 para 123.

[442] Johannsen (2009), 5 (6).

[443] Mortenson (2010), 257 (259–260) and (293); Johannsen (2009), 5 (6–7).

[444] ICSID [1997] ARB/96/3, 1378 (1384) para 28.

[445] Johannsen (2009), 5 (7).

[446] Borris and Hennecke (2008), 49 (55); Happ (2008), 19 (24); Idea mentioned, in: Gaillard (2009), 403 (409–410); Difficult to interpret what the author means, since there is no distinction between the liberal and objective approach, but there seems to be a favor for the liberal approach, in: Grubenmann (2010), 106 and 271.

[447] Yala (2005), 105 (106).

the test are only 'typical features' of an investment.[448] The Salini decision and therewith the 'double-barrelled test' is part of the liberal approach.[449] Even one arbitrator of the Salini tribunal mentioned 4 years later that he did not view the Salini test as a purely objective test.[450] He pointed out the risks and dangers which could result from an overly strict application of the Salini test.[451] No single criterion of the Salini test is truly objective as it can be compensated by other criteria.[452] Hence, it is not necessary for all the criteria to be present to form an investment, which means that the Salini criteria do not apply to all investments.[453]

In almost all cases of arbitration, the will of the party is of great importance.[454] However, investment arbitration is different due to the fact that BIT and the ICSID Convention are both relevant for a successful arbitration.[455] Both documents belong to international law and determine the jurisdiction of the ICSID tribunal. All other forms require "only" one international law document (e.g. the ICC Rules) for their jurisdiction.

3.6.4.4 Assessment of ICSID Case Law and the Literature

First and foremost, there is no consistent case law. Up to today, ICSID tribunals still use all of these approaches, making it difficult for an investor to foresee a possible outcome. There is no surprise that the Salini test is mostly used by the host state to argue against ICSID jurisdiction. Vice-versa, investors prefer the subjective approach, which gives them great freedom. Yet it is interesting that the utilization of the Salini test was contested by the claimant only in few cases.[456]

Based on the interpretation of the ICSID Convention, it is difficult to say whether the tribunals rightfully apply the Salini test or not. There are some arguments in favor of the liberal approach as it seems doubtable that the definition of investment and consequently ICSID jurisdiction solely depends on BITs. In both cases, in ICSID case law and the literature, the subjective approach seems to be favored. However, until recently, tribunals decided according to the objective and liberal approach, making it impossible to predict any outcome. It is advisable to include a specific clause in the investment contract which states that the project in question is an investment according to the parties.[457]

[448] Schöbener and Markert (2006), 65 (82); No strict requirements, but guidance, in: Schöbener et al. (2010), 315, Kap. 4 §19 para 384; Krishan (2009), 1 (13); Mortenson (2010), 257 (271–272).

[449] Yala (2005), 105 (106).

[450] Ho (2010), 633 (639–640).

[451] Ho (2010), 633 (639).

[452] Ho (2010), 633 (638).

[453] Ho (2010), 633 (638); Grubenmann (2010), 271.

[454] Yala (2005), 105 (106).

[455] Yala (2005), 105 (106–107).

[456] ICSID [2006] ARB/05/19, 21 paras 66–67.

[457] Schreuer (2009), 119 para 130.

One highly valued opinion must also be taken into account. Back in 2001, Christoph H. Schreuer mentioned duration, regularity of profit and return, risk, a substantial commitment and significance for the host state's development as typical features of an investment.[458] In his commentary, these features were under the section 'characteristics of an investment.'[459] Interestingly, shortly after this, the Salini tribunal picked out Schreuer's ideas and concluded the famous Salini decision, starting the discussion of investment definition all over again. In 2009, Christoph H. Schreuer released his second edition of the ICSID commentary. His remarks concerning the typical features of investment remained the same in the new edition.[460] However, he dropped the section's name 'characteristics of an investment' and renamed it 'A Test for the Existence of an Investment?.'[461] Later on in the same edition, Schreuer points out that it was not his intention to set up a mandatory test for investment.[462] Similar to the Salini tribunal, Schreuer did not understand the features as '(. . .) distinct jurisdictional requirements each of which must be met separately.'[463] Schreuer also points out that the features must be applied in conjunction to assess if there is an investment.[464] An isolated examination of each criterion does not help to strengthen ICSID tribunal's predictability.[465] Consequently, Schreuer seems to favor the liberal approach. Other legal author point out that the uncertainty concerning the notion of investment may lead to a reluctance of application of the ICSID Convention by investors.[466]

Since there is a risk that an ICSID tribunal might apply the objective approach, the project would be safe if it fulfilled the "strictest" requirements. Therefore, it is necessary to apply the objective approach to the Desertec Concept to find out whether the project constitutes an investment under all circumstances.

3.7 The Desertec Concept as an ICSID Investment According to Salini

As mentioned above, the Salini case relates to a dispute between *Salini Costruttori S.P.A. and Italstrade S.P.A. v the Kingdom of Morocco*. Since Morocco argued that the respective project was not an investment, there is an undeniable risk that this might happen again. Civil construction and infrastructure projects in particular have

---

[458] Schreuer (2001), 140 para 122.

[459] Schreuer (2001), 138.

[460] Schreuer (2009), 128 para 153.

[461] Schreuer (2009), 128 para 153.

[462] Schreuer (2009), 133 para 171.

[463] Schreuer (2009), 133 para 171.

[464] Schreuer (2009), 133 para 171.

[465] Schreuer (2009), 133–134 para 172.

[466] Reinisch (2014), 111 (145–146).

frequently been subject to ICSID arbitration because of its categorization as an investment.[467] Big scale projects, especially infrastructural ones, appear to be safe, although there remain uncertainties for projects like BOTs.[468] Some claim that BOTs fit perfectly under the Salini test.[469] Again others propose that the only way to avoid this risk is to abandon the ICSID Convention and rely on other arbitration methods.[470]

As mentioned above, the Desertec Concept is an investment according to the G/M-BIT and ICSID tribunals regard concession fees as an investment.[471] In one case, both parties agreed in the arbitration agreement that a PPA constitutes an investment.[472] The tribunal therefore assumed that an investment existed.[473] Additionally, it conferred that requirements of Art. 25(1) of the ICSID Convention and the BIT were met.[474] An ICSID tribunal pointed out that the concept of investment according to the ICSID Convention can differ from the projects economical concept of being an investment.[475] Due to the size and importance of the Desertec Concept and Morocco's previous challenge, an assessment is necessary. Consequently, the next section will assess the Desertec Concept under the Salini test. This means that (1) a contribution (in cash, in kind or in labor), (2) certain duration of performance, (3) participation in the risks of the transaction and (4) a contribution to the economic development of the host state must be present. In addition, the requirements (5) assets invested in accordance with the laws of the host State and (6) assets invested *bona fide* from the (2009) *Phoenix Action Ltd. v The Czech Republic* case will also be considered.

Even the application of the UNCITRAL Rules would not eradicate the problem of the Salini test. In recent years different ad hoc tribunals, which applied the UNCITRAL Rules, ruled differently concerning the application of the Salini test.[476] As a result it is not clear, if a tribunal (independent of the applicable arbitration rules) might apply the Salini test or not. Abovementioned examples (e.g. the PCA) illustrate that arbitration tribunals are willing to resort to other tribunals or courts decisions. This illustrates that the issue of the notion of investment even remains, if the G/M-BIT is replaced by an EU-BIT and the sole application of the UNCITRAL Rules. It remains unpredictable for an investor, if the Salini test is applied as the

[467] Reinisch (2014), 127–128 para 151.

[468] Mortenson (2010), 257 (278).

[469] Hunter (2007), 165 (169).

[470] Mortenson (2010), 257 (279).

[471] ICSID [2003] ARB/01/12—Decision, 34–35 para 64.

[472] ICSID [2008] ARB/04/19—Award, 36 para 148.

[473] ICSID [2008] ARB/04/19—Award, 36 para 148.

[474] ICSID [2008] ARB/04/19—Award, 36 para 149 and 41 para 166.

[475] ICSID [2008] ARB/05/22—Award, 45 para 100.

[476] Against the application of Salini, in: UNCITRAL [2011] *White Industries Australia Limited v The Republic of India* Award, 76 para 7.4.9; In favor of the Salini test, in: UNCITRAL [2011] *Alps Finance and Trade AG v The Slovak Republic* Award, 76–79 paras 239–247.

application of a certain arbitration process does not guarantee the non-applicability of the Salini test. Thus it would be best, if the Desertec Concept would meet all the above mentioned requirements, to reduce the risk of unwanted outcome.

3.7.1 Significant Contribution

Most tribunals agree on the necessity of assessing contribution.[477] Right at the beginning of this requirement, there is the question of whether the contribution must be significant or not. Some tribunals highlight that this is not important,[478] while others claim that "significant" is a vital part of the contribution requirement,[479] and some did not even mention anything concerning this fact.[480] Even in cases where ICSID tribunals denied the application of the Salini test, some judges dissented and specifically demanded a significant contribution to the local economy.[481] Nevertheless, no tribunal defines or indicates what the term significant means; they just consider it to be a given fact.

The Salini tribunal did not require a significant contribution as they only assessed a contribution.[482] Within the Salini case, the tribunal mentioned that the investor's contribution encompassed know-how, provision of the necessary equipment, and qualified personnel (including a secured payment of the workers).[483] The contribution can be in terms of know-how, equipment and personnel, and in

477 ICSID [2012] ARB/09/2—Award, 60 para 297; ICSID [2010] ARB/07/20—Award, 35–36 para 110; ICSID [2010] ARB/08/20, 28–29 para 80.

478 ICSID [2009] ARB/05/10—Annulment, 35–36 paras 80–82; Did not mention significance in their decision, in: ICSID [2001] ARB/00/4, 609 (623) para 57.

479 Talking about a substantial contribution, in: ICSID [2013] ARB/08/9—Decision on Jurisdiction and Admissibility, 163 para 483; ICSID [2008] ARB/05/22—Award, 85 para 310; Significant period of time, in: ICSID [2008] ARB/05/20, 37–38 para 125; ICSID [2007] ARB/05/10—Award, 47 para 138; Substantial work as investment, in: ICSID [2006] ARB/04/13, 29 para 92; ICSID [2005] ARB/03/29, 37 para 137; ICSID [2004] ARB/03/11—Award, 13–14 paras 57–58; Significant period of time, in: ICSID [2001] ARB/00/5, 39 para 101; ICSID [1997] ARB/96/3, 1378 (1387) para 43.

480 Not mentioning it as requirement, but later refer to a 'substantial contribution' and to a 'substantial economic value', in: ICSID [2012] ARB/09/2—Award, 60–61 paras 297–300; cf ICSID ARB/98/2—Award, 77 para 233; ICSID [2007] ARB/05/18, 32 paras 116–117; ICSID [2006] ARB/05/3, 17–19 para 73; ICSID [2005] ARB/02/17, 29 para 88; ICSID [2005] ARB/03/08—Award, 21–22 paras 13–14; ICSID [2001] ARB/00/6, 29–30 para 60; ICSID [1999] ARB/97/4, 251 (282–283) para 90.

481 ICSID [2009] ARB/05/10—Annulment—Dissenting Opinion (Shahabuddeen), 38 (38) para 2 and (48–50) paras 33–39.

482 ICSID [2001] ARB/00/4, 609 (622) paras 52 and 53.

483 ICSID [2001] ARB/00/4, 609 (622) para 53.

financial terms.[484] Assets (*en nature*) are possible, whereby no minimum amount is necessary because it is difficult to assess in cases of know-how or training of staff.[485]

The Desertec Concept certainly qualifies as a significant contribution. As mentioned above, an investor wants to share its ideas with the MENA region, which would be a transfer of know-how. In addition, an investor also desires to supply the MENA region with clean energy, which can be a contribution to all MENA countries. On a long-term scale, an investor wants the MENA countries to run the project themselves, which involves a training of qualified personnel. Hence, there can be no doubt that the Desertec Concept is a contribution on a significant scale.

3.7.2 Duration of the Investment

The contract duration within Salini was set to be at least between 2 and 5 years.[486] Within the drafting of the ICSID Convention, a duration of 5 years was discussed.[487] Some claim that the minimum duration should be 2 years minimum,[488] while others say that a project should take at least 5 years.[489] There is also the view that, under certain circumstances, the duration must not be exactly 2 years.[490] There are also proposals concerning calculation of the minimum investment duration.[491] Other tribunals view the duration of the investment in both qualitative and quantitative terms.[492] The quantitative duration relates to the duration of the investment (project), whereas the qualitative duration connects to the economic contribution.[493] The same tribunal proposed that the duration is closely related to the economic development of the state because the longer the project

[484] ICSID [2008] ARB/05/22—Award, 88–89 para 320; ICSID [2007] ARB/05/07, 29 para 100; ICSID [2007] ARB/05/10—Award, 36 para 109; ICSID [2006] ARB/05/19, 23–24 para 77; ICSID [2006] ARB/99/7—Annulment, 12 para 27; cf ICSID [2006] ARB/04/13, 29 para 92; ICSID [2005] ARB/03/29, 32–33 paras 115–120 and 35 para 131; ICSID [2001] ARB/00/4, 609 (622) para 53.

[485] Loncle (2006), 319 (326).

[486] Two years are sufficient, in: ICSID [2012] ARB/09/2—Award, 61 paras 300 and 303–304; ICSID [2007] ARB/05/10—Award, 36–37 para 110; 2 years are sufficient, in: ICSID [2007] ARB/05/07, 29 para 101; 3 years considered to be sufficient, in: ICSID [2006] ARB/04/13, 29–30 paras 95–96; ICSID [2001] ARB/00/4, 609 (622–623) para 54; Andreeva (2008), 161 (166).

[487] ICSID (1968) vol II(1), 116.

[488] ICSID [2007] ARB/05/07, 29 para 101.

[489] Johannsen (2009), 5 (15).

[490] ICSID [2006] ARB/04/13, 29–30 paras 94–96.

[491] Interruption and Suspensions are included, in: ICSID [2007] ARB/05/07, 29 para 102; Duration includes the tender stage as well, in: ICSID [2006] ARB/04/13, 29–30 paras 94–96.

[492] ICSID [2007] ARB/05/10—Award, 36–37 para 110.

[493] ICSID [2007] ARB/05/10—Award, 36–37 paras 110–111.

exists, the more it will contribute.[494] A determination must be on a case by case assessment,[495] but most tribunals agree on the necessity of a duration element.[496] The duration must be assessed in the light of the investor's overall commitment.[497] Older case law indicates that this prerequisite should not be too strict.[498]

An overall duration of at least 2 years seems to be sufficient. With regard to the Desertec Concept, the issue of duration does not cause any problems. An investor as well as Morocco probably plan to have a long-term relationship. The figures concerning possible profit (see Tables 1.1, 1.2 and 1.3) from the Desertec project also indicate that there must be a minimum cooperation of 30 years between an investor and Morocco/ONE. Consequently, the Desertec Concept satisfies the requirement of duration.

3.7.3 Risk

As mentioned above the Desertec Concept should encompass a PPA or subsidy or feed-in tariffs. This might cause problems with the Salini requirement "risk". All forms of support might breach the requirement of risk, and thus might not be an investment. Hence, it is necessary to analyze the Salini tribunal decisions. In addition, other tribunals which have dealt with the topic of risk and the literature will be assessed.

3.7.3.1 ICSID Case Law and the Literature's Approach

The Salini tribunal seems to differ between economic and political/legal risks.[499] The longer the investment takes, the more risk is involved.[500] This list included the burden to carry the change of contract, a potential increase of labor costs because of a change of respective Moroccan laws, an unforeseen incident (not force majeure), and an '(. . .) increase or decrease in volume of the work load not exceeding 20 % of the total contract price.'[501] Finally, the tribunal highlighted that it does not depend on whether the risk was taken freely.[502]

---

494 ICSID [2007] ARB/05/10—Award, 37 para 111.

495 Loncle (2006), 319 (326–327).

496 ICSID [2010] ARB/07/20—Award, 35–36 para 110; ICSID [2010] ARB/08/20, 28–29 para 80.

497 Peterson (2009b).

498 ICSID [2005] ARB/03/08—Award, 14 para 14(ii); cf ICSID [2005] ARB/03/29, 35–36 paras 132–133.

499 ICSID [2001] ARB/00/4, 609 (623) para 55.

500 ICSID [2001] ARB/00/4, 609 (623) para 56.

501 ICSID [2001] ARB/00/4, 609 (623) para 55.

502 ICSID [2001] ARB/00/4, 609 (623) para 55.

Other tribunals claimed that there is always a need to assess risk.[503] ICSID tribunals basically assume that investment risks are shared between the parties.[504] However, no-one has ever defined risk. In all cases (including the Salini decision), tribunals merely mentioned that the risk prerequisite exists and assessed the respective investment. Concerning construction contracts, one tribunal mentioned that the longer the construction contract takes, the more the criterion of risk is met.[505] Another ICSID tribunal said that providing retention money is an obvious risk in cases of long-term contracts.[506] One ICSID tribunal pointed out that there is a risk if the investor must keep a certain standard of the investment (project) over a certain amount of time according to the contract.[507] Two tribunals pointed out that the existence of a dispute already illustrates the existence of risk for an investor.[508] Others also claim that risk is a factor of jurisdiction, but not a legal requirement of investment.[509] An investment of 2.5 million US dollar was without a doubt accepted as a risk for the investor.[510] There is also a risk, if the host state intervenes, e.g. because of bankruptcy.[511]

Similar to ICSID tribunals, the literature does not define risk. Nevertheless, they try to offer a "kind of" catalog of what risk encompasses or might mean. Risk must go beyond mere commercial risks and '(...) entail some uncertainty as to the return to be achieved, and the ultimate outcome of the transaction.'[512] As some authors point out, '(...) the more the remuneration depends on the operating result, the higher the investor's risk.'[513] Risk also includes the risk of contractual non-performance.[514] Some authors mention that the Salini tribunal defined risk in a very broad and permissive way, which means that it includes day-to-day risks.[515] Also the fact that the tribunal does not link risk to remuneration illustrates that the tribunal favors a liberal approach over the classical understanding of risk.[516] Consequently, the literature does not see a problem to meet to criterion "risk" within the Salini test, because investments (partially) based on own funds almost imply risk for the investor.[517]

---

503 ICSID [2010] ARB/07/20—Award, 35–36 para 110; ICSID [2010] ARB/08/20, 28–29 para 80.

504 ICSID [2007] ARB/05/10—Award, 38–39 para 112.

505 ICSID [2009] ARB/07/12, 25–26 para 78; Grubenmann (2010), 117.

506 ICSID [2007] ARB/05/07, 30–31 para 109.

507 ICSID [2006] ARB/05/19, 23–24 para 77.

508 ICSID [2012] ARB/09/2—Award, 61 para 301; ICSID [1997] ARB/96/3, 1378 (1386) para 40.

509 ICSID [2006] ARB/03/16—Award, 59–60 para 317 and 61 para 325.

510 ICSID [2012] ARB/09/2—Award, 61 para 302.

511 ICSID [2013] ARB/08/9—Decision on Jurisdiction and Admissibility, 163 para 485.

512 Peterson (2009b); cf Grubenmann (2010), 117.

513 Loncle (2006), 319 (329).

514 Peterson (2009b).

515 Yala (2005), 105 (112).

516 Yala (2005), 105 (113).

517 Dolzer and Schreuer (2008), 68; Johannsen (2009), 5 (17); Grubenmann (2010), 118.

3.7.3.2 ICSID Tribunals' View on Guarantees

Since there is no real definition of risk, it is necessary to examine ICSID case law concerning the question of PPA, subsidies or feed-in tariffs. In all cases, there is no case law. Several ICSID cases encompassed PPAs; however, the PPA was never part of the discussion.[518] As mentioned above, one tribunal assessed a PPA in more detail, but it did not see any problems to categorize it as an investment. Thus, an investor and Morocco/ONE should also include a section in a contract (e.g. arbitration agreement) that the PPA is an investment or at least does not interfere with any definition of investment. By doing this, ICSID tribunals might not view them as infringing the requirement risk.

There are cases which deal with questions of "guarantees" of payment. Due to the close connection, these decisions might deliver a better understanding of the tribunals' views. First of all, one tribunal mentioned that guarantee payments do not exclude risk if they '(...) (have) been negotiated competitively and when unforeseen events may occur.'[519] Concerning a payment guarantee, another tribunal pointed out that there is no guarantee (but risk) if the price paid by the host state covers the actual costs of the investor's obligation of performance.[520] Furthermore, an ICSID tribunal decided that securing first demand bank guarantees is not a risk itself, but the danger of an unlawful call on these guarantees is a risk for the investor.[521]

Secondly, in the case of *MHS v. Malaysia*, the contract between the state and the private company contained a 'no find – no pay' clause, making MHS's search and salvage operation only profitable if it was successful.[522] A 'no find – no pay' agreement is:

> (...) [A] well established practice in marine salvage and, meant that all the costs of the, search and salvage operation (and its attendant risks) would be borne exclusively by (the investor). It also meant that the (investor) would recover its expenditure and make a profit only if both the salvage operation and the subsequent sale of the recovered items were successful.[523]

The tribunal mentioned in its award that the 'no find – no pay' contract was not a risk in the qualitative sense and was a mere commercial risk.[524] The tribunal did not mention what this exactly means. The decision was annulled later on.[525] Thus, the

[518] ICSID [2001], ARB/98/8, in: OECD (2006) Investor-to-state dispute settlement in infrastructure projects, 28–29; ICSID [2000], ARB/01/4, in: OECD (2006) Investor-to-state dispute settlement in infrastructure projects, 30–31; The Dabhol Power Project, in: OECD (2006) Investor-to-state dispute settlement in infrastructure projects, 31–33.

[519] ICSID [2009] ARB/07/12, 26 para 29.

[520] ICSID [2009] ARB/07/12, 28 para 86b).

[521] ICSID [2005] ARB/03/29, 36–37 paras 135–136.

[522] Andreeva (2008), 161 (163).

[523] ICSID [2007], ARB/05/10—Award, 2 para 10.

[524] ICSID [2007], ARB/05/10—Award, 38–39 para 112.

[525] ICSID [2009] ARB/05/10—Annulment, 35—36 paras 80–82.

only certain fact is that risk must be different from average risks of commercial contracts, e.g. termination.[526]

3.7.3.3 Risk and the Desertec Concept

It is difficult to evaluate what the tribunals think about guarantees. The decision concerning the "no find – no pay" clause does not apply to the Desertec Concept. Besides the fact that the decision was annulled, the first tribunal already stated that this is a typical clause in marine salvage. Marine salvage projects already bear a great risk if the parties find something at all. The clause tries to ease the risk for the ordering party. This clause transforms a service contract into a contract for work and services. Consequently, it deals with the transfer of risk, but not with guarantee related to the risk. Without a doubt, the mere fact that a subsidy or feed-in tariff exists does not exclude the requirement of risk.

Other decisions offer a type of guideline which the parties to the project should follow. First of all, there is the need for a competitive negotiation of PPAs, subsidies or feed-in tariffs between the parties of the Desertec Concept. Therefore, it is important that all parties negotiate in a transparent way to be able to illustrate competitiveness. Secondly, PPAs, subsidies or feed-in tariffs cannot cover the complete production costs of the energy. The main idea of them is to support the company and not to pay all its bills. In the latter case, there would be no risk because the state would always step in for the company. In addition, an investor also underlines that PPAs, subsidies or feed-in tariffs are only important at the start of the production. Otherwise there would be the risk that desert sun is not competitive with other technologies. ICSID tribunals also did not view any of these supportive measures as being contradictory to investment as long as these tools meet certain requirements, e.g. not guaranteeing all investor costs. There is a great possibility that ICSID tribunals are going to accept the supportive tools of the Desertec Concept. This seems to support the idea that a project might need help at the beginning, but not total coverage. To avoid problems with the requirement of risk, the amount and percentage of total production costs concerning PPAs, subsidies or feed-in tariffs should be transparent.

As mentioned above, the Desertec Concept faces a lot of different procedural risks and project risks which are connected to its size and "prototype" status. Most of these risks do not even relate to the issue of production costs and sale of services. This also underlines the importance of certain guarantee tools concerning some risks. Due to the great investment sum, it is "doubtful" whether the project can be completely covered. There is a great chance that ICSID tribunals view the requirement of "risk" as met. However, the above-mentioned cases did not deal with the Salini test, which leaves a residual risk, as risk did not form part of the notion of

[526] ICSID [2007] ARB/05/10—Award, 38–39 para 112; ICSID [2004] ARB/03/11—Award, 13 para 57.

investment. If an investor and Morocco/ONE comply with the above-mentioned requirements, there is no doubt that they meet the requirement of risk.

3.7.4 *Contribution to the Economic Development of the Host State*

The most widely discussed element of the Salini test is the contribution to the economic development of the host state. As mentioned above, the liberal approach often entails three out of four Salini elements, excluding this element. A lot of tribunals which favor the liberal approach deny the application of the contribution to the economic development of the host state element, due to its uncertainties and its complex nature.

3.7.4.1 The Salini Tribunal

The Salini tribunal only briefly addressed the question of contribution to the economic development of the host state. In its case, the tribunal only assessed the respective investment, without mentioning any further details concerning the element. It pointed out that construction of a highway cannot be contested as not contributing to the development of the host state.[527] Furthermore, the Salini tribunal considered the provision of know-how and concluded that the Italian company shared its knowledge.[528]

3.7.4.2 Tribunals Supporting Contribution to the Economic Development Element

The Salini tribunal insisted on its importance, but assessed it the "contribution to the economic development element" with a great amount of flexibility.[529] As referred to earlier, even in cases where the application of the Salini test was denied, dissenting opinions of arbitrators supported its utilization. There were several tribunals dealing with this prerequisite and assessing the individual investment.[530]

[527] ICSID [2001] ARB/00/4, 609 (623) para 57.

[528] ICSID [2001] ARB/00/4, 609 (623) para 57.

[529] ICSID [2009] ARB/06/5—Award, 33 para 84.

[530] Funds, but without a further assessment, in: ICSID [2013] ARB/08/9—Decision on Jurisdiction and Admissibility, 164 para 487; A highway in Morocco, in: ICSID [2001] ARB/00/4, 609 (623) para 57; A highway in Pakistan, in: ICSID [2005] ARB/03/29, 6–7 paras 10–12; A legal consulting firm in Congo, in: ICSID [2006] ARB/99/7—Annulment, 15–16 para 39; A five star hotel in Egypt as contribution to tourism, in: ICSID [2006] ARB/05/19, 23–24 para 77; A banking infrastructure contributed in Slovakia, in: ICSID [1999] ARB/97/4—Decision, 251 (252) para 1 and (282) para

The *MHS v Malaysia* tribunal was one of the few applying the requirement of contribution to the economic development in a very strict and decisive way.[531] It claimed that ICSID tribunals must ignore every political and cultural benefit resulting from the project, except when these benefits result in economic benefits.[532] Hence, there should be a difference between benefits for the business (investor) and benefits for the local economy.[533] A frequent argument supporting the application of this element rests in the Preamble of the Convention, which already mentions the need for "economic development".[534] In addition, it is sometimes mentioned that the ICSID Convention was concluded under the auspices of the IBRD.[535] Since one of the main tasks of the IBRD is the support of developing countries, investment must contribute to the local economy. One ICSID tribunal claims that all ICSID tribunals take the criterion of contributing to the economic development into account, explicitly or implicitly.[536] Other tribunals just simply assessed the criterion without further explanation.[537] Another ICSID tribunal pointed out that the ICSID Convention does not protect 100 %- owned foreign property, but only projects which fulfill the Salini test and hence contribute to the economic development.[538] Then, the tribunal mentioned that:

> (. . .) [A]n essential – although not sufficient – characteristic or unquestionable criterion of the investment, (which) does not mean that this contribution must always be sizable or successful (. . .).[539]

The same tribunal highlighted that there can be an expression in various forms in the economic development which varies from case to case.[540] In a dissenting opinion it is highlighted that the criterion should be part of the notion of investment, but it is part of the other Salini criteria and does not need a separate assessment.[541]

88; The bank operated in the area of host state and investor's home state (however both countries used to be one), in ICSID [1999] ARB/97/4—Decision, 251 (252) para 2; Widening and deepening southern stretches of the Suez Canal in Egypt, in: ICSID [2006] ARB/04/13, 6 para 9 and 29 para 92; BOT/concession power plant in Turkey, in: ICSID [2004] ARB/02/5—Decision, 8–9 paras 18–20.

[531] ICSID [2007] ARB/05/10—Award, 47 para 138.

[532] ICSID [2007] ARB/05/10—Award, 47 para 138.

[533] ICSID [2007] ARB/05/10—Award, 47 para 138.

[534] ICSID [2007] ARB/05/10—Award, 20 para 66; ICSID [2006] ARB/99/7—Annulment, 12–13 paras 28–30; ICSID [2005] ARB/03/29, 37 para 137; ICSID [1999] ARB/97/4—Decision, 251 (273) para 64.

[535] ICSID [2006] ARB/99/7—Annulment, 12 para 28.

[536] ICSID [2006] ARB/99/7—Annulment, 12 para 29; E.g. implicitly checked, in: ICSID [1999] ARB/94/2—Award, 197 (228–229) paras 119–125.

[537] ICSID [2010] ARB/08/20, 28–29 para 80.

[538] ICSID [2006] ARB/99/7—Annulment, 15 para 38.

[539] ICSID [2006] ARB/99/7—Annulment, 14 para 33.

[540] ICSID [2006] ARB/99/7—Annulment, 14 para 33.

[541] ICSID [2012] ARB/09/2—Dissenting Opinion of Makhdoom Ali Khan, 17–18 paras 46–48.

Finally, tribunal emphasized the fact that a contribution to the economic development cannot be merely speculative (unlikely in cases of public infrastructure).[542]

3.7.4.3 Tribunals Opposing Contribution to the Economic Development Element

Especially within the liberal approach, there are some tribunals opposing the requirement of contribution to the economic development of the host state.[543] These tribunals share the opinion that the criterion of economical contribution is too vague and that the other characteristics of the Salini test encompass economic contribution.[544] One tribunal pointed out that the contribution of an investment to the development is almost impossible to assess and hence a less ambiguous approach like the contribution to the economy should be taken.[545] Such a contribution '(. . .) is (. . .) normally inherent in the mere concept of investment as shaped by the elements of contribution/duration/risk, and should therefore in principle be presumed.'[546] Another tribunal mentioned that this element may be a result of the investment, but is not a prerequisite.[547] Nevertheless, it viewed it as a given in the present case.[548] Another tribunal supported this, as contribution to the development of the local economy is not a mandatory criterion for an investment.[549] Contribution to the local economy is an objective of the ICSID Convention, but there is no interpretation of the preamble in the way which shows that it is a prerequisite for an investment.[550] It is difficult to predict how an investment will evolve over time as it may be fruitful at the beginning, but turn into a disaster at the end.[551] If the other Salini elements exist, the contribution does not deliver any new aspects.[552] Vice versa, if there is a only a partial fulfillment of the other Salini requirements, the contribution to the economic development of the host state is necessary.[553] In 2012

---

[542] ICSID [2007] ARB/05/10—Award, 48–49 para 144; This decision was later on annulled, in: ICSID [2009], ARB/05/10—Annulment, 1 (35–36) para 80–82.

[543] ICSID [2006] ARB/05/3, 16–17 para 72; ICSID [2005] ARB/03/08—Award, 426 (450) para 13 (iv); Arbitration under UNCITRAL Rules, in: Peterson (2009b).

[544] ICSID [2006] ARB/05/3, 16–17 para 72; ICSID [2005] ARB/03/08—Award, 426 (449–450) para 13.

[545] ICSID [2009] ARB/06/5—Award, 34 para 85.

[546] ICSID [2009] ARB/06/5—Award, 34 para 85.

[547] ICSID [2008] ARB/98/2—Award, 76 para 232.

[548] ICSID [2008] ARB/98/2—Award, 77 paras 233–234.

[549] ICSID [2010] ARB/07/20—Award, 36 para 111.

[550] ICSID [2010] ARB/07/20—Award, 36 para 111; ICSID [2008] ARB/98/2—Award, 76–77 para 232.

[551] ICSID [2010] ARB/07/16—Award, 110 para 312; ICSID [2010] ARB/07/20—Award, 36 para 111; ICSID [2008] ARB/98/2—Award, 76–77 para 232.

[552] ICSID [2010] ARB/07/16—Award, 110 para 312.

[553] ICSID [2007] ARB/05/10—Award, 34 para 124; '(. . .) often included (. . .)', in: ICSID [2005] ARB/03/29, 37 para 137.

one tribunal dealt with this issue while applying the Salini test and pointed out that it '(...) the commitment of the investor (is important) and not whether he positively contributed to the economic and social development of the host state'.[554]

Overall, the contribution to the economic development should be handled in a flexible, not too strict way. Especially since the last tribunal using Salini elements in 2009 also preferred a more flexible approach. It is difficult to argue that there is no contribution to the economic development of the host state if all three elements are present. Based on the different opinions, to assess the commitment of the investor seems to be the best solution, if the contribution to the economic development of the host state should be regarded as a binding requirement. Therefore, there is a lot of evidence that the liberal approach is more advisable.

3.7.4.4 Case Law Analysis

Even ICSID tribunals admit that the contribution to the economic development of the host state element is the most widely discussed element of the Salini test.[555] Since no tribunal has been able to deliver a definition, it is necessary to review the case law concerning the application of this element.

One tribunal said that monetary magnitude can never be a prerequisite of an investment.[556] Another tribunal viewed a dedication of over a half million Euro before the actual start of the contract as 'financially meaningful.'[557] Some tribunals concluded that infrastructure projects or the service of general interest frequently constitute a contribution to the local economy.[558] In one case of a BOT/Concession power plant, the ICSID tribunal just mentioned that the project contributes to the development of the host state.[559] However, one tribunal underlined that monetary profit for the host state was not enough because the investor must prove 'the spread of knowledge.'[560]

Only one ICSID tribunal faced a case involving a power plant set up as a BOT. The *PSEG case* dealt with a BOT/concession concerning a power plant.[561] The tribunal assessed the concession contract and if it is valid.[562] With the existence of the contract, the tribunal decided that there is an investment.[563] Consequently, the

[554] ICSID [2012] ARB/09/2—Award, 62 paras 306–307.

[555] ICSID [2009] ARB/06/5—Award, 33–34 para 84.

[556] ICSID [2009] ARB/07/21—Award, 11 para 45.

[557] ICSID [2010] ARB/08/8, 61 para 133.

[558] ICSID [2005] ARB/03/29, 37 para 137; Verifying the decision of ICSID [2001] ARB/00/4, in: ICSID [2004] ARB/03/11—Award, 15 para 62; ICSID [2001] ARB/00/4, 609 (623) para 57.

[559] ICSID [2008] ARB/05/12, 40 para 132.

[560] ICSID [2006] ARB/99/7—Annulment, 15–16 para 39.

[561] ICSID [2004] ARB/02/5—Decision, 8–9 paras 18–20.

[562] ICSID [2004] ARB/02/5—Decision, 25–31 paras 79–104.

[563] ICSID [2004] ARB/02/5—Decision, 31 para 104.

tribunal did not apply the Salini test as it categorized the investment as a 'readily recognizable investment'[564]—a concept introduced during the draft of the Executive Directors' Report.[565] However, another tribunal incidentally applied the Salini test on the *PSEG* case to assess its own case. The total investment sum of 804.8 million USD satisfies the criterion of contribution.[566] Concerning risk, the tribunal concluded that despite an adjustment article in the concession contract, the investor bears the risk that the request of adjustment is denied by the state.[567] It also approved the requirement duration, as the project was planned for 38 years.[568] Concerning the element of contribution to the economic development of the host state, it briefly stated that the construction of a power plant is sufficient.[569]

3.7.4.5 The Literature's Opinion

Without a doubt, the definition of contribution to the economic development is of high relevance as it can be the deciding factor of the Salini test.[570] Nevertheless, some authors oppose this criterion.[571] The criterion is a purpose of the investment and not a prerequisite, and the ICSID Convention's Preamble does not offer a different interpretation.[572] In addition, it is impossible to illustrate how the investment actually contributes to the local economy.[573] By definition, 'the economic transaction constituting an investment (. . .) contribute(s) to the economic development.'[574] The Salini test disregards the fact that a contribution to the development of the host state can happen years later.[575] There are several problems connected to contribution to the economic development. Taking the wording of ICSID tribunal's decision into account, by implication this means that all forms of development, except economic development, are excluded to define investment.[576] The proposal that a tribunal should ignore every political and cultural benefit is quite problematic because it the ICSID tribunal to predict future economic developments.[577] Furthermore, ICSID tribunals do not seem to consider tax payments as a contribution to the

[564] ICSID [2007] ARB/05/10—Award, 40–41 para 119.
[565] ICSID [2007] ARB/05/10—Award, 40–41 para 119.
[566] ICSID [2007] ARB/05/10—Award, 41 para 120a).
[567] ICSID [2007] ARB/05/10—Award, 41–42 para 120b).
[568] ICSID [2007] ARB/05/10—Award, 42 para 120c).
[569] ICSID [2007] ARB/05/10—Award, 42 para 120d).
[570] Hamida (2007), 287 (295).
[571] Grubenmann (2010), 123.
[572] Krishan (2009), 1 (15); With further references, in: Grubenmann (2010), 123.
[573] Krishan (2009), 1 (15); cf Loncle (2006), 319 (328).
[574] Krishan (2009), 1 (15).
[575] Krishan (2009), 1 (21).
[576] Hamida (2007), 287 (296).
[577] Johannsen (2009), 5 (20).

local economy, as one tribunal pointed out that besides not contributing anything to the local economy, the investor also removed his income from the tax system.[578] There is also the problem that this term is so undefined that almost every investment of a certain size fulfills the requirement.[579] Thus, there is the claim that the contribution to development criterion should be abandoned.[580] The objective or liberal approach to narrow down investment might lower the attractiveness of the whole ICSID system.[581]

Nevertheless, some authors propose different solutions to the problem. To assess if an investment contributes to the economic development, there should be a double test, including an 'economic test' and a 'development test.'[582] The 'economic test' assesses whether the investment is more likely to lead to an increase in the real per capita income of the host country, and the 'development test' assesses whether the project is likely to contribute to public goods (e.g. better education, better health standards, less poverty, and a cleaner environment).[583] Others propose that the only analysis should be if the project is according to the local laws, because ICSID tribunals are not able to fully assess the presumable economic development.[584]

Finally, there is Schreuer, who proposes a completely different approach. The Convention's object and purpose is to support the host state's development. However, if an international transaction is designed to promote development, it can be viewed as an investment.[585] It would be wrong to look at it from the other way and claim that an international transaction which does not obviously contribute is automatically excluded from ICSID Convention's protection.[586] Schreuer mentions that:

> [I]t should not be restricted to measureable contribution to GDP but should include development of human potential, political and social development and the protection of the local and global environment.[587]

Overall, the literature is as diverse on this topic as the ICSID tribunals. The only well established similarity is that the element of contribution to the economic development of the host state is difficult to handle. The best recommendation concerning the presumption is offered by Schreuer. Nevertheless, the fact remains that some tribunals still use this Salini element. Hence, there is a risk that the Desertec Concept will face this criterion. Therefore, the next section offers a

---

578 ICSID [2006] ARB/99/7—Annulment, 17 para 44.

579 Grubenmann (2010), 123.

580 Loncle (2006), 319 (328).

581 Hamida (2007), 287 (304).

582 Hamida (2007), 287 (297).

583 Hamida (2007), 287 (297).

584 Hamida (2007), 287 (297).

585 Schreuer (2009), 134 para 173.

586 Schreuer (2009), 134 para 173.

587 Schreuer (2009), 134 para 174.

catalog of what contribution to the economic development of the host state entails, and a guideline for the Desertec Concept to comply with this requirement.

3.7.4.6 Contribution to the Economy in Other Treaties and Conventions

There also other conventions or treaties requiring a contribution to the local economic and development. First of all, there are treaties like the 1993 Framework Cooperation Agreement between the EEC and the Republics of Costa Rica, El Salvador, Guatemala, Honduras, Nicaragua and Panama, which includes provisions on the contribution to the development.[588]

Next, there is the MIGA Convention. Art. 12(d) of the MIGA Convention limits MIGA Convention application to investments which contribute to the economic and social development of the host country.[589] In the context of MIGA practice, there is a difference between a subjective and an objective contribution. The 2009 Operational Regulations of MIGA stipulate this difference. The objective contribution encompasses the potential to generate revenue for the host state, maximizing the host state's productive potential, producing exports or import substitutes and reducing vulnerability to external economic changes, the diversification the economic activities, impact on employment and income, the amount of knowledge transfer, the effect on the social infrastructure and the environment.[590] The scope of the subjective contribution includes compliance with the host states laws and projects consistency with its declared local development objectives.[591]

Since MIGA is part of the World Bank, like the ICSID, there is a possibility that either side may make use of each other. In 2013 an ICSID tribunal actively used parts of the MIGA Convention for the first time.[592] Although they did not discuss Art. 12(d) MIGA, this illustrates that the possibility of future usage is significantly higher. The protection according to MIGA is desirable for the investor. This protection is only possible if the investor complies with Art. 12(d) of the MIGA Convention. The investor might not be faced with the issue of contribution to the economic development of the host state within ICSID arbitration, but if the MIGA Convention applies. This also underlines that the need for the Desertec Concept to comply with these standards. Yet neither the ICSID nor MIGA Convention offer proper definitions.

---

588 As the treaty regulates the obligations to e.g. train staff, in: Zampetti and Sauvé (2007), 211 (239–241).

589 Commentary on Convention – Establishing the Multilateral Investment Guarantee Agency (2010), para 21.

590 MIGA (2012), 47 para 3.06.

591 MIGA (2012), 48–49 para 3.09.

592 ICSID [2013] ARB/07/30—Decision on Jurisdiction, 66–67 para 246.

3.7.5 Desertec Concept and Contribution to the Economic Development Element

There can be a lot of tensions between the state and a private operator in the field of the service of general interest, as it is a core element of the state. This is mainly due to the fact that the transfer of operation to a private operator includes a "loss of sovereignty." Since a lot of cooperation agreements are set up as concessions, this difficult situation often applies to concession treaties.[593] ICSID case law indicates that there seems to be no problem with large scale infrastructure projects and power plants. This also finds some support within the literature.[594]

3.7.5.1 Intentions of the Parties

First of all, the intentions of the parties might be relevant. Both parties have different intentions if it comes down to an investment. The state party normally desires as much contribution to its own economy, whereas the investor desires as much profit as possible. Thereby, the investor is not as interested in the contribution to the state, as long as the investment pays off.

As mentioned above, there have been some indicators that Dii plans to primarily supply Europe and not the MENA region. Besides the existing figures, which only deal with export to Europe, there are some Dii documents which can be interpreted this way. Such an intention could already exclude the fulfillment of the requirement contribution to the economic development of the host state. If the investor is not planning to primarily contribute to the host state market, it is not considered to be an investment according to the Salini test. There is a chance that this might also apply to other investors. Any investor's main goal would be to refinance its investment and it is likely that this would be achieved by focusing on the European market.

However, the intention of the investor is not to evaluate whether the primary source meets the Salini requirements. The state will always maintain its ground that the primary goal of the investment must be to contribute to its economy. On the other hand, the investor will try everything to keep its investment (e.g. power plant) running. In the case of a solar thermal power plant, like the Desertec Concept, long-term success is only possible if Europe is integrated into the grid. Therefore, the intention of Dii was not as odd as it might seem. Furthermore, it can be very difficult to prove the real intentions of the investor as it is rare to find a document evincing it. The only real proof of the intention of the parties can be the concluded contracts (e.g. concession and state contract). This is the "only" contract clarifying the relationship between the investor and the state relating to all investment matters. There might be other indicators, but these might be only hearsay evidence.

[593] Karl (2003), 37 (44).

[594] Hauschka (2005), 1550 (1555).

Intentions of the parties can only a secondary tool to determine, if there is a contribution to the economic development of the host state.

3.7.5.2 Desertec Concept as a Contribution and Its Benefits

The primary goal of Dii was to supply Europe and it is most likely that any investor pursues the same target. This means that MENA countries would not be the prime beneficiaries of the project. It would instead serve the energy needs of Europe. Nevertheless, MENA and therewith Morocco would benefit from the project. Even if Morocco is not the primary market of the Desertec Concept, there would be still positive side effects.

Previously, ICSID tribunals accepted BOT/Concession power plants as a contribution to the economic development of the host state. Similarly the provision of a service of general interest was also regarded as a contribution. However, if the Desertec Concept does not primarily supply the MENA region, there could be doubt about the applicability of both decisions. One tribunal dealt with the issue of transit highways and the profit of third parties. In this case, there is a possibility that the neighboring countries benefit more than the host state. The ICSID tribunal mentioned that facilitation of land transportation and strengthening a country's position as a transit country within a region is a contribution to the economic development.[595]

Due to the fact that there is still some time before HVDC transmissions lines exist all over Europe, there should be a focus on the supply of the MENA region.[596] Therefore, it would be best for an investor to keep its promises (e.g. open technology) and implement them.[597] Furthermore, there might be a legal obligation for an investor to supply Morocco (and MENA countries). As mentioned above, the contracts between the parties might be evidence of a contribution to the economic development. The three main contracts are the concession contract, the state contract and the PPA. Therewith, an investor legally binds itself to its promises concerning MENA supply, if this obligation is part of these contracts. Vice versa, these contracts (especially a PPA) create a legal obligation for Morocco/ONE to purchase the produced energy. In addition, there are the concession fee and tax payments of the Desertec Concept. As mentioned above, the best way to set up the Desertec Concept would as a BOT/Concession. This would include a concession fee paid to the public party. Due to the concession fee, the state of Morocco would also benefit even if the energy is primarily produced for Europe. Both parties can also agree within the concession agreement that the prime function of the project is the supply of the MENA (e.g. Morocco) region. Therewith, the ICSID decisions concerning BOT/Concession power plants and service of general interest would

[595] ICSID [2009] ARB/07/12, 28 para 86d).

[596] Deutsch and Hobohm (2010), 54 (58).

[597] Deutsch and Hobohm (2010), 54 (58).

also apply to the Desertec Concept. Although there are taxes which must be paid by the investor so that it may run the project, these were not considered by former ICSID tribunals.

3.7.5.3 Tests According the Literature

According to Schreuer's idea, it is important that the design of the investment includes development promotion. So far, the intention of an investor can be understood in both ways, which makes it important to clarify in the concession and state contract that the investment's main purpose is the promotion of development. The project's setup should be adjusted to the needs of Morocco (and the MENA region). If the setup already indicates that the main focus is Europe, a project might not be considered to be an investment.

According to the double test, which performs an 'economic test' and a 'development test', the Desertec Concept is also an investment. Due to the concession fee, Morocco is likely to increase its real per capita income and the application of solar thermal energy leads to a reduction in conventional power plants. Consequently, a cleaner environment and better health would be the result. This is under the condition that the Desertec Concept produces energy for Morocco and not only Europe. The last proposed test only demands that the project be in accordance with the local laws. At this point of time, there can be no statement concerning the consideration of contribution.

3.7.5.4 Art. 12(d) of the MIGA Convention

As mentioned above, the MIGA Convention is not directly applicable, but it offers comprehensive guidance. According to Art. 12(d) of the MIGA Convention, contribution to the economic development of the host state encompasses an objective and a subjective element.

The Desertec Concept can fulfill several requirements of the objective element. It has the potential to generate revenue for the host state. Due to the concession fee and tax payments, the Desertec Project will produce revenue for the state of Morocco. The utilization of sun power also helps to reduce vulnerability to external economic changes and the diversification. Down to the present day, Morocco's economy is still heavily reliant on few products (e.g. fabric industry), and energy creation mainly depends on oil. Due to the increasing oil price, which is frequently subject to changes due to the international crisis, Morocco's economy relies on external factors. The same applies to the economy; because of cheap textile products from Asia, the main export branch is under pressure. A new solar thermal power plant and the possibility to export energy after the end of the concession period would help to diversify products. Furthermore, the Desertec Concept would also have an impact on employment and income. Studies predict that there could be up to 120,000 jobs created during construction, and during manufacturing parts of the plant (e.g. mirrors) up to 80,000 people per 20 GW.[598] The number of jobs created because of a joint venture is significantly higher if production of all the parts is within the host state.[599] In addition, it is vital to train and educate African engineers.[600] The training of African engineers would also be a part of the knowledge transfer, which is frequently demanded by ICSID tribunals. Finally, the Desertec Concept could also have positive effects on the environment. There is no question that the replacement of conventional power plants by solar thermal plants takes pressure off the environment. The cutback on CO_2 emissions and thus cleaner air are just two examples of the benefits for the environment. The support of environmental protection through new technology is also within the interest of the developing country and the international community.[601]

Secondly there is the subjective element. The scope of the subjective contribution includes compliance with the host state's laws and compliance with declared local development objectives. As mentioned above, compliance with the law cannot be assessed at the moment. However, the second requirement is important as it requires the investor to keep to its promises and declarations. Since the Desertec Concept is a different kind of project, this requirement might be more relevant than in other investments. It underlines the need that an investor primarily supplies Morocco and not Europe. The concentration throughout the whole project must be on the MENA region, respectively the host states. If an investor fails, there is a risk that it will not be considered an investment anymore. This might also happen despite the concession fee. Nevertheless, if an investor keeps it promises, the Desertec Concept would also fulfill the subjective requirement. Assessment of these elements by ICSID tribunals might be difficult, as economic development is mostly a political term and difficult to assess. The ICSID Convention focuses on legal questions and does not deal with national policies. Nevertheless, it offers a guideline in a field where legal political issues overlap.

3.7.5.5 Cost Efficiency

One problem remains, and this is connected to the concession fee. As mentioned above, the produced energy from the Desertec Concept will cost more at the beginning than conventional energy. Hence, it is necessary to support solar thermal energy with feed-in tariffs or subsidies. In both cases, the state must pay this money itself, which could lead t the fact that the benefits Morocco has from the concession fee and taxes could be consumed by the need to subsidize the project. The fact that the need for subsidies exists prior to the start of the project is rather unusual because it indicates that the product is not at all competitive. This is almost contrary to the

[598] Komendantova (2010).
[599] Komendantova (2010).
[600] Saß (2009), 27 (30).
[601] Herdegen (2003), 13 (29).

purpose and idea of an investment because no host state would purposely accept a project which needed public financial aid.

This might already exclude the requirement of contribution to the economic development, which means that the Desertec Concept is something completely new in the field of investment law. This is a particular problem in Morocco because the price of solar thermal energy is twice as high as the average price.[602] Without these subsidies, there is only a small chance that solar thermal energy could be competitive. If Morocco/ONE agree to buy the produced energy (e.g. PPA), they will have an obligation to purchase. This may be a problem for both parties, especially at the beginning of the Desertec Concept. Since there are no transmission lines between MENA countries and Europe at the moment, an investor cannot export the energy. Hence, an investor must focus at first on the MENA region. The same problem arises for the economy and export. The domestic economy must also pay higher prices, which leads to increasing production costs and consequently to more expensive products, which are difficult to sell and export. In addition, the solar thermal energy price is also considered to be expensive in other countries. These countries also need to subsidize the energy to make it compatible. This is another example of how Morocco depends on external factors, creating a vicious circle. If Morocco does not subsidize the energy, it will harm its own economy and if it does subsidize it, it creates a burden on its public finance. In both cases, it is difficult to talk about a contribution to the economic development of the host state if there is nothing left of the concession fee.

One possible solution to the problem would be if third countries supported the implementation of the Desertec Concept. As mentioned above, Morocco and the EU have a lot of cooperation agreements which might offer additional financial aid. The EU has a lot of programs devoted to assist developing countries and supports external ones as well, e.g. the UM. Once it is up and running, it could financially assist MENA countries. Therewith, a combination of these tools might offer a way to subsidize the Desertec Concept within the first years. Morocco also applied for funds of the UNFCCC and cooperates closely with the World Bank, as mentioned above. Furthermore, most European actors have declared that they are willing to support the Desertec Concept. Different parts of the World Bank have also showed their support for clean technology and the Desertec Foundation and the newly founded Desertec University Network could be helping to make the project public and create additional support. External financing would help to relieve the Moroccan finances and thus turn the Desertec Concept into a contribution for the host state. Nevertheless, it is difficult to predict how the project will develop and whether all the third parties will keep their financial aid promises. Recent negotiations concerning a follow-up protocol for Kyoto have illustrated again that there is a difference between making and keeping a promise.

[602] Since Morocco has the highest energy prices at the moment within the MENA region, costs in other MENA countries would be even higher.

A second approach is a different structuring of the Desertec Concept, excluding a concession contract. This might open a new box of discussions concerning the contribution to the economic development. If there is no concession, there would be no concession fee and a major benefit for Morocco would be missing. Consequently, the question of "who is the primary benefiter" would rise again. This includes uncertainties like "must the host state benefit from an investment directly or indirectly?" Overall, a missing concession would make the whole situation even more complicated for an investor.

3.7.5.6 Evaluation of the Desertec Concept

The topic of contribution to the economic development of the host state is difficult to assess, especially as there are no data available and neither Dii, as an example of an investor, nor the Desertec Foundation were willing to offer any information. Due to the lack of material the following evaluation is based on own considerations and evaluations.

It is no surprise that a lot of ICSID tribunals and most of the literature have rejected this element. Even tribunals supporting it did not have a consistent case law concerning this Salini element. Based on case by case decision practice, most of the tribunals accepted the criterion contribution to the economic development of the host state without a detailed observation. Cases of a denial were frequently subject to successful annulments. Due to the superficial assessment of ICSID tribunals, it is almost impossible to predict a possible outcome. Taking the benchmark of the ICSID tribunal's decision into account, there is a great chance that the Desertec Concept is a contribution to the economic development of the host state. This is also supported by the idea that a project's contribution increases equivalent to its duration. The same applies for the proposed tests by the literature. So far, ICSID tribunals have not referred to Art. 12(d) of the MIGA Convention in any verdict. Therefore it is not clear whether they do not accept it as an indicator, or just did not explicitly consider it. According to the recent case law, the chances that ICSID tribunals might consider this article are rather small. The fact that the Desertec Concept differs from earlier projects could be the reason for taking different aspects into account. However, there is no certainty. It must be assumed that the ICSID tribunals do not refer to Art. 12(d) of the MIGA Convention. This article only serves as possible guidance to solve the problem, which deserves attention and the Desertec Concept is a contribution to the economic development of the host state at the moment. The issue of cost efficiency might be problematic as it is uncertain which way the parties will head. A financial support by third parties would still be helpful, to prevent a possible consumption of whole concession fee.

The question remains at what point the drawbacks outweigh the benefits. One might say that if subsidies consume the concession fee completely, there is no contribution to the economy anymore. However, this overlooks the fact that Morocco will gain a new power plant which produces clean energy. Since Morocco is heavily on oil for energy production, this is an advantage. Besides less CO_2

emissions, Morocco will also save money which it needs to import oil and production of energy will be much safer and more reliable. A big scale plant or several small scale plants create new jobs. It comes down on the issue of how much extra money (besides the concession fee) Morocco must spend to keep the Desertec Concept profitable. This is difficult to answer as individual preferences prevail. Some might say that 10 % of the concession fee on top of the fee excludes a contribution, and others might claim 50 %. Taking the importance and potential of the Desertec Concept into account, these figures should be set at the higher end of the percentage scale. The biggest gain from this project could be the replacement of nuclear energy as the main energy source of the (developed) world. There are other forms of renewable energy, but only solar thermal energy has the capability to replace nuclear efficiently. The fact that the project will have a breakeven point in a couple of decades also supports this assumption. Consequently, the figures should be at 50–75 % of the concession fee as an additional subsidy. Beyond this point, it is difficult to argue that the project is a contribution to the local economy.

If there is a financial support by third parties, these figures could be adjusted. So far, ICSID tribunals have not been faced with this issue. Since ICSID tribunals do not (officially) support renewable energies, their decision is likely to be "purely" objective. This means that they are not going to rule in the favor of the project just because of its status as renewable energy. ICSID case law shows no signals that subsidies or any other support tool of the host state exclude a categorization as an investment. The only difference is the certainty that the Desertec Concept is going to need subsidies right from the start. Normally an investor builds a project and later on finds out that it needs financial support. However, as long as the subsidies do not completely consume all benefits from the concession fee and tax payments, there is a great chance that ICSID tribunals will accept it as an investment, because it also contributes to the economic development of the host state. Referring back to the issue of public policy, the Desertec Concept would contribute financially and economically to Morocco. In the case that an investor sticks to its plan to primarily supply the MENA region; the Islamic community would profit. Consequently, Islamic states would have the chance to generate an infinite amount of clean energy. In summary, the Desertec Concept would not be put at risk due to public policy.

3.7.6 Assets Invested in Accordance with the Laws of the Host State

The requirement of 'in accordance of the laws of the host state' has been around for a long time within ICSID case law. However, it was mostly a question of the merits. The *Phoenix Action Ltd. v The Czech Republic* tribunal was the first one to claim that it is part of the definition of investment. Its relevance is also underlined by the fact that Art. 12(d) of the MIGA Convention considers this as a part of the subjective element of contribution.

3.7.6.1 Requirement as Part of Art. 25(1) of the ICSID Convention

Based on this decision, it would be part of Art. 25(1) of the ICSID Convention and the jurisdiction of the ICSID tribunals. The *Phoenix Action Ltd. v The Czech Republic* ICSID tribunal pointed out that the ICSID convention does not protect investments which violate host state laws.[603] The laws of the host, which were in force at the moment of the establishment of the investment, are relevant.[604] However, the tribunal mentions that:

> [T]he State is not at liberty to modify the scope of its obligations under the international treaties on the protection of foreign investments, by simply modifying its legislation or the scope of what it qualifies as an investment that complies with its own laws.[605]

There are also other tribunals which seem to support the approach that this is part of jurisdiction,[606] because '(a) failure to comply with the national law to which a treaty refers will have an international legal effect.'[607] One ICSID tribunal denied jurisdiction because of a violation of the criterion 'in accordance with the law.'[608] However, the tribunal restricted its approach. If a violation of law takes place after the investment was initiated and the process of investing was according to the law, this does not lead to a denial of ICSID jurisdiction.[609] To violate national law, there must be certain prerequisites according to the ICSID tribunal.[610] Another example is a flight terminal project in the Philippines. After a lot of complaints by local enterprises, the government became convinced that the contract with Fraport violated public policy.[611] In 2003, the Philippine Supreme Court concluded that the contract did violate the public policy and therefore was nugatory *ex tunc*.[612] The ICSID tribunal came to the decision that a project which violates national law is not an investment according to the respective BIT.[613]

3.7.6.2 Requirement as an Issue of the Merits

The question of whether the requirement of 'in accordance with the laws of the host state' is an issue of jurisdiction is highly disputed. The ICSID Convention is neutral

603 ICSID [2009] ARB/06/5—Award, 39 para 100.

604 ICSID [2009] ARB/06/5—Award, 40–41 para 103.

605 ICSID [2009] ARB/06/5—Award, 40–41 para 103.

606 ICSID [2010] ARB/07/20—Award, 39 para 121; Concerning violation of the BIT, in: ICSID [2007] ARB/03/25—Award, 141 para 306 and 153 para 323.

607 ICSID [2007] ARB/03/25—Award, 187 para 394.

608 ICSID [2007] ARB/03/25—Award, 188–190 paras 396–404.

609 ICSID [2007] ARB/03/25—Award, 164–165 para 345.

610 ICSID [2007] ARB/03/25—Award, 185 para 387 and 188–189 para 396.

611 Borris and Hennecke (2008), 49 (51).

612 Borris and Hennecke (2008), 49 (51).

613 Cabrol (2008), 796 (798); Borris and Hennecke (2008), 49 (51).

on the question of 'legality' of the investment.[614] Conferring to domestic law is not a qualification of an investment as such, but a '(. . .) condition of its validity under domestic law.'[615] Similarly, the Salini ICSID tribunal ruled, as it pointed out that 'accordance with the law' is not part of the notion of investment.[616] Thus, the question of violation of national law is an issue related to the merit of the decision and not an issue of ICSID jurisdiction.[617] Other tribunals decided differently on this issue. Some pointed out that such a case concerns the legitimacy of the investment, but not the jurisdiction of the tribunal.[618] Again others decided if the investment is against the law of the state, there is no consent of the state to participate in an ICSID proceeding, and thus no jurisdiction.[619] Additionally, it is also highly disputed whether a violation of host state law would lead to a denial of BIT investment protection.[620] To exclude every investment from BIT protection which is does not completely conform in a legal sense could result in an unwanted erosion of the BIT application.[621]

The requirement of 'in accordance with the law' is also beneficial because it helps to "limit" the effects on sovereignty.[622] If the investment is in accordance with the local laws, there is no infringement of state sovereignty. Nevertheless, problems remain related to the correctives of host state law violation. First, the good faith of the investor is hard to prove and the severity of the violation is irrelevant.[623] Second, the state can pass non-discriminatory laws as 'emergency reserve' prior to the investment and use them later on to influence the investment.[624] Third, the fact that the deliberate toleration is a subjective approach can cause problems as the pure impartial enforcement of a law might not be covered.[625] Finally, the issue of a violation while initiating an investment is difficult to assess, as there is no agreement when an investment starts.[626]

There is no doubt that a state cannot invoke its own breach of the law to avoid BIT jurisdiction.[627] There is a legitimate expectation of the investor because of the

---

[614] ICSID [2010] ARB/07/20—Award, 37 para 114.

[615] ICSID [2010] ARB/07/23, 53 para 140.

[616] ICSID [2001] ARB/00/4, 609 (620–621) para 46.

[617] ICSID [2007] ARB/03/25—Award—Dissenting Opinion of Bernardo M. Cremades, 22 para 38; Borris and Hennecke (2008), 49 (55); Tietje (2010), 5 (16–17).

[618] ICSID [2005] ARB/03/29, 30–31 paras 109–114; ICSID [2001] ARB/00/4, 609 (620–621) para 46.

[619] ICSID [2006] ARB/03/26—Award, 50–52 paras 168–173 and 100–101 para 332

[620] Tietje (2010), 5 (10).

[621] Restraining investment could be against the will of parties (according to the BIT), in: Hobe and Müller (2009), 65 (68); Borris and Hennecke (2008), 49 (56).

[622] Maier (2010), 95 (104).

[623] Borris and Hennecke (2008), 49 (56–57).

[624] Borris and Hennecke (2008), 49 (57).

[625] Borris and Hennecke (2008), 49 (57–58).

[626] Borris and Hennecke (2008), 49 (58).

[627] ICSID [2007] ARB/05/18, 49–50 para 182.

performance of governmental authorities.[628] Hence, the principle of estoppel is applied by the ICSID tribunals. Estoppel in international law means that a state is bound to the raised expectations due to its behavior, and the opposite party can rely on it according to good faith.[629]

Overall evaluation of this requirement is quite difficult. This is also due to the fact that other international conventions also include this prerequisite concerning investment. The most prominent example is Art. 12(d) of the MIGA Convention.[630] Since the MIGA is a very useful tool to secure investments, the investor should make a considerable effort to meet all requirements. There can be no doubt that the violation of national law does not matter. Nevertheless, the issue does not need to be determined, since a violation of national law causes problems for the investor. Concerning the Desertec Concept, it is important that there is a previous assessment of the national law. In Art. 1(1) of the G/M-BIT it is stipulated that the investment must be in accordance with the law of the contracting state. Despite the fact that this supports social acceptance of the project, it prevents serious mistakes. It would be beneficial to include local lawyers or legal professionals to secure that there is enough knowledge concerning the local laws. If an investor violates host state law, there is a high risk of it not being included under the ICSID protection umbrella.

3.7.7 *Assets Invested* **Bona fide**

The Salini decision did not include this requirement. In 2009, another tribunal introduced it. The main idea is that ICSID jurisdiction cannot apply to investments not made in good faith.[631] International investment arbitration cannot take place if this would be against general principles of international (e.g. good faith), whereby good faith means '(. . .) to deal honestly and fairly with each other, to represent their motives and purposes truthfully, and to refrain from taking unfair advantage (. . .).'[632] Concerning good faith there is a dual approach, as it is said that a violation of good faith automatically violates national and international law.[633] However, another tribunal highlighted that Art. 25(1) of the ICSID Convention does not

[628] ICSID [2007] ARB/05/18, 52 para 191; ICSID [1992] ARB/84/3—Award [1992] ICSID Review, 328 (352) paras 82–83; Verdross and Simma (2010), 801–803 paras 393–394 para 615; cf ICJ [1998] Land and Maritime Boundary between Cameroon and Nigeria—Preliminary Objects, 275 (303) para 57.

[629] Verdross and Simma (2010), 801–803 paras 393–394 para 615; cf ICJ [1998] Land and Maritime Boundary between Cameroon and Nigeria—Preliminary Objects, 275 (303) para 57.

[630] Commentary on Convention – Establishing the Multilateral Investment Guarantee Agency (2010), para 21.

[631] ICSID [2009] ARB/06/5—Award, 42 para 106.

[632] ICSID [2009] ARB/06/5—Award, 42 paras 106–107.

[633] ICSID [2009] ARB/06/5—Award, 44 para 112; ICSID [2008] ARB/03/24—Award, 42 paras 143–144.

differentiate between legal and illegal, good faith or not concerning an investment.[634] There is no answer to this requirement because no investment has taken place so far. Therefore, it is impossible to say whether the investment is in done *bona fide*.

3.7.8 *Regularity of Profit and Return*

Lastly, there is the requirement of regulatory of profit and return, introduced by the ICSID tribunal in the case *Joy Mining v Egypt*.[635] The Salini decision did not mention this element within its decision. However, another ICSID tribunal pointed out that this criterion is not essential in every case.[636] Another tribunal also checked it in its decision.[637] One tribunal explicitly rejected this criterion, although it applied the liberal approach.[638]

The contractual setup as a BOT enables the investor to get regular profits and returns because he can charge an appropriate tariff.[639] Since no investment has taken place so far, it is impossible to check if regularity of profit and return exist. Based on the amount of money and the intention of an investor, it can be expected that this requirement is met. In addition, the contractual setup of the Desertec Concept, as proposed above, also supports the existence of regular profits and returns.

3.7.9 *In the Territory of the Host State*

Although 'territory of the host state' is not mentioned as a requirement of an investment according to Art. 25(1) of the ICSID Convention, it is an important requirement of investment. The basic logic of investment says that the investment must be within the territory of the host state.

3.7.9.1 Case Law Concerning Territory Requirement

In the case of pre-shipment inspections, Pakistan claimed that the investment did not take place in its territory.[640] These pre-shipment inspections were partly

---

[634] ICSID [2010] ARB/07/20—Award, 36 para 112.

[635] ICSID [2004] ARB/03/11—Award, 12 para 53.

[636] ICSID [2007] ARB/05/10—Award, 35–36 para 108.

[637] ICSID [2010] ARB/08/20, 28–29 para 80.

[638] ICSID [2012] ARB/09/2—Award, 62 para 305.

[639] ICSID [2007] ARB/05/10—Award, 42 para 120 e).

[640] ICSID [2003] ARB/01/13, 307 (321–322) para 46.

conducted outside of Pakistan territory.[641] Although the investor's expenditures were small, the ICSID tribunal concluded that they were in connection with the territory of Pakistan.[642] Another tribunal in a similar cases mentioned that 'in accordance with normal principles of treaty interpretation, investments made outside the territory of the Respondent State, however beneficial to it, would not be covered by the BIT.'[643] One tribunal also highlighted that nothing prevents an investment from being connected to the investor's home state, as long as they are allocated to the project to be carried out abroad.[644] In a case, where an immovable property is located within the territory of the host state, it is not really controversial if it is within the territory.[645] Concerning issues of financial business, ICSID tribunals have a different point of view. In these cases, it is not unusual that the funds involved are not physically in the host country.[646]

3.7.9.2 Requirements of the G/M-BIT

It is also important to check the respective BIT concerning the territoriality requirement.[647] If a BIT includes such a requirement, the ICSID tribunal lacks jurisdiction when the investment was done outside of state's territory.[648] As one tribunal highlighted, '(...) an injection of funds into the territory of the host State in connection with an investment will typically satisfy the requirement (...).'[649] Art. 1(1) G/M-BIT also emphasizes this requirement. Otherwise the BIT does not regard it as an investment.

There is no doubt that an investor plans to build a solar thermal power plant in the MENA countries. Morocco is the first country to be considered. Unlike the case of pre-shipment inspection or financial businesses, a power plant is immovable construction. Consequently, the Desertec Concept fulfils the requirement of physically being in the territory of the host state.

---

641 ICSID [2003] ARB/01/13, 307 (310) para 11.

642 cf for pre-shipment inspection in the Philippines, in: ICSID [2004] ARB/02/6, 43 paras 111–112; ICSID [2003] ARB/01/13, 307 (348) para 136.

643 ICSID [2004] ARB/02/6, 38 para 99.

644 ICSID [2006] ARB/05/3, 17–18 para 73(i).

645 Knahr (2009), 42 (52).

646 ICSID [2013] ARB/08/9—Decision on Jurisdiction and Admissibility, 167–171 paras 496–510; ICSID [2010] ARB/08/8, 56–57 paras 123–124; cf ICSID [1999] ARB/97/4—Decision, 251 (276–277) paras 75–77; ICSID [1997] ARB/96/3, 1378 (1386) para 41.

647 ICSID [2010] ARB/07/16—Award, 97–98 para 279; ICSID [2010] ARB/08/8, 51 para 114; ICSID [1997] ARB/96/3, 1378 (1386) para 41; Permanent Court of Arbitration [2009] PCA Case No. AA280, 60–61 para 237.

648 ICSID [2010] ARB/08/8, 54 para 120.

649 ICSID [2010] ARB/08/8, 56 para 123.

3.7.10 Assessment Results of Investment Protection

Overall, the Desertec Concept is within the scope of the ICSID Convention and its protection of investments. This also includes the applicability of the G/M-BIT as a basis to formulate any claims. Both parties must negotiate and conclude their contracts in a careful way. The state contract and the concession contract in particular need a lot of work. In both cases, it is important to clarify the parties to it. Application of the umbrella clause and the question of contract claims make it necessary to know the exact parties of each contract. Furthermore, it is of great relevance that both contractors include the exact obligations of an investor. As shown above, this forms a part of the acceptance as an investment. It would be wrong to claim that a previous review of all possibilities can happen. In 2010, one ICSID tribunal pointed out that there is still no universal approach so far concerning the notion of investment.[650] This remains one of the biggest problems of investment law besides the question of stare decisis.

Although there seems to have been a decrease in the application of a strict Salini test, the risk remains. Thereby the risk is not the test itself, but the fact that no-one really knows the prerequisites for the different elements. The study above illustrates that the Desertec Concept is an investment according to the strictest application of the Salini test. It is not possible to identify and solve all of the remaining uncertainties. This is mostly due to the unpredictability of the ICSID tribunals' setup and their interpretation of the BIT and the ICSID Convention. Hence, a lot of facts are in favor of the Desertec Concept.

One of the main points which can influence the outcome of any ICSID proceeding is the behavior of the investor and the state. The state, in this case Morocco, should avoid using purely nationalist reasons to "influence" the project. ICSID case law revealed that an abuse never pays off for any party. Concerning the Desertec Concept, the investor has an even greater potential to influence arbitration. Art. 12 (d) of the MIGA Convention as well as the parts of the literature mentioned that the behavior of the investor can be decisive for the element of contribution to the economic development of the host state. Therefore, an investor must stick to its promises. Besides the fact that different behavior could cause serious problems related to the acceptance of the project, it can be decisive for arbitration. Since the Desertec Concept is a new and unknown project in ICSID history, there is great chance that the ICSID might assess it in a more detailed way. Hence, it is up to the investor how the tribunal might treat the investment. In summary, the Desertec Concept is an investment according to all approaches taken by ICSID tribunals. In the case of a strict objective approach, it is within the investor's sphere of influence, if the investment is according to the BIT and the ICSID Convention ICSID jurisdiction. Nevertheless, the project still depends on a lot of support from third countries or institutions like the EU or the UM.

---

[650] ICSID [2010] ARB/07/20—Award, 32 para 97.

References

Abraham DC (2009) Arbitration of investment disputes: a Malaysian perspective. Arbitration 75:206–212

Alvarez JE (2003) The new dispute settlers: (half) truths and consequences. Tex Int Law J 38:405–444

Andreeva Y (2008) Salvaging or sinking investment? MHS v. Malaysia revisited. Law Pract Int Courts Tribunals 7:161–175

Aust A (2007) Modern treaty law and practice, 2nd edn. Cambridge

BDI (2011) Wirtschaftsreformen müssen im südlichen Mittelmeerraum oberste Priorität haben. http://www.bdi.eu/dbwt2013/Mittlerer-Osten_Interview-mit-dem-EUAbgeordneten-Caspary.htm. Accessed 24 Apr 2014

Bhala R (1999) The myth about stare decisis and international trade law (part one of a trilogy). Am Univ Int Law Rev 14:845–956

Binder C (2009a) Changed circumstances in the investment law: interfaces between the law of treaties and the law of state responsibility with a special focus on the Argentine crisis. In: Binder C, Kriebaum U, Reinisch A, Wittich S (eds) International investment law for the 21st century – essays in honour of Christoph Schreuer. Oxford, pp 608–630

Binder C (2009b) Die Veränderung innerstaatlicher Verhältnisse als Nichterfüllungsgrund von völkerrechtlichen Vertragspflichten. Archiv des Völkerrechts 47:187–219

BMWi (2014) Übersicht über die bilateralen Investitionsföderungs- und schutzverträge (IFV) der Bundesrepublik Deutschland. http://www.bmwi.de/BMWi/Redaktion/PDF/B/bilaterale-investitionsfoerderungs-und-schutzvertraege-IFV,property=pdf,bereich=bmwi,sprache=de,rwb=true.pdf. Accessed 10 Apr 2014

Borris C, Hennecke R (2008) Das Kriterium der Einhaltung von Vorschriften nationalen Rechts in ICSID-Schiedsverfahren – Anmerkungen zum Schiedsspruch in der Sache Fraport v Philippines. SchiedsVZ 49–58

Broches A (1966) The convention on the settlement of investment disputes: some observations on jurisdiction. Columbia J Transnat Law 5:263–280

Brower CN, Ottolenghi M, Prows P (2009) The saga of CMS: res judicata, precedent, and the legitimacy of ICSID arbitration. In: Binder C, Kriebaum U, Reinisch A, Wittich S (eds) International investment law for the 21st century – essays in honour of Christoph Schreuer. Oxford, pp 843–864

Cabrol E (2008) International investments - law and practice. Int Bus Law J 796–816

Commentary on Convention – Establishing the Multilateral Investment Guarantee Agency (2010) III. Operations. http://www.miga.org/documents/commentary_convention_november_2010.pdf. Accessed 05 May 2014

Commission JP (2007) Precedent in investment treaty arbitration. J Int Arbitr 24:129–158

Crawford J (2008) Treaty and contract investment arbitration. Arbitr Int 24:351–374

Delaume GR (1966) Convention on the settlement of investment disputes between states and nationals of other states. Int Lawyer 1:64–80

Desbordes R, Vicard V (2009) Foreign direct investment and bilateral investment treaties: an international political perspective. J Comp Econ 37:372–386

Deutsch M, Hobohm J (2010) Strom aus der Wüste – Ja, aber aus welcher Quelle. Energiewirtschaftliche Tagesfragen 54–59

Deutsche Industrie- und Handelskammer in Marokko (2008) Marktprofil Marokko. http://www.een-bayern.de/een/inhalte/Laender/Anhaenge/Marktprofil-Marokko.pdf. Accessed 17 July 2010

Diehl AN (2007) Juris-conference "investment treaty arbitration: a debate and discussion" und ICSID/ECT/SCC-Konferenz in Washington, D.C. SchiedsVZ 268–273

Dolzer R, Schreuer C (2008) Principles of international investment law. Oxford

Dörr O, Schmalenbach K (2012) Vienna convention on the law of treaties. Berlin

Ebner T (2005) Streitbeilegung im Welthandelsrecht. Tübingen

Egonu MI (2007) Investor-state arbitration under ICSID: a case for presumption against confidentiality? J Int Arbitr 24:479–489

Europäische Kommission (2010) Vorschlag für eine Verordnung des Europäischen Parlaments und des Rates zur Einführung einer Übergangsregelung für bilaterale Investitionsabkommen zwischen Mitgliedstaaten und Drittländern KOM(2010) 344 endgültig. Brüssel

European Commission (2010) Towards a comprehensive European investment policy – COM (2010) 343 final. http://trade.ec.europa.eu/doclib/docs/2011/may/tradoc_147884.pdf. Accessed 10 Apr 2014

European Commission (2011) Follow up to the European Parliament resolution on the future European international investment policy, adopted by the commission on 5 July 2011. www.europarl.europa.eu/oeil/spdoc.do?i=19829&j=0&l=en. Accessed 11 Apr 2014

European Commission (2012a) Proposal for a regulation of the European Parliament and the Council establishing a framework for managing financial responsibility linked to investor-state dispute settlement tribunals established by international agreements to which the European Union is party – COM(2012) 335 final. http://trade.ec.europa.eu/doclib/docs/2012/june/tradoc_149567.pdf. Accessed 11 Apr 2014

European Commission (2012b) Press release. http://europa.eu/rapid/press-release_IP-12-677_en.htm. Accessed 10 Apr 2014

European Commission (2013a) List of bilateral investment agreements referred to in Article 4 (1) of regulation (EU) No 1219/2012 of the European Parliament and of the council establishing transnational arrangements for bilateral investment agreements between member states and third countries. Off J Eur Union C 131:2–98

European Commission (2013b) Investitionsschutz und Beilegung von Investor-Staat-Streitigkeiten in EU-Abkommen. http://trade.ec.europa.eu/doclib/docs/2013/december/tradoc_151995.pdf. Accessed 11 Apr 2014

European Commission (2014) Commission to consult European public on provisions in EU–US trade deal on investment and investor-state dispute settlement. http://europa.eu/rapid/press-release_IP-14-56_en.htm. Accessed 10 Apr 2014

European Parliament (2011) European Parliament resolution of 6 April 2011 on the future European international investment policy (2010/2203(INI)). http://www.europarl.europa.eu/sides/getDoc.do?type=TA&reference=P7-TA-2011-0141&language=EN. Accessed 10 Apr 2014

Fiebiger G (2010) Vienna arbitration days 2010. SchiedsVZ 269–271

Franck SD (2005) The legitimacy crisis in investment treaty arbitration: privatizing public international law through inconsistent decisions. Fordham Law Rev 73:1521–1625

Gaillard E (2009) Identify or define? Reflections on the evolution of the concept of investment in ICSID practice. In: Binder C, Kriebaum U, Reinisch A, Wittich S (eds) International investment law for the 21st century – essays in honour of Christoph Schreuer. Oxford, pp 403–416

Gallus N (2008) An umbrella just for two? BIT obligations observance clauses and the parties to a contract. Arbitr Int 24:157–169

Gantz DA (2004) The evolution of FTA investment provisions: from NAFTA to the United States – Chile free trade agreement. Am Univ Int Law Rev 19:679–767

Gill J, Gearing M, Birt G (2004) Contractual claims and bilateral investment treaties. J Int Arbitr 21:397–412

Griebel J (2008) Internationales Investitionsrecht. München

Griebel J (2009) Überlegungen zur Wahrnehmung der neuen EU-Kompetenz für ausländische Direktinvestitionen nach Inkrafttreten des Vertrages von Lissabon. RIW 469–474

Griebel J, Kim Y-I (2007) Zwischen Aufbruch und Rückschritt – Überlegungen zur Zukunkt des internationalen Investitionsrechts. SchiedsVZ 186–195

Grubenmann B (2007) Aktuelle Probleme und Entwicklungen im internationalen Investitionsrecht. http://public.univie.ac.at/fileadmin/user_upload/int_beziehungen/Assistententagung/thesen/ref_uebers_grubenmanparapdf. Accessed 02 Nov 2010

Grubenmann B (2010) Der Begriff der Investition in Schiedsgerichtsverfahren in der ICSID-Schiedsgerichtsbarkeit. Basel

Hamida WB (2007) Two nebulous ICSID features: the notion of investment and the scope of annulment control. J Int Arbitr 24:287–306

Happ R (2007) Die praktische Bedeutung des Energiechartavertrages – eine EG-feste Grundrechtscharta für die deutsche Energiewirtschaft? Energiewirtschaftliche Tagesfragen 129–131

Happ R (2008) Rechtsprechung der ICSID-Schiedsgerichte 2005–2007. SchiedsVZ 19–30

Harten G (2007) Investment treaty arbitration and public law. Oxford

Harten G, Loughlin M (2006) Investment treaty arbitration as a species of global administrative law. Eur J Int Law 17:121–150

Hauschka C E (2005) Bauprojekte im Ausland – Durchsetzung der Ansprüche des Bauunternehmers gegen den Staat vor ICSID-Schiedsgerichten. BauR 1550–1560

Hellmann V (2009) Der Vertrag von Lissabon. Heidelberg

Herdegen M (2003) Rechtsprobleme des internationalen Konzessionswesens. In: Baur JF, Hobe S (eds) Rechtsprobleme von Auslandsinvestitionen. Baden-Baden, pp 13–35

Herdegen M (2010) Völkerrecht, 9th edn. München

Herrmann C, Weiß W, Ohler C (2007) Welthandelsrecht, 2nd edn. München

Hilf M, Oeter S (eds) (2010) WTO-Recht, 2nd edn. Baden-Baden

Ho J (2010) The meaning of 'investment' in ICSID arbitrations. Arbitr Int 26:633–647

Hobe S, Müller A (2009) Die Schiedsgerichtsbarkeit des International Centre for Settlement of Investment Disputes. WiVerw 1:65–76

Hobér K (2008) State responsibility and investment arbitration. J Int Arbitr 25:545–568

Hunter R (2007) Investitionsschutz und Streitbeilegung bei BOT-Projekten im Ausland. In: Nicklisch F (ed) Betreibermodelle – BOT/PPP-Vorhaben im In- und Ausland. München, pp 165–174.

IBRD (2009) Energizing climate-friendly development. Available via The World Bank Group. http://www.ifc.org/ifcext/sustainability.nsf/AttachmentsByTitle/p_EnergizingClimateFriendly Investment-WBG/$FILE/476190WP0Box331010RE1EE1Report12008.pdf. Accessed 26 Oct 2010

ICSID (1968) Convention on the settlement of investment disputes between states and nationals of other states – analysis of documents concerning the origin and the formulation of the convention, vol II(1), Washington D.C.

ICSID (1968) Convention on the settlement of investment disputes between states and nationals of other states – analysis of documents concerning the origin and the formulation of the convention, vol II(2), Washington D.C.

ICSID (1970) Convention on the settlement of investment disputes between states and nationals of other states – analysis of documents concerning the origin and the formulation of the convention, vol I, Washington D.C.

ICSID (2014) List of contracting states and other signatories of the convention. https://icsid.worldbank.org/ICSID/FrontServlet?requestType=ICSIDDocRH&actionVal=ShowDocument&language=English. Accessed 05 May 2014

IFC (2010) Annex 1: institutional support for renewable energy and energy efficiency. http://www.ifc.org/ifcext/enviro.nsf/AttachmentsByTitle/p_CatalyzingPrivateInvestment_Annex1/$FILE/Annex1.pdf. Accessed 26 Oct 2010, pp 47–50

Jagusch S, Gearing M, Sinclair A (2010) ICSID. In: Rowley JW (ed) Arbitration world, 3rd edn. London, pp 75–100

Johannsen SLE (2009) Die Kompetenz der Europäischen Union für ausländische Direktinvestitionen nach dem Vertrag von Lissabon. In: Tietje C, Kraft G, Lehmann M (eds) Beiträge zum Transnationalen Wirtschaftsrecht, vol 90. Halle(Saale), pp 5–37

Karl J (1994) Neue Entwicklungen beim internationalen Investitionsschutz. RIW 809–817

Karl J (2003) Aktuelle Probleme des Investitions- und Enteignungsschutzes am Beispiel des Internationalen Konzessionsvertrages. In: Baur JF, Hobe S (eds) Rechtsprobleme von Auslandsinvestitionen. Baden-Baden, pp 37–61
Kaufmann-Kohler G (2007) Arbitral precedent: dream, necessity or excuse? Arbitr Int 23:357–378
Kemper R (2003) Der Energiecharta-Vertrag – multilateraler Rahmen für Energiezusammenarbeit in Eurasien. Energiewirtschaftliche Tagesfragen 504–507
Kim HW (2007) Internationale Investitionsordnung und Umweltschutz. Göttingen
Knahr C (2009) Investments in the territory of the host state. In: Binder C, Kriebaum U, Reinisch A, Wittich S (eds) International investment law for the 21st century – essays in honour of Christoph Schreuer. Oxford, pp 42–53
Krajewski M (2005) External trade law and the constitution treaty: towards a federal and more democratic common commercial policy? Common Mark Law Rep 42:91–116
Krajewski M (2009) Wirtschaftsvölkerrecht, 2nd edn. Heidelberg
Krishan D (2009) A notion of ICSID investment. Received via email from author
Loncle J-M (2005) The arbitration option in protection of investment treaties: treaty claim vs. contract claims. Int Bus Law J 1:3–12
Loncle J-M (2006) The investment notion under ICSID convention. Int Bus Law J 3:319–331
Lörcher T (2005) ICSID-Schiedsgerichtsbarkeit. SchiedsVZ 11–21
Maier B (2010) How has international law dealt with the tension between sovereignty over natural resources and investor interests in the energy sector? Is there a balance? Int Energy Law Rev 95–109
Marauhn T (1996) Beyond pollution control: energy efficiency instruments in a liberalised international energy market. In: Wolfrum R (ed) Enforcing environmental standards: economic mechanisms as viable means? Berlin, pp 301–335
Matz N (2005) Wege zur Koordinierung völkerrechtlicher Verträge. Berlin
McIlwrath M, Savage J (2010) International arbitration and meditation. Alphen aan den Rijn
Metje TM (2008) Der Investitionsschutz im internationalen Anlagenbau. Tübingen
MIGA (2012) Operational regulations. http://www.miga.org/documents/Operations-Regulations.pdf. Accessed 05 May 2014
Mortenson JD (2010) The meaning of investment: ICSID's travaux and the domain of international investment law. Harv Int Law J 51:257–318
Moses ML (2008) The principles and practice of international commercial arbitration. Cambridge
Muchlinski P (2014) Corporations in international law. Available via Max Planck encyclopedia of public international law. http://opil.ouplaw.com/view/10.1093/law:epil/9780199231690/law-9780199231690-e1513#law-9780199231690-e1513-div2-8. Accessed 07 May 2014
Nariman F (2004) East meets west: tradition, globalization and the future of arbitration. Arbitr Int 20:123–138
Nathan KVSK (1995) Submissions to the International Centre for Settlement of Investment Disputes in breach of the convention. J Int Arbitr 1:27–52
Neumann J (2001) Die materielle und prozessuale Koordination völkerrechtlicher Ordnungen. ZaöRV 61:529–576
Nmehielle VOO (2001) Enforcing arbitration awards under the international convention for the settlement of investment disputes. Annu Surv Int Comp Law 7:21–48
OECD (2006) Investor-to-state dispute settlement in infrastructure projects. http://www.oecd.org/dataoecd/51/45/37475044.pdf. Accessed 30 Apr 2014
Office national d'électricité (2010) L'ONE au service du développement du pays. http://www.one.org.ma/. Accessed 18 Nov 2010
Peterson LE (2009a) Analysis: tribunal grapples with definition of investment under ICSID convention in RSM v Grenada case. Available via IA reporter. http://www.iareporter.com.ezproxy.auckland.ac.nz/articles/20090922_19. Accessed 12 Dec 2010
Peterson LE (2009b) Analysis: Uzbek case hinges on definition of investment under BIT. Available via IA reporter. http://www.iareporter.com.ezproxy.auckland.ac.nz/articles/20091222_10. Accessed 12 Dec 2010

Peterson LE (2010) In parallel decisions, tribunals diverge on admissibility of umbrella clause claims. IA Report 3:1–29

Potestà M (2009) Bilateral investment treaties and the European Uniopara recent developments in arbitration and before the ECJ. Law Pract Int Courts Tribunals 8:225–245

Prawoko P (2005) An analysis of ICSID case Law on the state of nationality of foreign investors. Griffin's View 6:143–158

Rang H (2011) MENA-Region – Chancen und Risiken. Available via GTAI. http://www.agaportal.de/pdf/dia_ufk/berichte/dia_exkurs_mena.pdf. Accessed 24 Apr 2014

Reed L, Paulsson J, Blackaby N (2004) Guide to ICSID arbitration. The Hague

Reinisch A (2008) The proliferation of international dispute settlement mechanisms: the threat of fragmentation vs. the promise of a more effective system? Some reflections from the perspective of investment arbitration. In: Buffard I, Hafner G (eds) International law between universalism and fragmentation – Festschrift in honour of Gerhard Hafner. Boston, pp 107–125

Reinisch A (2014) The EU on the investment path – Quo Vadis Europe? The future of EU BITs and other investment agreements. Santa Clara J Int Law 12:111–157

Saß KU (2009) Wissenschaftlich-Technologische Zusammenarbeit mit Afrika. APuZ 27–32

Schöbener B, Markert L (2006) Das International Centre for Settlement of Investment Disputes (ICSID). ZVglRWiss 65–116

Schöbener B, Herbst J, Perkams M (2010) Internationales Wirtschaftsrecht. Heidelberg

Schreuer C (2001) The ICSID convention – a commentary. Cambridge

Schreuer C (2008) What is a legal dispute? In: Buffard I, Hafner G (eds) International law between universalism and fragmentation – Festschrift in honour of Gerhard Hafner. Boston, pp 959–979

Schreuer C (2009) The ICSID convention – a commentary, 2nd edn. Cambridge

Schreuer C (2010) Denunciation of the ICSID convention and consent to arbitration. In: Waibel M, Kaushal A, Chung K-H, Balchin C (eds) The backlash against investment arbitration. Alphen aan den Rijn, pp 353–368

Schulz S (2012) Nach dem Frühling? Chancen nachhaltiger Entwicklung in der MENA-Region. Available via Heinrich Böll Stiftung. https://www.boell.de/de/2012/11/08/nach-dem-fruehling-chancen-nachhaltiger-entwicklung-der-mena-region. Accessed 24 Apr 2014

Schwarze J (eds) (2009) EU-Kommentar, 2nd edn. Baden-Baden

Seattle to Brussels Network (2010) EU investment agreements in the Lisbon Treaty Era: a reader. http://www.s2bnetwork.org/fileadmin/dateien/downloads/eu_investment_reader.pdf. Accessed 12 April 2014

Shany Y (2005) Contract claims vs treaty claims: mapping conflicts between ICSID decisions on multisourced investment claims. Am J Int Law 99:835–851

Shaw MN (2008) International law, 6 edn. Cambridge

Sinclair A (2009) Bridging the contract/treaty divide. In: Binder C, Kriebaum U, Reinisch A, Wittich S (eds) International investment law for the 21st century – essays in honour of Christoph Schreuer. Oxford, pp 92–104

Sureda AR (2009) Precedent in investment treaty arbitration. In: Binder C, Kriebaum U, Reinisch A, Wittich S (eds) International investment law for the 21st century – essays in honour of Christoph Schreuer. Oxford, pp 830–842

Tams C (2014) International arbitration: heating up or under pressure. Available via EJIL: talk. http://www.ejiltalk.org/international-arbitration-heating-up-or-under-pressure/. Accessed 10 Apr 2014

Tietje C (2003) Grundstrukturen und aktuelle Entwicklungen des Rechts der Beilegung internationaler Investitionsstreitigkeiten. In: Tietje C, Kraft G (eds) Arbeitspapiere aus dem Institut für Wirtschaftsrecht, vol 10. Halle (Saale), pp 5–20

Tietje C (2005) Die Beilegung internationaler Investitionsstreitigkeiten. In: Marauhn T (ed) Streitbeilegung in den internationalen Wirtschaftsbeziehungen. Tübingen, pp 47–62

Tietje C (2010) Internationales Investitionsschutzrecht im Spannungsverhältnis von staatlicher Regelungsfreiheit und Schutz wirtschaftlicher Individualinteressen. In: Tietje C, Kraft G,

Lehmann M (eds) Beiträge zum Transnationalen Wirtschaftsrecht, vol 93. Halle (Saale), pp 5–22

UNCTAD (2009) Selected recent developments in IIA arbitration and human rights – IIA MONITOR No. 2. http://www.unctad.org/en/docs/webdiaeia20097_en.pdf. Accessed 05 May 2014

United Nations (2001) Draft articles on responsibility of states for internationally wrongful acts, with commentaries 2001. Yearb Int Law Comm 31–143

Universität Frankfurt (2014) Investition und Finanzierung. http://www.wiwi.uni-frankfurt.de/~elsas/ir.pdf. Accessed 24 Apr 2014

Verdross A, Simma B (2010) Universelles Völkerrecht, 3rd edn. Berlin

Vinuesa RE (2005) Jurisdictional objections to ICSID arbitration under bilateral investment treaties. In: Bröhmer J, Bieber R, Calliess C, Langenfeld C, Weber S, Wolf J (eds) Internationale Gemeinschaft und Menschenrechte – Festschrift für Georg Ress. Köln, pp 331–355

Wälde TW (2009) Interpreting investment treaties: experiences and examples. In: Binder C, Kriebaum C, Reinisch A, Wittich S (eds) International investment law for the 21st century – essays in honour of Christoph Schreuer. Oxford, pp 724–781

Walter A (2006) Die Reichweite von Schirmklauseln in Investitionsschutzabkommen nach der jüngsten ICSID-Spruchpraxis. RIW 815–824

Yala F (2005) The notion of "Investment" in ICSID case Law: a drifting jurisdictional requirement? J Int Arbitr 22:105–126

Zampetti AB, Sauvé P (2007) International investment. In: Guzman A, Sykes AO (eds) Research handbook in international economic law. Cheltenham, pp 211–270

Conference Speeches

Komendantova N (2010) Investment into Large-Scale Deployment of Renewable Energy Sources in North Africa. Presented at the SEPA 2010 (Desert Sun, Yours, Mine, Ours?) conference, University of Giessen, Giessen, November 2010

Cases

ECJ

Case C-162/96 *A. Racke GmbH & Co. v Hauptzollamt Mainz* [1998] ECR I-3655

Schlussanträge des Generalanwaltes Antonio Tizzano C-466/98, *Kommission der Europäischen Gemeinschaft* v *Vereinigtes Königreich Großbritannien und Nordirland* [2002] ECR I-9431

German Courts

BVerfG, *Treaty of Lisbon* [2009] 2 BvE 2/08 [2009] NJW, 2267

PCIJ

The Government of the Greek Republic v The Government of His Britannic Majesty [1924] The Mavrommatis Palestine Concessions, Series A – No. 2, 7

WTO

Appellate Body, *India, Malaysia, Pakistan and Thailand v United States of America* [1998] United States - Import Prohibition of Certain Shrimp and Shrimp Products - Recourse to Article 21.5 of the DSU by Malaysia WT/DS58/AB/RW

International Court of Justice

Advisory Opinion [1950] Competence of the General Assembly for the Admission of a State to the United Nations [1950] ICJ Reports, 4

Advisory Opinion [1950] Interpretation of Peace Treaties with Bulgaria, Hungary and Romania [1950] ICJ Reports, 65

Bosnia and Herzegovina v Serbia and Montenegro [2007] Case concerning the application of the Convention on the Prevention and Punishment of the Crime of Genocide – Judgement [2007] ICJ Reports, 60

Botswana v Namibia [1999] Case concerning Kasikili/Sedudu Island – Judgment [1999] ICJ Reports, 1045

Cameroon v Nigeria (Equatorial Guinea intervening) [1998] Land and Maritime Boundary between Cameroon and Nigeria – Preliminary Objections Judgment [1998] ICJ Reports, 275

Cameroon v Nigeria (Equatorial Guinea intervening) [2002] Land and Maritime Boundary between Cameroon and Nigeria – Judgement [2002] ICJ Reports, 303

Germany v United States of America [2001] LaGrand case – Judgement [2001] ICJ Reports, 466

Indonesia v Malaysia [2002] Case concerning Sovereignty over Pulau Ligitan und Pulau Sipadan – Judgement [2002] ICJ Reports, 625

Libyan Arab Jamahiriya v Chad [1994] The Territorial Dispute – Judgment [1994] ICJ Reports, 6

Spain v Canada [1998] Fisheries Jurisdiction – Judgement [1998] ICJ Reports, 432

United Kingdom of Great Britain and *Northern Ireland v Albania* [1949] The Corfu Channel – Judgment [1949] ICJ Reports, 4

Arbitration courts

Ad hoc Arbitration, Eureko B.V. v Republic of Poland [2005] Partial Award

LCIA, *EnCana Corporation v Republic of Ecuador* [2006] Date of Award 3 February 2006

Permanent Court of Arbitration, *Romak S.A. (Switzerland) v The Republic of Uzbekistan* [2009] PCA Case No. AA280 – Award

Permanent Court of Arbitration, *Saluka Investments BV (The Netherlands) v The Czech Republic* [2006] Partial Award (under the UNICTRAL Arbitration Rules 1976)

SCC, *Nykomb Synergetics Technology Holding AB v The Republic of Latvia* [2003] Date of Award 16 December 2003

Sergei Paushok, CJSC Golden East Company and CJSC Vostokneftegaz Company v Mongolia [2011] Award – Proceeding pursuant to UNCITRAL Arbitration Rules

UNCITRAL, *White Industries Australia Limited v The Republic of India* [2011] Award

UNCITRAL, *Alps Finance and Trade AG v The Slovak Republic* [2011] Award

ICSID Tribunals

Abaclat and others v Argentine Republic [2011] ARB/07/5 – Decision on Jurisdiction and Admissibility

ADC Affiliate Limited and ADC & ADMC Management Limited v The Republic of Hungary [2006] ARB/03/16 – Award

AES Corporation v The Argentine Republic [2005] ARB/02/17

Alpha Projektholding GmbH v Ukraine [2010] ARB/07/16 – Award

Ambiente Ufficio S.p.A. and others v Argentine Republic [2013] ARB/08/9 – Decision on Jurisdiction and Admissibility

Ambiente Ufficio S.P.A. and Others v The Argentine Republic [2013], ARB/08/9 – Dissenting Opinion of Santiago Torres Bernárdez

American Manufacturing and Trading, Inc. v Zaire [1997] ARB/93/1 – Award

Antoine Goetz and others v Republic of Burundi [1999] ARB/95/3 – Award (1999) ICSID Review, 458

ATA Construction, Industrial and Trading Company v The Hashemite Kingdom of Jordan [2010] ARB/08/2 – Award

Autopista Concesionada de Venezuela, C.A. v Bolivarian Republic of Venezuela [2001] ARB/00/5

Autopista Concesionada de Venezuela, C.A. v Bolivarian Republic of Venezuela [2003] ARB/00/5 – Award
Azurix Corp. v The Argentine Republic [2003] ARB/01/12 – Decision
Azurix Corp. v The Argentine Republic [2006] ARB/01/12 – Award
Azurix Corp. v The Argentine Republic [2009] ARB/01/12 – Annulment
Bayindir Insaat Turizm Ticaret Ve Sanayi A.S. v Islamic Republic of Pakistan [2005] ARB/03/29
Bernardus Henricus Funnekotter and others v Republic of Zimbabwe [2009] ARB/05/6 – Award
Biwater Gauff (Tanzania) Ltd. v United Republic of Tanzania [2008] ARB/05/22 – Award
Bosh International, Inc. and B&P, LTD Foreign Investments Enterprise v Ukraine [2012] ARB/08/11 – Award
Bureau Veritas, Inspection, Valuation, Assessment and Control, BIVAC B.V. v The Republic of Paraguay [2009] ARB/07/9
Burlington Resources Inc. v Republic of Ecuador [2010] ARB/08/5
Cable Television of Nevis, Ltd. and Cable Television of Nevis Holdings, Ltd. v Federation of St. Kitts and Nevis [1997] ARB/95/2 – Award (1997) ICSID Review, 329
Camuzzi International S.A. v The Argentine Republic [2005] ARB/03/2
Ceskoslovenska Obchodni Banka, A.S. v The Slovak Republic [1999] ARB/97/4 – Decision, [1999] ICSID Review, 251
CMS Gas Transmission Company v The Republic of Argentina [2003] ARB/01/8, (2003) 42 ILM, 788
CMS Gas Transmission Company v The Republic of Argentina [2007] ARB/01/8 – Annulment
Compañiá de Aguas del Aconquija S.A. and Vivendi Universal S.A. v Argentine Republic [2002] ARB/97/3 – Annulment [2002] ICSID Review, 89
ConocoPhillips Petrozuata B.V., ConocoPhillips Hamaca B.V. and ConocoPhillips Gulf of Paria B.V. v Bolivarian Republic of Venezuela [2013] ARB/07/30 – Decision on Jurisdiction
Consortium R.F.C.C. v Kingdom of Morocco [2001] ARB/00/6
Consorzio Groupement L.E.S.I. – DIPENTA v People's Democratic Republic of Algeria [2005] ARB/03/08 – Award [2005] ICSID Review, 426
Continental Casualty Company v The Argentine Republic [2006] ARB/03/9
Deutsche Bank AG v Democratic Socialist Republic of Sri Lanka [2012] ARB/09/2 – Award
Deutsche Bank AG v Democratic Socialist Republic of Sri Lanka [2012] ARB/09/2 – Dissenting Opinion of Makhdoom Ali Khan
Duke Energy Electroquil Partners & Electroquil S.A. v Republic of Ecuador [2008] ARB/04/19 – Award
El Paso Energy International Company v The Argentine Republic [2006] ARB/03/15
Elsamex, S.A. v Republic of Honduras [2012] ARB/09/4 – Award
Emilio Agustín Maffezini v The Kingdom of Spain [2000] ARB/97/7 – Decision
Emmis International Holding, B.V., Emmis Radio Operating, B.V., and MEM Magyar Electronic Media Kereskedelmi és Szolgáltató Kft. v Hungary [2014] ARB/12/2 – Award
Enron Corporation and Ponderosa Assets, L.P. v Argentine Republic [2004] ARB/01/3 - Decision on Jurisdiction (Ancillary Claim)
Enron Corporation and Ponderosa Assets, L.P. v Argentine Republic [2004] ARB/01/3 – Decision
Enron Corporation and Ponderosa Assets, L.P. v Argentine Republic [2008] ARB/01/3 – Annulment
Enron Corporation and Ponderosa Assets, L.P. v Argentine Republic [2010] ARB/01/3 – Annulment
Fedax N.V. v The Republic of Venezuela [1997] ARB/96/3 (1998) 38 ILM, 1378
Fraport AG Frankfurt Airport Services Worldwide v Republic of the Philippines [2007] ARB/03/25 – Award
Fraport AG Frankfurt Airport Services Worldwide v Republic of the Philippines [2007] ARB/03/25 – Award – Dissenting Opinion of Bernardo M. Cremades
Gas Natural SDG, S.A. v The Argentine Republic [2005] ARB/03/10
GEA Group Aktiengesellschaft v Ukraine [2011] ARB/08/16 – Award

Generation Ukraine, Inc. v Ukraine [2003] ARB/00/9 – Award
Giovanni Alemanni and others v Argentine Republic [2014] ARB/07/8 – Decision
Helan International Hotels A/S v The Arab Republic of Egypt [2006] ARB/05/19
Helan International Hotels A/S v The Arab Republic of Egypt [2008] ARB/05/19 – Award
Iberdrola Energía, S.A. v Republic of Guatemala [2012] ARB/09/5 – Award
IBM World Trade Corporation, Actor v Republic of Ecuador [2003] ARB/02/10
Impregilo S.p.A. v Argentine Republic [2011] ARB/07/17 – Award
Impregilo S.p.A. v Islamic Republic of Pakistan [2005] ARB/03/3
Inceysa Vallisoletana S.L. v Republic of El Salvador [2006] ARB/03/26 – Award
Inmaris Perestroika Sailing Maritime Services GmbH and others v Ukraine [2010] ARB/08/8
Ioannis Kardossopoulos and Ron Fuchs v The Republic of Georgia [2007] ARB/05/18, 32 para 116
Jan de Nul N.V. Dredging International N.V. v Arab Republic of Egypt [2006] ARB/04/13
Joseph Charles Lemire v Ukraine [2010] ARB/06/18
Joy Mining Machinery Limited v The Arab Republic of Egypt [2004] ARB/03/11 – Award
KT Asia Investment Group B.V. v Republic of Kazakhstan [2013] ARB/09/8 – Award
L.E.S.I., S.p.A. and Astaldi, S.p.A. v People's Democratic Republic of Algeria [2006] ARB/05/3
Lanco International Inc. v The Argentine Republic [1998] ARB/97/6 – Preliminary Decision (2001) 40 ILM, 457
LG&E Energy Corp., LG&E Capital Corp. and LG&E International Inc. v Argentine Republic [2006] ARB/02/1 – Decision on Liability
loan Micula, Viorel Micula, S.C. European Food S.A, S.C. Starmill S.R.L. and S.C. Multipack S.R.L. v Romania [2008] ARB/05/20
M.C.I. Power Group L.C. and New Turbine, Inc. v Republic of Ecuador [2007] ARB/03/6 – Award
Malaysian Historical Salvors, SDN, BHD v The Government of Malaysia [2007] ARB/05/10 – Award
Malaysian Historical Salvors, SDN, BHD v The Government of Malaysia [2009] ARB/05/10 – Annulment
Malaysian Historical Salvors, SDN, BHD v The Government of Malaysia [2009] ARB/05/10 – Annulment – Dissenting Opinion (Shahabuddeen), 38
Middle East Cement Shipping and Handling Co. S.A. v Arab Republic of Egypt [2002] ARB/99/6 – Award
Mihaly International Corporation v Democratic Socialist Republic of Sri Lanka [2002] ARB/00/2 – Award (2002) ICSID Review, 142
Millicom International Operations B.V. and Sentel GSM SA v The Republic of Senegal [2010] ARB/08/20
Mobil Corporation and others v Bolivarian Republic of Venezuela [2010] ARB/07/27
Mr. Franck Charles Arif v Republic of Moldova [2013] ARB/11/23 – Award
Mr. Patrick Mitchell v The Democratic Republic of Congo [2006] ARB/99/7 – Annulment
Mr. Saba Fakes v Republic of Turkey [2010] ARB/07/20 – Award
Niko Resources (Bangladesh) Ltd. v Bangladesh Petroleum Exploration & Production Company Limited ("Bapex") and Bangladesh Oil Gas and Mineral Corporation [2013] ARB/10/11 and ARB/10/18 – Decision on Jurisdiction
Noble Energy, Inc. and MachalaPower Cia. Ltda. v The Republic of Ecuador and Consejo Nacional de Electricidad [2008] ARB/05/12
Noble Ventures, Inc. v Romania [2005] ARB/01/11 – Award
Pac Rim Cayman LLC v Republic of El Salvador [2012] ARB/09/12 – Decision on Jurisdiction
Pantechniki S.A. Contractors & Engineers (Greece) v The Republic of Albania [2009] ARB/07/21 – Award
Parkerings-Compagniet AS v Lithuania [2007] ARB/05/8 – Award
Philip Morris Brand Sàrl (Switzerland), Philip Morris Products S.A. (Switzerland) and Abal Hermanos S.A. (Uruguay) v Oriental Republic of Uruguay [2013] ARB/10/7 – Decision
Philippe Gruslin v Malaysia [2000] ARB/99/3 (2002) 5 ICSID Rep., 484

Phoenix Action Ltd. v The Czech Republic [2009] ARB/06/5 – Award
Plama Consortium Limited v Bulgaria [2008] ARB/03/24 – Award
PSEG Global Inc., The North American Coal Corporation, and Konya Ilgın Elektrik Üretim ve Ticaret Limited Şirketi v Republic of Turkey [2004] ARB/02/5 – Decision
Railroad Development Corporation (RDC) v Republic of Guatemala [2010] ARB/07/23
Saipem S.p.A. v The People's Republic of Bangladesh [2007] ARB/05/07
Salini Costruttori S.P.A. and Italstrade S.P.A. v Kingdom of Morocco [2001] ARB/00/4 (2003) 42 ILM, 609
Salini Costruttori S.p.A. and Italstrade S.p.A. v The Hashemite Kingdom of Jordan S.p.A. [2004] ARB/02/13
Sempra Energy International v The Argentine Republic [2007] ARB/02/16 – Award
SGS Société Générale de Surveillance S.A. v Islamic Republic of Pakistan [2003] ARB/01/13 [2003] ICSID Review, 307
SGS Société Générale de Surveillance S.A. v Republic of Paraguay [2012] ARB/07/29 – Award
SGS Société Générale de Surveillance S.A. v Republic of the Philippines [2004] ARB/02/6
Siemens v The Argentine Republic [2004] ARB/02/8 – Decision
Siemens v The Argentine Republic [2007] ARB/02/8 – Award
Société Ouest Africaine des Bétons Industriels v Senegal [1984] ARB/82/1, [1984] ICSID Review, 217
Société Ouest Africaine des Bétons Industriels v Senegal [1988] ARB/82/1 – Award [1988] ICSID Review, 125
Southern Pacific Properties (Middle East) Limited v Arab Republic of Egypt [1992] ARB/84/3 – Award [1992] ICSID Review, 328
Standard Chartered Bank v United Republic of Tanzania [2012] ARB/10/12 – Award
Suez, Sociedad General de Aguas de Barcelona S.A., and InterAguas Servicios Integrales del Agua S.A. v The Argentine Republic [2006] ARB/03/17
Tanzania Electric Supply Company Limited v Independent Power Tanzania Limited [2001] ARB/98/8 – Award
Teinver S.A., Transportes de Cercanías S.A. and Autobuses Urbanos del Sur S.A. v Argentine Republic [2012] ARB/09/1 – Decision on Jurisdiction
Telenor Mobile Communications A.S. v Republic of Hungary [2006] ARB/04/15
The Loewen Group, Inc. and Raymond L. Loewen v United States of America [2001] ARB(AF)/98/3
The Rompetrol Group N.V. v Romania [2008] ARB/06/3
Tidewater Investment SRL and Tidewater Caribe, C.A. v Bolivarian Republic of Venezuela [2013] ARB/10/5 – Decision on Jurisdiction
Tokios Tokelės v Ukraine [2004] ARB/02/18, [2004] ICSID Review, 205
Tokios Tokelės v Ukraine [2004] ARB/02/18 – Dissenting Opinion (Weil), [2004] ICSID Review, 245
Toto Costruzioni Generali S.P.A. v The Republic of Lebanon [2009] ARB/07/12
Tradex Hellas S.A. v Republic of Albania [1999] ARB/94/2 – Award, [1999] ICSID Review, 197
Venezuela Holdings B.V. and others v Bolivarian Republic of Venezuela [2014] ARB/07/27 – Award
Victor Víctor Pey Casado and President Allende Foundation v Republic of Chile [2008] ARB/98/2 – Award

Chapter 4
Concluding Remarks

After assessing all of the core legal aspects of the Desertec Concept, there is no doubt that the project can be implemented. This thesis serves as a guideline to avoid difficulties closely connected to the start of the project which may arise. Depending on which country hosts the Desertec Concept or parts of it, other questions of domestic law and its application will come up. If an investor or any other investor chooses to build a decentralized network in different MENA states, several domestic laws and their pitfalls will come into play. This will automatically lead to an even more complex contractual setup.

Nevertheless, recent events like the nuclear catastrophe in Fukushima (Japan) illustrate the dangers of nuclear energy. Furthermore, the ongoing climate change, which is closely connected to CO_2 emissions, illustrates that other forms of conventional energy are not a realistic option for the future. One of the few positive effects of this tragic event is that it has forced the world to start to focus more on alternative energies. Thus, it is necessary to explore alternatives such as the Desertec Concept in greater depth. Only this way can a dream become reality. The next sections will summarize the main theses of this thesis.

4.1 Only Certain Actors Come into Consideration as Contractual Partners

After analyzing the different actors involved in the implementation of the Desertec Concept, there can be no doubt that only a few fulfill the prerequisites of a contractual partner. There were several exclusion criteria, e.g. the legal incompetence to act, distrust by third parties and political instability. Although a lot of people would like to see the EU take a leading role in this project, this is unrealistic at the moment, as there is no legal basis for the EU to be an active actor within the Desertec Concept. It is only possible for the EU to take on a supporting role.

R. Gömmel, *Investing into North African Solar Power*,
DOI 10.1007/978-3-319-15756-6_4

The often hailed directives 2003/54/EC and 2009/28/EC cause a lot of difficulties as they are built on a very weak legal basis. The fact that leading members of the Desertec Foundation still invoke these directives illustrates that there is a need to deal with this material in greater depth. It would be highly risky to build this project with these directives as they are subject to a lot of discussion.

Furthermore, the foundation of the UM highlighted that even European countries still have issues with the leading roles of what are known as "close partners". The French answer to the "German" Desertec, embodied in the foundation of Transgreen, is a perfect example that it would be completely unacceptable for Germany to take a leading role. This is also connected to the fear that the Desertec Concept might provide a service of general interest. Thus, it is unlikely that any European country could take on a leading role in the implementation of the Desertec Concept. Based on this, there is only a private initiative left to implement the project. Within Europe, Dii was the company which has dealt the most with the topic of solar thermal energy from North Africa. Since Dii's failure, there is no other investor in sight that might be able to facilitate this kind of project.

On the other hand, there is the MENA region. There are no international or supranational organizations which could be contractual partners. In particular, the AU does not offer any help as it still lacks internal acceptance and the necessary setup. It is going to be important to include the AU in all relevant negotiations as a third party (vice versa to the EU on the European side), but at the moment it cannot take over any leading role. Although all parties stress that the MENA region must be understood as one continuum, this is window dressing. There is a clear line between Middle Eastern and North African countries. At the moment, ongoing revolutions or internal conflicts are shaking the whole region at the grassroots level and creating unforeseeable situations for any Western investor. The outcomes of these movements are unpredictable. Nevertheless, countries of the Middle East face even more severe problems than an unstable or unjust government. The utilization of the Shari'a and the problem of terroristic attacks hamper this region tremendously. On the contrary, North African countries have a more European-influenced background, which also reflected in their domestic laws. Morocco in particular has a close cooperation with Europe, which makes it an interesting partner for the implementation of the Desertec Concept. The fact that King Mohammed VI initiated democratic reforms long before the revolutions in Tunisia, Egypt and Libya created a more stable political system. During the revolutions, the Moroccan government was never at risk of being overthrown. Even recent terroristic attacks did not harm the political stability. Political reforms, as happened in 2011, also help to stabilize the system. Without a doubt, there are also problems connected to Morocco, e.g. the Western Sahara Conflict. Furthermore, the judicial system is not as independent as Moroccan politicians claim. The internal setup of legal competences within Morocco is rather complicated, but overall it seems that an investor must deal with ONE as a contractual partner for some of the Desertec Concept contracts. Nevertheless, it is the best suited country at the moment.

4.2 PPPs Are a Valid Contractual Tool of International Law

There is a lot of discussion about PPPs. Most of them revolve around their categorization, benefits and drawbacks. First of all, PPPs are nothing new anymore and are part of international and national law. Although some lawyers do not accept this, it does not change the fact that PPPs have been around for almost over 30 years.

It is rather difficult to categorize PPPs as there is no real definition. This is mainly due to their complexity, which however offers various types of contractual setup. Therefore, only certain attributes can help us understand PPPs. These attributes are: (1) a long duration of the contract, (2) one private and one public party (3), a privately funded project, and (4) an allocation of risks.

Furthermore, cooperation in the case of PPPs can vary with regard to the percentage of involvement. This is similar to the cooperation model, where the state remains an influential actor. Other forms of cooperation like concessions or joint ventures exclude almost all state influence. Unlike substantive privatization, PPPs do not include a complete transfer of duties to the private sector. In the case of PPPs, the public sector is still in charge to provide the respective service. Even in the case of a concession PPP or a joint venture, the state still has some control over the project. The main idea of PPPs is not to completely exclude the public influence.

There are positive and negative experiences with PPPs, which are mostly connected to the different applications and expectations of the contracting parties. Without a doubt, it is impossible to include all possible risks and events within a PPP. A PPP cannot offer a complete legal framework, covering all possible future situations. PPPs need a prior assessment of the respective situation. In the best case scenario, there are PPP guidelines or governmental departments dealing exclusively with PPPs. In most cases, PPPs failed because of missing or wrong feasibility studies or exceptions. PPPs are not the solution to all financial problems of the state. Every state must understand that a PPP is a long-lasting and complicated form of cooperation which offers, due to its flexibility, an excellent chance to successfully realize the Desertec Concept. Yet, it requires a lot of commitment and work. In addition, the fact that PPPs are complex is not a valid argument against their application. In the case of the Desertec Concept, it would be foolish to believe that any other contract form would be less complex concerning its setup. In the case of big scale projects and Greenfield investments, every contract requires a long and properly implemented negotiation. If both parties are willing to cooperate as the term PPP already indicates, the project will be successful.

4.3 Institutional PPPs to Successfully Set Up the Desertec Concept

The amount of investment needed for the implementation of the Desertec Concept illustrates that all other forms of contractual cooperation are overcharged. State budgets in the MENA region are not capable to come up with the required sum of around 400 billion euros, which also applies to a decentralized project. The only functional solution is the application of a PPP. International PPPs also illustrate that this form of cooperation can be very successful. In the field of energy production in particular, the positive experiences are far-reaching, irrespective of the cultural and religious background.

The best way to set the PPP up for the implementation of the Desertec Concept is as an institutional PPP with a joint venture at the core. This means that the newly formed SPV is a joint venture of the public and private sector. The setup as a cooperation model will not be accepted by the private sector as it must provide most of the money, but would only get max. 49.9 % of shares—i.e., not a majority. In the case of a joint venture, the share allocation can be in any desired way.

Nevertheless, to secure government's rights, the public party should get a minimum amount of shares, which guarantees them minority rights. The set-up of the joint venture must be according to national company laws, which in the case of the Desertec Concept should be according to German law and GmbH should be the legal form of the SPV. Moroccan law does not forbid a GmbH to operate within Morocco. In addition the question of SPV nationality would be easily answered, since the main seat theory (as prevailing in international law) would determine the SPV's nationality as German. Due to the manner in which numerous international arbitration institutions, including the ICSID, go about determining nationality, it is paramount that there is no dispute related to the Desertec Concept. The unpredictable practice creates a risk. In today's world, a US company is not necessarily held by US citizens. It is quite common that shareholders stem from all over the world. If a dispute resolution institution takes this into account to determine investor nationality, the outcome might be surprising. Within the GmbH, Morocco should have at least 25 % of all shares so as to have influential rights. This is comparable to the Moroccan S.A.R.L., which is similar to the French S.A.R.L.

The SPV should conclude a BOT/concession with either Morocco or ONE concerning the actual performance of service. It is very important that an investor transfers the project to Morocco at the end of contractual period, since its main idea is to provide energy for MENA countries. Furthermore, it offers new industrial perspectives for MENA countries, as they can export energy after the contractual period. During the contractual period, they have the concession fee as a payoff. The combination of BOTs and concessions are a possible option within PPP frameworks. In fact, they have already served as cooperation forms in different projects all over the world. These also included energy production projects. Hence, it is suitable for the Desertec Concept as well. Besides these two contracts, it is also necessary to conclude a PPA. Therefore, the an investor has a certain guarantee that its product is going to be purchased.

4.4 Subsidies or Feed-Ins Are Necessary to Implement the Desertec Concept

All figures illustrate that the project will not sustain itself at the beginning. The low energy prices in all MENA countries highlight the importance of additional financial aid. There are doubts that the Moroccan people can easily pay twice the price for energy than before. This also relates to the local industry. The effect of high energy prices are increasing production costs, which makes the product itself more expensive. In the case of Morocco, the country relies greatly on its textile industry export. This branch has been under enormous pressure for several years because of cheap Chinese products. An increase in energy prices could lead to a collapse of this export branch. In other MENA countries, the change would be even worse, since Morocco already has a very high energy price compared to its neighbors. Combined with a PPA, there would be an obligation to purchase the energy. Hence, they would "force" their customers to buy the new and more expensive energy.

Nevertheless, it is highly unlikely that prices are going to drop in the long term. Especially in the case of solar thermal energy, there is going to be a break-even point within the first few decades, making the energy cheaper. Some even claim that the costs could decrease from 22 Euro cents per kWh to 5–6 Euro cents per kWh, making it cheaper than today's energy price in Morocco. However, based on the average price of energy (22 Euro cents per kWh), the contract period should be around 30–40 years before it pays off for the investor. This long lifespan is not uncommon in international law. In international law, a contractual lifespan (also set up as a PPP) can be up to 99 years. Yet, it depends on each country if the domestic law has special requirements concerning the duration of contracts. In the case of Morocco, the maximum lifespan is 25 years. It is unrealistic to expect that Morocco will change its laws to allow longer projects. Even German law recognizes a limitation of contract periods. Therefore, the PPP should include a renewal option after the first 25 years. A renewal option is a common tool in PPPs. Moroccan law also does not forbid such a renewal option. Nevertheless, there must be financial support until this point is reached.

There are two main ways to achieve effective financial support, namely subsidies and feed-in tariffs. The latter is complicated in the case of Morocco as national law prohibits the utilization of feed-in tariffs. Subsidies might come from the host state itself or from third parties, e.g. AU, EU or European states. The host state might use some of the concession money to support the Desertec Concept at the beginning, risking a zero-sum situation. Until the first plant is set and ready, it is difficult to predict how much subsidization is necessary to successfully operate the project. Besides the above-mentioned organizations or states, there are also projects of the World Bank or under the Kyoto Protocol which might offer additional money. Unfortunately, the latter ends in 2012 and no new protocol is in sight. Thus, all the financial tools like the CDM are going to end in 2012 as well. Even if these tools apply to the Desertec Concept, it is questionable whether they have the capability to effectively support the Desertec Concept. Those projects currently being supported are not close to the size and complexity of this project. Overall, it comes down to whether European states are willing to support this project or just pay it lip service. Further funding could stem from other national institutes like the GTZ, DEG, and KfW. As a guarantee for all investments, institutes like MIGA offer good options (even for third parties) to secure their financial aid.

4.5 Commercial Arbitration Should Take Place at the CAM or ICC

The question of arbitration is essential for the success of the project. Due to the long contractual time, it is clear that there are going to be disputes about contract interpretation or obligations. Therefore, a dispute settlement must be at hand.

First of all, it is very important to avoid any kind of arbitration. If possible, both parties should try to resolve the dispute amicably. Taking someone to "court" makes it difficult to cooperate later on. Consequently, ADRs are interesting ways of preventing arbitration. Secondly it is necessary to include certain regulations in the contract. Both parties should conclude a dispute resolution agreement. This agreement must be divided into at least two sections—one dealing with international commercial disputes and the other one with international investment disputes. At the beginning it was unclear whether national or international courts should deal with the Desertec Concept, but an examination revealed that this is not an option.

No international court has the necessary competence to deal with the project. There are also a lot of problems in relation to national courts. The independence of Moroccan courts is questionable due to the king's influence, and European investors are not in favor of "foreign" national courts. The worst start for a dispute resolution is distrust in the independence of the respective court in charge. As a result, the only possible and accepted form of dispute resolution is international arbitration.

When comparing international arbitration institutions, it is noticeable that most arbitration rules are similar, apart from the CRCICA, which is rather complicated. Hence, it comes down to the regional focus of each institution and its experiences. With regard to commercial disputes, the CAM and the ICC offer the best possibility to settle disputes. First of all, there is the ICC—the oldest and most experienced arbitration institution in the world. Its expertise is tremendous, especially with regard to big-scale projects. The ICC is also a well-known institution on the African continent. Its arbitration rules offer a lot of benefits, like the review of the award and the possibility to assign the arbitration costs to each party independently of the outcome. On the other hand, there is the CAM. Unlike the ICC, the CAM is a national arbitration institution which is located in Italy. It has been dealing with international cases for quite some time. Its main focus is on the Mediterranean area. It cooperates closely with other regional arbitration institutions, like the CRCICA, to exchange knowledge and arbitrators. One of the major drawbacks is that the CAM has no experience with international big-scale projects.

Overall, it depends on how the Desertec Concept is implemented. If an investor decides to build one centralized plant, the ICC is the better option for the commercial arbitration. Vice versa, if they want to build several decentralized plants, the ICC is also the best choice, but the CAM offers a great arbitration possibility due to its local focus. The argument that the CAM has not arbitrated a large-scale project and therefore should not arbitrate one right now is not a persuasive one. As a result, the CAM would never get the chance to arbitrate any large-scale project. However, it is doubtful that this argument would be significant for an investor. The CAM is a great solution to combine and solve European interests and MENA resentments against European supremacy. Due to the focus on the MENA region and the cooperation with "Arab" institutions, the CAM might have the most experience of the cultural specifics of the MENA region, which might lead to it being more widely accepted. Both contractual parties might be more willing to accept CAM rulings as they represent both of them. Yet, this does not mean that the ICC is not highly respected, even in the MENA region. The CAM might be a new solution for a new project, but based on the new arbitration rules of the ICC, it is advisable to choose it. Due to the importance of the project and the wide impact investors tend to resort to a "safe harbor". In the case of commercial arbitration there is no institution as prominent as the ICC. The application of the ICC could therefore help to find and enthrall more and new investors.

Finally, arbitration is also a widely accepted form of dispute resolution in Arab countries. A historical assessment reveals that arbitration has a long tradition within Islam. The fact that Moroccan laws deal with international arbitration and Moroccan courts accept arbitration verdicts are other indicators that arbitration is accepted in the region. Although there are some inconsistencies within Moroccan laws on international arbitration, due to its historical acceptance and importance, there are no doubts that arbitration is the most effective form of dispute resolution for the Desertec Concept.

4.6 Investment Arbitration Must Be According to the ICSID Convention

No other arbitration institution offers the same amount of investment protection as the ICSID Convention. First, ICSID arbitrators have a lot of experience with investment disputes. Since 1965, ICSID tribunals have dealt with investment issues.

Unlike other arbitration frameworks, the ICSID Convention was especially designed to solely host international investment disputes. The growing amount of cases pending at ICSID tribunals also highlights the wide acceptance of these investment institutions. The two main facts supporting this are the enforceability of awards and the option of annulments. Unlike others, the ICSID Convention does not leave enforcement to the NY Convention. Second, there is a high voluntary execution rate of ICSID awards in almost every ICSID member state. In addition, the ICSID Convention offers special provisions dealing with interpretation, revision and annulment of awards (Art. 50 to 52 ICSID Convention). Art. 54(1) of the ICSID Convention is the great benefit of the ICSID Convention compared to other arbitration facilities, because it leads to a direct enforceability of ICSID awards. However, Art. 55 of the ICSID Convention limits its application, as it excludes immunity of execution. This is a general problem which all international arbitration institutions face and nothing specific to the ICSID Convention. It is almost remarkable that the ICSID Convention limits execution problems.

There are other options concerning international investments, as the ICC or UNCITRAL offer ways to arbitrate investment disputes. Nevertheless, only the ICSID Convention offers this unparalleled system of enforcement, award acceptance and annulment, which makes it unique. The fact that the ICSID is part of the World Bank also underlines the relevance of the ICSID Convention. States face more repressions in cases of non-compliance than in the case of ICC or CAM arbitrations. Disobeying an ICSID award might automatically lead to problems with the World Bank. Acceptances of ICSID awards serve to some extent as a reputation indicator of the respective state for the World Bank. A state constantly refusing to implement ICSID awards runs the risk of losing World Bank trust and support. Although many authors complain about the ICSID system, it is the best investment arbitration system at the moment.

4.7 Domestic Law Should Be the Applicable Law Combined with a Stabilization Clause

International law only has a gap filling or corrective function. This is mainly due to its insufficient regulatory density compared to national legal systems. Thus, the contracts should not include an internationalization clause which regulates that international law is the applicable law of the contract. A comparable clause leaves a great interpretation possibility for any arbitration court dealing with the issue. This creates a tremendous risk for both parties because the outcome is highly unpredictable. Arbitration courts might be completely free to decide due to missing regulations or guidelines. The risk is outweighs any positive effect for the investor or the host state.

This is different if it comes down to a stabilization clause, which "freezes" the status quo concerning the project. In this case, the investor has a certain guarantee that the host state will not introduce new and discriminatory laws after the start of the investment. There is no doubt that a stabilization clause might interfere with state sovereignty. However, this is acceptable, as the state may freely to conclude such a clause or not. If the clause does not touch core elements of the state, it is a valid legal tool. The latest revolutions in Egypt, Tunisia and Libya illustrate the importance of a stabilization clause. Until the beginning of 2011, an overthrow of these regimes was almost unthinkable in Egypt and Tunisia. Due to the ongoing internal conflicts in Libya and NATO involvement, the outcome can still not be predicted. This new legal situation is especially underlined by the fact that in Egypt the revolution peaked in a "new" constitution. All countries belonging to the MENA region have a more European legal background. Besides the legal changes and the reform of the political system, the revolution has caused people to fear a greater influence of the Shari'a in this region. In Egypt in particular, the opposition favors a more traditional approach. Comparable to the fear of "foreign" national courts, investors also often fear the Shari'a as well. Especially since the Shari'a might become an important document of the judicial system in the Mediterranean region again. Its imprecise wording makes Western investors feel uncomfortable because it can be interpreted in numerous ways. Hence, a stabilization clause could guarantee the application of the old, more European influenced law. If both parties are willing to include a stabilization clause, they should declare within the contract that a stabilization clause is not subject to the applicable law of the contract, but to international law.

Although most MENA countries were shaken up by the so-called "Arab Spring", Morocco seems to have had the least problems. There have been demonstrations in Morocco as well demands for democracy, but unlike in other countries these did not end in a revolution or civil war. Authors mentioned that this is closely connected to the political reforms by the king a couple of years ago. During the revolutions, the Moroccan system proved to be more stable than others. Hence, it is more interesting for investors than other countries in the region.

Moroccan law, as illustrated in laws for arbitration and energy, was reformed a few years ago. It is now more open to international projects and is also capable of successfully "supporting" large scale investments. Furthermore, the constitution of Morocco underlines that Moroccan law is not Islamic law in the classical sense as the Saudi Arabian or Iranian law. Over the past few years, Moroccan law was constantly reformed by the king, which means it is one of the most modern legal systems in the North African region. Consequently, Morocco and an investor should apply Moroccan law to the Desertec Concept, combined with a stabilization clause freezing today's status quo. Morocco is also an international actor that knows about the consequences of investor discrimination as it was the first country sued under the ICSID Convention. In all cases including Morocco, the country obeyed all the awards filed. Therefore, it is doubtful that Morocco would willingly disobey obligations, arbitration awards or purposely change its national law adversely. The stabilization clause serves as a "compromise" for the application of Moroccan law for the investor. Finally, both parties should agree that the stabilization clause applies to all contracts.

4.8 Public Policy and State Immunity Illustrate the Need of Cooperation

Public policy and state immunity are two of the main reasons why arbitration awards are not enforced in the in MENA countries. To date, arbitration courts, states and sometimes even their national courts have tried to limit the application of state immunity as tool to prevent foreign award implementation. This is different concerning public policy. Due to the definition problem and hence the scope of public policy, it is almost impossible to constrain it. Public policy might cause problems in countries where the Shari'a has a lot of influence on the legal system.

The biggest problem is that both tools are frequently misused to avoid execution of unwanted arbitration awards. Nevertheless, there is a great acceptance of arbitration within the Arab world and only a few arbitration awards have been subject to discussion. Both state tools emphasize the importance that both partners (e.g. an investor and Morocco/ONE) cooperate. The best way to avoid utilization of these tools is to have a good partnership prior to this event. Both parties must accept and tolerate each other to successfully complete the project. This includes the avoidance of stereotypes like European supremacy and Arab states being less developed. As the term PPP already includes partnership, it seems to be forgotten from time to time that partnerships are a mixture of "give-and-take."

In the case of Morocco, national and international public policy might apply. Morocco does not have the reputation of resorting to public policy. The experiences with previous oil concessions in the Middle East also illustrate that the objection to public policy was not an issue. As long as the project contributes to the economy, there is only a low risk that someone will raise the issue of public policy. This is closely connected to the importance of financial support to make the project beneficial right from the start. Then a basic rule might be set up: the more the Shari'a influences the national legal system, the higher the risk is that public policy claims will be used to avoid recognition of foreign awards. With regard to state immunity, an investor should demand a waiver of immunity of execution. This is especially important since the normal arbitration agreement does not encompass such a waiver. French national courts in particular viewed an ICC arbitration clause as a waiver. However, this decision was highly controversial and thus cannot directly apply to other cases. Moroccan courts have not dealt with this issue yet. The ICSID Convention faces fewer problems with state immunity, but the issue of immunity from execution also remains. This underlines its importance, but it is going to be difficult for an investor to negotiate a comparable waiver.

It is important both parties should try to avoid arbitration in the first place. This can only happen if they are willing to accept that it is a partnership. Hence, both sides must benefit from the cooperation. Only if there is a positive atmosphere of talks and cooperation can issues like public policy and state immunity be circumvented. Negotiations must be free of any stereotypes an investor should not try to get as much benefit out of the project as possible and Morocco should not try to offload as many obligations as possible due to resentments towards Europe.

It is within the interests of both contractual parties that the project does not fail. While the investor carries the financial risk, the state faces a loss of reputation and consequently fewer investments.

4.9 The Desertec Concept Fulfills all Investment Requirements

Since the definition of investment is a widely discussed topic within investment arbitration, it was necessary to assess if the Desertec Concept fulfills all possible requirements. When the Salini tribunal set up these requirements—(1) a contribution (in cash, in kind or in labor), (2) certain duration of performance, (3) participation in the risks of the transaction, and (4) a contribution to the economic development of the host state)—different approaches developed. These are the subjective, objective and liberal approaches. The subjective approach claims that only the definition of the BIT is relevant to determine the definition of investment. The objective approach views the requirements of Art. 25(1) of the ICSID Convention, the Salini criteria and the BIT definition as relevant. Finally, the liberal approach fits in the middle of the two other approaches as it claims that only certain parts of the Salini test are necessary besides the basic requirements of Art. 25(1) of the ICSID Convention and the BIT definition. Although the subjective approach seems to be favored, this did not exclude the ongoing application of what are known as the Salini elements. One of the latest decisions in 2009 illustrated that some tribunals even added more requirements to the definition of investment according to Art. 25(1) of the ICSID Convention. Furthermore, one ICSID tribunal mentioned in 2010 that the question of investment definition is still unanswered. The different approaches taken by ICSID tribunals and the literature emphasize the uncertainty which is closely related to the issue of "what is an investment?" Due to the fact that it is difficult to predict which approach a possible ICSID tribunal might follow, the Desertec Concept must meet the strictest requirements. If the project is in accordance with the objective approach, the risk that it is not considered to be an investment is significantly minimized.

The Desertec Concept meets all the necessary requirements of the objective approach. First of all, it is an investment according to the G/M-BIT. Since the definition within the BIT is not very narrow, the Desertec Concept easily fulfills this prerequisite. With regard to the requirement of Art. 25(1) of the ICSID Convention, this is a bit more difficult. Without a doubt, the Desertec Concept is a significant contribution and has a certain duration. Besides the fact that the complete project costs around 400 billion euros, it will take at least 30–40 years before the investor withdraws from the project. However, the element risk is already causing some difficulties. As mentioned above, there is a need to financially support the project. The PPP should include a PPA to secure the investor's money. Both tools could exclude the Salini requirement of "risk" because the investor does not want to run the risk of losing its money. Due to the PPA, the state has an obligation to pay, and due to the subsidies there is always enough money. ICSID tribunals have accepted PPAs and subsidies in different cases without excluding the requirement "risk". It would be wrong to believe that risk stops at the question of remuneration. In the case of the Desertec Concept, there is a lot of risk inherent to the project because there have been no similar projects up to now. ICSID tribunals must accept that and should refrain from taking an overly narrow approach. Due to the size, uniqueness and importance of the Desertec Concept, ICSID tribunals should be aware that risks besides remuneration might occur. Hence, they should not exclude risk.

There are also some requirements left, namely "in the territory of the host state", "regularity of profit and return", assets invested *bona fide* and assets invested "in accordance with the laws of the host state." The requirements of invested *bona fide* and assets invested in accordance with the laws of the host state also highlight the importance of a partnership and mutual consideration. Cases involving German companies have stressed that the investor should not try to circumvent local laws just for its benefit. Although it is an international project, an investor must play by the rules of the host state. Vice versa the host must obey international rules and its laws. Finally, there is the problem of contribution to the economic development of the state. Even within the different approaches within investment law, this is the most widely discussed element of the Salini test.

4.10 The Desertec Concept Is a Contribution to the Economic Development of the Host State

As the most diverse element of the Salini test, the criterion contribution to the economic development of the host state has been the subject of a lot of controversial discussion within ICSID case law and the literature. The main argument against this element is that it is almost impossible to define. In cases where ICSID tribunals have applied it, they assessed every investment individually, without setting any standards or guidelines. Therefore, the only way to find out whether the respective investment will contribute to the economic development of the host state depends solely on the view of the tribunal. This leads to an enormous unpredictability, and it also shows why this element is such a hot topic.

With regard to the Desertec Concept, there seems to be no initial problem. It is a major investment sum which will go into Morocco. The investment will take at least 30 years. New technology is coming into the Morocco. There is a great chance of new jobs and the transfer of knowledge. In addition, it will replace conventional energy production. This will lead to a cleaning of the environment, which will automatically raise living standards. It will also lift Morocco to the level of a global player. Due to the importance of the project and Europe's interest, it offers a great chance for "average" countries to gain more international influence.

However, the project also has drawbacks. The biggest problem is that the produced energy is very expensive at the beginning. For this reason, the project must be financially supported. This support must be primarily done by Morocco, because they are going to be under the obligation to buy the energy (due to the PPA). The low average energy price will force the state to counteract against the expensive new solar energy. Morocco might use the concession fee to "refinance" the energy. Yet, this leaves the question of whether there is any contribution to the economic development of the host state if there is no concession fee left. As mentioned above, the project will not be transferred very soon. As a result, the only benefit for Morocco will be the concession fee. If this fee is completely consumed by subsidies, there will be nothing left for the state. In the worst case scenario, the concession fee will not be enough. This might be an argument against a contribution to the economic development of the host state, as nothing would be left of the concession fee, which is itself a vital part of the initial "acceptance" of the contribution requirement.

It is almost impossible to set abstract guidelines if the situation turns negative. The importance of the Desertec Concept for the future makes it necessary to adjust the terms of the positive and negative effects. It would be wrong to define these terms too narrowly, as they are connected to the length of the actual implementation of the project. E.g. short-term negative effects might be outweighed by long-term positive effects. Due to the importance of the project, the approach towards this problem should not be too strict. Looking at the positive "side-effects" of the Desertec Concept, like a cleaner environment, less CO_2, no risk of nuclear power plants in the MENA region, and an increase in the number jobs and knowledge, it is easy to see how the project would contribute to the economic development of the host state. Furthermore, there must be a separation between the short-term positive effects and long-term positive effects. On a short-term scale, there is the concession fee and an increase in jobs and knowledge. If the concession fee is missing, the amount of positive short-term effects on the economic development of the host state will be greatly reduced. However, the long-term positive effects also require consideration. The fact that these benefits will be visible in years to come or that they cannot be valued in money does not make them irrelevant. On the long-term scale, a cleaner environment and a reduced risk of nuclear catastrophes combined with the knowledge to run this project and the possibility to export energy to Europe is clearly a contribution to the economic development of the host state. Hence, the overall assessment reveals that the Desertec Concept would contribute to the economic development of the host state.

4.11 A Certain Amount of Uncertainty Remains

Finally, a lot of uncertainties remain. The Desertec Concept is a unique project, so it is impossible to predict all the risks or challenges it will eventually face. Therefore, it is important that the PPP is set up in a flexible way if it is to survive. Other projects have been important lessons, e.g. the desire to regulate all possible situations still to come frequently ended in the failure of the whole project. Flexibility can be achieved by including adjustment clauses and with ongoing communication between all relevant contract parties. Nevertheless, there are already risks which are known but cannot be changed. The major one is the unpredictability of the arbitration institutions' decision. The ICC, the CAM and the ICSID are not bound by precedents. Thus, it is almost impossible to predict an outcome of an arbitration process. The best example is the Salini decision itself. Before this decision, no-one had ever thought of actually defining investment or setting up any requirements. This makes it very difficult for lawyers to advise their clients. As another example, the umbrella clause and its treatment shows how difficult it is to predict an outcome. The almost exact same case was decided in two different ways by different ICSID tribunals. The different outcomes in the arbitration process can also be illustrated by the Salini test. Since 2001, several ICSID tribunals have dealt with this issue and come up with numerous different decisions. There have been cases decided in accordance with the Salini which were annulled later on, and vice versa annulment processes which supported the Salini application, unlike the initial court. There is a categorization into three main groups, but even the liberal approach does not agree on how to handle all the criteria. Therefore, it was important to assess the Desertec Concept according to the most restrictive approach within ICSID case law. There is still a chance that another tribunal will lift the Salini test to a new level, but the only way to predict this would be with a crystal ball. However, the shifting of law and opinions is part of the lawyer's job. Consequently, it is important to pay attention to the development of case law and adjust to it. The best solution would be to avoid any confrontation in front of an arbitration institution. This always ends with a winner and loser, which is not helpful in cooperation. As the PPP already indicates, it must be a partnership, which means that all interests must be balanced at all times. There are risks, but the term "investment" encompasses risk. This thesis offers a guideline to avoid the most severe risks. Consequently, there is a great chance that the Desertec Concept will be a success for year.

MIX
Papier aus verantwortungsvollen Quellen
Paper from responsible sources
FSC® C105338

If you have any concerns about our products,
you can contact us on
ProductSafety@springernature.com

In case Publisher is established outside the EU,
the EU authorized representative is:
Springer Nature Customer Service Center GmbH
Europaplatz 3, 69115 Heidelberg, Germany

Printed by Libri Plureos GmbH
in Hamburg, Germany